Machine Learning and IoT

A Biological Perspective

Machine Learning and IoT

A Biological Perspective

Edited by
Shampa Sen
Leonid Datta
Sayak Mitra

CRC Press
Taylor & Francis Group
Boca Raton London New York

CRC Press is an imprint of the
Taylor & Francis Group, an **informa** business

CRC Press
Taylor & Francis Group
6000 Broken Sound Parkway NW, Suite 300
Boca Raton, FL 33487-2742

Printed on acid-free paper

International Standard Book Number-13: 978-1-138-49269-1 (Hardback)

Library of Congress Cataloging-in-Publication Data

Names: Sen, Shampa, editor.
Title: Machine learning and IoT : a biological perspective / [edited by] Shampa Sen, Leonid Datta, Sayak Mitra.
Description: Boca Raton : Taylor & Francis, 2019. | Includes bibliographical references and index.
Identifiers: LCCN 2018014315 | ISBN 9781138492691 (hardback)
Subjects: LCSH: Biology--Data processing--Methodology. | Bioinformatics--Methodology.
Classification: LCC QH324.2 .M34 2019 | DDC 570.285--dc23
LC record available at https://lccn.loc.gov/2018014315

Visit the Taylor & Francis Web site at
http://www.taylorandfrancis.com

and the CRC Press Web site at
http://www.crcpress.com

I would like to dedicate this book to my dearest niece and light of my life, Neha

Shampa Sen

To my beloved late grandma Bimalprabha, my father Haranath, my mother Mallika, my uncle Ramanath, and aunt Karuna Datta, who have always been my inspirations in life.

Leonid Datta

Dedicated to all the teachers, my mother and my grandmother, who continue to have a profound effect on my life. Also dedicated to my late grandfather Dr. Bimalendu Majumdar, who would have been very happy to see me reach for the stars

Sayak Mitra

Contents

Foreword

Machine learning and IoT are presently two of the most rapidly developing aspects in the field of computer science. Machine learning deals with the ability of the computers to learn and predict an output without being explicitly programmed. IoT, on the other hand, deals with connecting everything and/or everybody to each other using sensors. Both machine learning and IoT find numerous interdisciplinary applications in various fields, especially biology. The present book deals with how machine learning and IoT complement research in the field of biotechnology and health care. At present, biology is the foremost field generating enormous amounts of data needing to be sorted. These data also needs to be analyzed for a better understanding of the research at hand.

It is often hard for budding researchers to visualize how to bring together the two very contrasting fields of computer science and biology. This book beautifully unifies these two fields and provides an outlook on the ways biology has and can be improved when computer science technique intervenes. While there are many publications in this regard, most of them focus on the already established field of bioinformatics. But there is more to this interdisciplinary wedding than bioinformatics, and those aspects have been covered by the book. The topics chosen in the book are present day hot topics, thus making them relatable to undergraduate students and researchers of both fields alike.

What adds to the glamour of the book is how two of the editors—Leonid Datta and Sayak Mitra—both being undergraduate students, with hardly any experience in their respective core backgrounds of computer science and biology have effectively merged the two fields, while being guided by the senior editor Dr. Shampa Sen. Even the list of editors showcases a colorful array of undergraduate and postgraduate students guided by eminent researchers, along with industrial experts who are constantly updated on the current technologies used in industries through their technological demands.

I highly recommend this valuable and timely book. I heartily congratulate the editors and contributing authors for their immense efforts in producing this book.

Prof. Sarit K. Das
Director
Indian Institute of Technology Ropar

Preface

The present era of research and development is all about interdisciplinary studies attempting to better comprehend and model our understanding of this vast universe. The fields of biology and computer science are no exception. This book discusses some of the innumerable ways in which computational methods can be used to facilitate research in biology and medicine—from storing enormous amounts of biological data to solving complex biological problems and enhancing the treatment of various diseases.

In 1959, Arthur Samuel defined machine learning as the "field of study that gives computers the ability to learn without being explicitly programmed". Since then, this field of study has undergone many advancements to make machine learning a combination of several disciplines such as statistics, information theory, theory of algorithms, probability, and functional analysis. Another aspect of machine learning is the extraction of common patterns from sets of data, known as data mining. Hence, a more inclusive definition of machine learning would be design and testing of algorithms facilitating pattern recognition, classification, and prediction based on existing data.

Internet of Things (IoT) is based on the concept of various hardware devices, including sensors, being embedded with software and apps able to transmit and process data. These new easy-to-use devices have innumerable applications, especially in the field of biotechnology and medicine. Any object, outfitted with the right sensors, can observe and interact with its environment, allowing hard-working individuals to control their personal and professional responsibilities from far away, using only a voice command in smartphone.

The relationship between biology and machine learning has existed for decades. The present status of research in the field of biotechnology calls for sophisticated methods to store the large amounts of data generated, and to ease the task of the researcher by going so far as to predict certain outcomes. In short, biological research has become much dependent on computational algorithms. Likewise, the management of the health care system also becomes easier with the intervention of machine learning and other computational tools.

In this book, we have started our discussion with how various machine learning algorithms can be used as a powerful tool for aiding biologists in their research work. In the first half of the book, we have talked about how different aspects of machine learning can complement bioprocess engineering, bioinformatics, and biofouling. We have also discussed how health care facilities and treatments can benefit from machine learning, covering a wide range of aspects such as autoimmune diseases, nosocomial infections, brain lesions, and even theranostics. The latter half of the book talks about how IoT can facilitate many aspects of biology, starting from health care systems to analytical labs, even revolutionizing traditional farming practices.

In conclusion, this book gives a comprehensive overview of some of the innumerable ways in which computational methods can aid research in biology and medicine to allow us to live a better life.

Shampa Sen
Leonid Datta
Sayak Mitra

About the Editors

Dr. Shampa Sen earned her PhD in environment from Indian Institute of Technology, Guwahati, India. She is the Associate Professor at School of Bio-Sciences and Technology, VIT, Vellore, India. With extensive experience in academia, she has 60 publications in the fields of biotechnology, drug design, nanobiotechnology, and nutraceuticals. She was actively involved in many professional development activities. Her research interests include biosynthesis of metallic nanoparticles, nanoparticles in biomedical and environmental applications, metabolic engineering, drug design, and computational biology. She is a life member of the Biotech Research Society, India (BRSI) and the Environmental Mutagen Society of India (EMSI), as well as a member of the International Neural Network Society (INNS). She is also a fellow of the Royal Society of Biology.

Leonid Datta has earned his BTech degree in computer science and engineering from VIT, Vellore, India. He has various publications in reputed peer-reviewed international journals and books. His research interests include data mining and machine learning. His most notable project works include designing truncation techniques for a search engine, developing a system that runs based on RFID scanning for attendance purposes or maintaining other records, and automating the processing of big data analysis through development of a portal. He is also a student member of INNS, USA.

Sayak Mitra is currently pursuing his BTech degree in Biotechnology at VIT, Vellore, India. At present, he has publications in the fields of nanoremediation and metabolic engineering in international peer-reviewed journals and conferences. His research interests include biochemical engineering, bioprocess engineering and metabolic engineering. He was awarded Best Oral Presentation Award at the 7th International Science Congress conference (ISCA-2017) held in Bhutan. He is also a fellow under the Khorana Program for Scholars (an Indo-U.S. Science and Technology collaborative research scheme). His avid interest in research motivates him to pursue a PhD in the upcoming years.

Contributors

Neha Agnihotri
School of Bio Sciences and Technology
Vellore Institute of Technology
Vellore, Tamil Nadu, India

Mohanapriya Arumugam
School of Bio Sciences and Technology
Vellore Institute of Technology
Vellore, Tamil Nadu, India

Srijita Banerjee
School of Bio Sciences and Technology
Vellore Institute of Technology
Vellore, Tamil Nadu, India

Gaurav Bansal
Bank of America
Hyderabad, India

Shishir K. Behera
Department of Chemical Engineering
Vellore Institute of Technology
Vellore, Tamil Nadu, India

Adrish Bhattacharya
Department of Computer Science and Technology
Indian Institute of Engineering Science and Technology
Shibpur, India

and

Department of Electronics and Telecommunication Engineering
Indian Institute of Engineering Science and Technology
Howrah, West Bengal, India

Sweta Bhattacharya
School of Information Technology and Engineering
Vellore Institute of Technology
Vellore, Tamil Nadu, India

Vibhash Chandra
Department of Computer Science and Technology
Indian Institute of Engineering Science and Technology
Howrah, West Bengal, India

Ashmita Das
School of Bio Sciences and Technology
Vellore Institute of Technology
Vellore, Tamil Nadu, India

Denim Datta
CIDSE (School of Computing Informatics and Decision Systems Engineering)
Arizona State University
Tempe, Arizona

Emilee Datta
R G Kar Medical College and Hospital
Kolkata, India

Leonid Datta
Department of Software Systems
Vellore Institute of Technology
Vellore, Tamil Nadu, India

Prajit Kumar Datta
Bank of America
Hyderabad, India

Gayathri M.
Department of Biotechnology
School of Bio Sciences and Technology
Vellore Institute of Technology
Vellore, Tamil Nadu, India

Karthikeya Srinivasa Varma Gottimukkala
Department of Biotechnology
Sreenidhi Institute of Science and Technology
Hyderabad, Telangana, India

Prerna Grover
School of Bio Sciences and Technology
Vellore Institute of Technology
Vellore, Tamil Nadu, India

Madhurima Gupta
School of Bio Sciences and Technology
Vellore Institute of Technology
Vellore, Tamil Nadu, India

Nitu Joseph
School of Mechanical and Building Sciences
Vellore Institute of Technology
Vellore, Tamil Nadu, India

Shreyasi Kundu
School of Bio Sciences and Technology
Vellore Institute of Technology
Vellore, Tamil Nadu, India

Sajitha Lulu
School of Bio Sciences and Technology
Vellore Institute of Technology
Vellore, Tamil Nadu, India

Deepa Madathil
Department of Sensor and Biomedical Technology
School of Electronics Engineering
Vellore Institute of Technology
Vellore, Tamil Nadu, India

Diptesh Mahajan
School of Bio Sciences and Technology
Vellore Institute of Technology
Vellore, Tamil Nadu, India

Debayan Mandal
School of Mechanical and Building Sciences
Vellore Institute of Technology
Vellore, Tamil Nadu, India

Saroj K. Meher
Systems Science and Informatics Unit
Indian Statistical Institute
Bangalore, India

Bishwambhar Mishra
Department of Biotechnology
Sreenidhi Institute of Science and Technology
Hyderabad, Telangana, India

Avitaj Mitra
Department of Biomedical Engineering
School of Electronics Engineering
Vellore Institute of Technology
Vellore, Tamil Nadu, India

Sayak Mitra
School of Bio Sciences and Technology
Vellore Institute of Technology
Vellore, Tamil Nadu, India

Abhishek Mukherjee
School of Information Technology and Engineering
Vellore Institute of Technology
Vellore, Tamil Nadu, India

Sharmila Nageswaran
Department of Sensor and Biomedical Technology
School of Electronics Engineering
Vellore Institute of Technology
Vellore, Tamil Nadu, India

Harshit Nanda
Department of Biotechnology
School of Bio Sciences and Technology
Vellore Institute of Technology
Vellore, Tamil Nadu, India

Ramkumar Lakshmi Narayanan
Department of Information Technology
Sur College of Applied Sciences
Ministry of Higher Education
Sultanate of Oman

Abanish Roy
Department of Biotechnology
School of Bio Sciences and Technology
Vellore Institute of Technology
Vellore, Tamil Nadu, India

Subhrodeep Saha
School of Bio Sciences and Technology
Vellore Institute of Technology
Vellore, Tamil Nadu, India

Soumyadipto Santra
School of Bio Sciences and Technology
Vellore Institute of Technology
Vellore, Tamil Nadu, India

Avipsha Sarkar
School of Bio Sciences and Technology
Vellore Institute of Technology
Vellore, Tamil Nadu, India

Shampa Sen
School of Bio Sciences and Technology
Vellore Institute of Technology
Vellore, Tamil Nadu, India

Sourish Sen
School of Bio Sciences and Technology
Vellore Institute of Technology
Vellore, Tamil Nadu, India

Sombuddha Sengupta
School of Bio Sciences and Technology
Vellore Institute of Technology
Vellore, Tamil Nadu, India

Aditya Shah
School of Bio Sciences and Technology
Vellore Institute of Technology
Vellore, Tamil Nadu, India

Sehaj Sharma
GE Healthcare
Bangalore, India

Bharti Singh
Abbott
California

Shakti Singh
Los Angeles Biomedical Research Institute Harbor-UCLA Medical Center
Torrence, California

Riddhi Srivastava
Department of Biotechnology
School of Bio Sciences and Technology
Vellore Institute of Technology
Vellore, Tamil Nadu, India

Sudharsana Sundarrajan
School of Bio Sciences and Technology
Vellore Institute of Technology
Vellore, Tamil Nadu, India

Gaurav K. Verma
EBS University of Business and Law
Wiesbaden, Germany

S. Vidhya
Department of Sensor and Biomedical Technology
School of Electronics Engineering
Vellore Institute of Technology
Vellore, Tamil Nadu, India

Mohd Zafar
Department of Applied Biotechnology
Sur College of Applied Sciences
Ministry of Higher Education
Sultanate of Oman

1 Machine Learning

A Powerful Tool for Biologists

Mohd Zafar, Ramkumar Lakshmi Narayanan, Saroj K. Meher, and Shishir K. Behera

CONTENTS

1.1 MACHINE LEARNING (ML): AN OVERVIEW

1.1.1 What is ML?

ML is a method of data analysis dealing with the construction and evaluation of algorithms. It is the science that gives computers and computing machines the ability to act without being explicitly being programmed. It is defined by the ability to choose effective features for pattern recognition, classification, and prediction based on the models derived from existing data (Tarca et al. 2007). A suitably programmed computing machine is required to perform the aforementioned tasks with the help of an efficient data analysis algorithm in such a way that the classifier itself is highly mechanized without the involvement of human input.

1.1.2 ML versus Other Computing Environments

Data mining is an advanced technique used for data exploration in these domains and is broadly categorized into descriptive and predictive approaches. Under the descriptive approaches, the interesting patterns (i.e., relations) in the dataset can be identified and clustered into meaningful groups. The predictive approaches refer to supervised learning which establishes the relationship between input (independent) variables and target (dependent) variables through a structural presentation, that is, a model. Among different ML algorithms, a supervised learning algorithm includes a set of mathematical instructions which can derive a typical, high-dimensional model in the form of input-output relationships. The developed supervised learning model help in identification of mapping of input variables to output variables based on a given example of joint observations of the values of these variables (Geurts et al. 2009).

Data science compares and overlaps with many related fields such as ML, deep learning, artificial intelligence, statistics, operations research, and applied mathematics. ML is largely a hybrid field, taking its motivation and practices from both computational science and mathematics. It is the core part of both data mining and predictive analytics.

1.1.2.1 ML versus Deep Learning

The two main supervised models of ML are classification and regression. The regression model is used to predict the demand of a given product in relation to its characteristics. However, the classification model maps the input variables into predefined classes. The widely used classifiers under ML approaches are support vector machine (SVM), artificial neural network (ANN), and decision trees (DT). Deep learning is referred to as the intersection between ML and artificial intelligence. It is a subfield of ML focusing narrowly on a subset of ML tools and techniques supported by neural network inspired algorithms.

1.1.2.2 ML versus Statistics

The main focus of ML is the study and design of systems which can learn from data. On the other hand, statistical modeling represents the mathematical equation revealing the relationships between variables of a data set. ML is all about predictions, classification, regression, forecasting, clustering, and association models, whereas statistics is concerned about sample, population, and hypothesis testing. ML techniques and approaches completely rely on computing power. On the other hand, statistical techniques can be developed where computing power is not visualized as an option.

1.2 VARIOUS APPROACHES TO ML

1.2.1 Decision Tree (DT)

The application of DTs is very popular in biological data mining because of its simplicity and transparency (Table 1.1). The constructed DT is self-explanatory in nature and can efficiently explain and interpretate the output result without the help of a data mining expert.

The structure of a DT has much resemblance in real life, as well as in artificial intelligence life. It is an intersect between artificial intelligence and statistics, which can be used to build the predictive model from the observation of a system (Geurts et al. 2009). In its structure, the input and output variables are known as attributes or features, while joint observations of these values are called objects. The given example of objects, which is supposed to be used as a model, is known as a learning sample.

A DT is made up of various nodes, which can appear in the form of rooted tree having no incoming edges. Besides, the other nodes comprise exactly one incoming edge. A node with outgoing edges is called an "internal" or "test" node, whereas all other nodes are known as "leaves" (also called "terminal" and "decision" nodes) (Figure 1.1). As per the specific discrete function of the input variables, each internal node can split the case space into two or more subspaces.

TABLE 1.1
Application of DT in Biological Science

Type of DT	Study Focus	References
Genetic algorithm based DT and neural network	Prediction of benthic macroinvertebrates	D'heygere et al. (2003); D'heygere et al. (2006)
Genetic algorithm based DTs	Prediction of microRNA target genes	Behzad et al. (2015)
Wavelets and genetic algorithm	Extraction of cancer data classification using mass spectrometry	Nguyen et al. (2015)
DT	DT for binding of dipeptides to the thermally fluctuating surface of cathepsin K	Nishiyama (2016)
Genetic algorithm based DTs	Ensemble model (iACP-GAEnsC) proposed for classification of anticancer peptides	Akbar et al. (2017)
Discrete bacterial algorithm based DTs	Selection of feature in classification of microarray gene expression cancer data	Wang et al. (2017b)

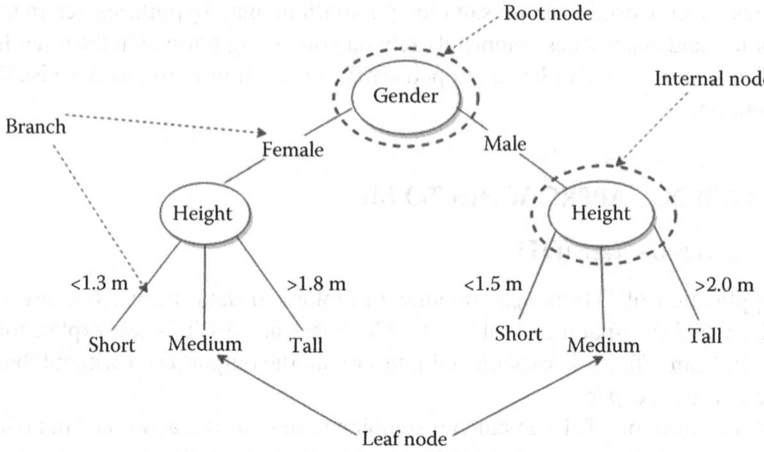

FIGURE 1.1 Typical diagram of DT.

The complexity of a DT depends on available variables in data and the total number of nodes, leaves, and tree-depth. It can be controlled by using a suitable pruning method and stopping criteria (Geurts et al. 2009).

The conventional greedy inducers such as ID_3, C4.5, and CART (Rokach and Maimon 2008) are widely used in DT, which help in construction of a DT in recursive manner, that is, divide and conquer. The greedy nature of DT induction usually fails to elucidate underlying relationships between the datasets and endpoints (output). However, using evolutionary programing, the induction of a DT is more significant than the induction by recursive portioning. It has been reported that the evolutionary programming induced DTs to exhibit a significant accuracy on previously unseen data (DeLisle and Dixon 2004). They have produced less complex classifiers with an increase in predictive accuracy of 5% to 10% over the traditional method (DeLisle and Dixon 2004). Evolutionary Algorithm (EA) is a search heuristic mimicking the process of natural biological evolution. During induction using EA, a collective learning process within the selected population of individuals is randomly initiated. Using the EAs, the problem of local minima in the search space can be overcome by randomization process (Zafar et al. 2010). Multiobjective algorithm comprises different objectives in aggregated form and combined into one objective using a fixed weight. Ant colony optimization (ACO) algorithm is another heuristic-based algorithm successfully applied to extract classification rules, and DT induction in unexplored research fields. Bacterial-based optimization algorithms with embedded weighted feature selection strategy has been developed with the aim to reduce the feature dimension in classification. The feature selection algorithm-based optimization processes possess low computational complexity and improve search ability, even in discrete optimization problems (Wang et al. 2017a).

1.2.2 Artificial Neural Networks

The inspiration for computing with ANNs came from various studies on biological neurons. The major components of a neuron are shown in Figure 1.2. They consists of

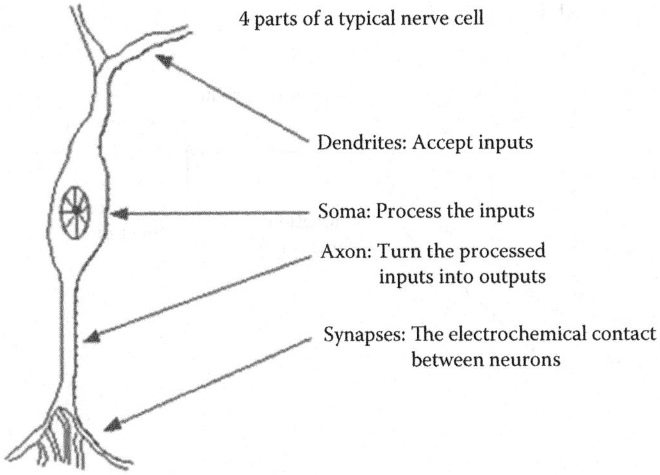
4 parts of a typical nerve cell
Dendrites: Accept inputs
Soma: Process the inputs
Axon: Turn the processed inputs into outputs
Synapses: The electrochemical contact between neurons

FIGURE 1.2 Typical biological neuron. (From Rene, E. R., "Removal of volatile organic compounds using compost biofiters." PhD Thesis, Submitted to the Dept Chem. Eng., IIT Madras, 2005.)

dendrites, soma, axon, and synapses. While dendrites are the receptive zones (input surface of the neuron), the axon is the transmission (output) line. Synapses connect the axon of one neuron to various parts of other neurons. When the input signals come in contact with these synapses, it results in local changes in the input potential in the cell body of receiving neurons, which eventually spread through the main body of the cell, and are then weighted.

ANNs possess neurons joined together by variable strength connections (synaptic weights) to form a highly interconnected information system, which helps in the modeling and control of nonlinear systems (Hornik et al. 1989).

The input signals from an artificial neuron (Figure 1.3) flow through a gain or weight called synaptic weight, whose function is analogous to that of the synaptic junction in a biological neuron. The weights can be positive or negative depending on the acceleration or inhibition of the flow of signals in a biological cell. The summing node accumulates all the input weighted signals, adds a bias signal, and then passes to the output through the activation function. The most commonly used activation function is the sigmoidal function also called the log-sigmoid or logistic function, which is defined by:

$$f(x) = \frac{1}{1+e^{-x}} \tag{1.1}$$

A generalized example of a sigmoid function with inputs ranging between −1 to +1 is shown in Figure 1.4.

The interconnections of these neurons results in an ANN, and its prime objective is to emulate the human brain and to solve scientific, engineering, and many other real life problems. Various neural network models have been developed in the

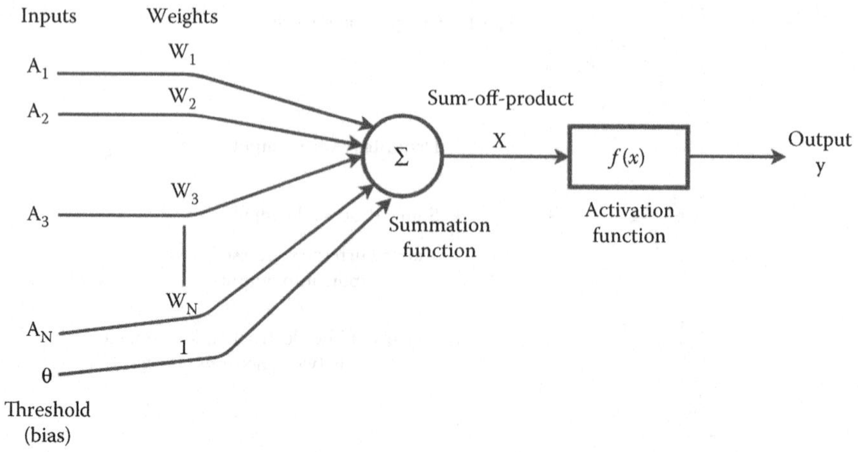

FIGURE 1.3 Schematic of an artificial neuron.

recent past due to the fuzzy understanding about the interconnections between the biological neurons. These networks are generally classified as feed forward and recurrent types. Of considerable interest is the feed forward network, where signals from one neuron to another flows only in the forward direction. Some of the common network models developed are perceptron, back propagation networks, radial bias function networks, modular neural networks, etc. (Haykin 1994). Many ANN topologies have been proposed (Widrow and Sterns 1985; Rumelhart et al. 1986; Kohonen 1988).

A feed forward multilayer perceptron (MLP) consists of two or more layers of processing elements, which are linked by weighted connections. Neural networks can be thought of as "black box" devices, which accept inputs and produces the desired outputs. The architecture of a three layered ANN with four input and two output parameters is shown in Figure 1.5. It consists of (1) an input layer receiving

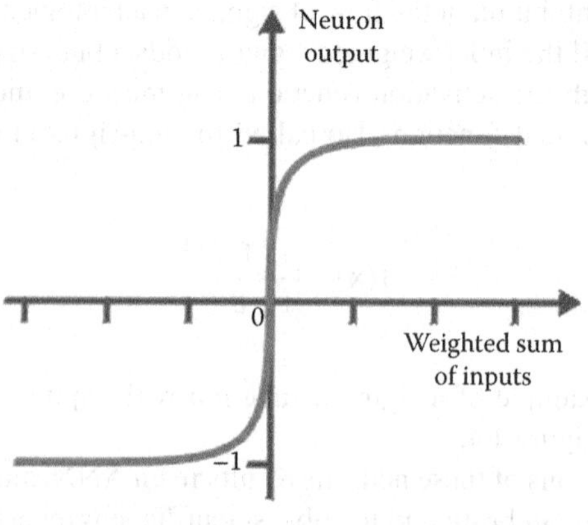

FIGURE 1.4 Sigmoid transfer function.

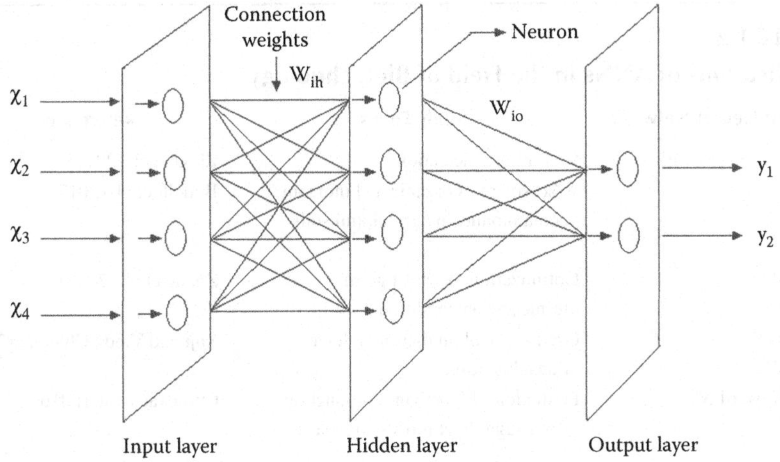

FIGURE 1.5 Basic structure of a multilayered perceptron (χ_1–χ_4: input vector; y_1, y_2: output vector).

information from external sources, and passing this information to the network for processing; (2) a hidden layer receiving information from the input layer and processing them in a hidden way; (3) an output layer receiving processed information and signaling the desired output; and (4) a bias acting on a neuron like an offset. The function of the bias is to provide a threshold for the activation of neurons. The bias input is connected to each of the hidden and output neurons in a network. The network is operated in two distinct phases called training and testing.

More commonly, ANNs have been widely studied and successfully employed in diverse applications, such as pattern recognition (Mah and Chakravarthy 1992), fault detection (Hoskins and Himmelblau 1988; Sharma et al. 2004), weather forecasting (Foster et al. 1990; Ramírez et al. 2005), and image analysis (Nattkemper 2004). In addition to this, ANNs have also been widely used for prediction purposes (Maier and Dandy 2000; Al-Shayji and Liu 2002; Zhang et al. 2002; Sözen et al. 2005).

For simplicity, and since the connection weights are the most influential variable, Rumelhart et al. (1986) proposed the back error propagation (BEP) training algorithm, which uses the gradient descent technique to adjust the weights in which the global error function, E, is minimized by modifying the weights using the Equation 1.2:

$$\Delta W_{ji} = -\eta \frac{\partial E}{\partial W_{ji}} \tag{1.2}$$

where, ΔW_{ji} is the weight increment from node i to node j; and η is the learning rate, by which the size of the step taken along the error surface is determined.

The weights are then updated by adding the delta weight, ΔW_{ji}, to the corresponding previous weight as follows:

$$W_{ji}(n+1) = W_{ji}(n) + \Delta W_{ji}(n+1) \tag{1.3}$$

TABLE 1.2
Applications of ANNs in the Field of Biotechnology

Type of Neural Network	Study Focus	References
Deep neural network	Cell mitosis detection	Zhou et al. (2017)
ANN	Assessment of organic and inorganic contaminants in agricultural soils	Bonelli et al. (2017)
ANN-GA	Optimization of wheat germ fermentation condition	Zheng et al. (2017)
MLP NN	Graft survival prediction in liver transplantation	Raji and Vinod Chandra (2016)
Feed forward NN	Prediction of bioethanol production from sugar beet processing waste	Grahovac et al. (2016)
MLP NN	Prediction of volatile organic compound removal in bioreactors	López et al. (2017)
Back propagation NN	Prediction of methane percentage in biogas recovered from a landfill	Behera et al. (2015)

Where, $W_{ji}(n)$ = the value of a weight from node i to node j at step n (before adjustment); and $W_{ji}(n + 1)$ = the value of the weight at step (n + 1) (after adjustment). First of all, the weights between the hidden layer and the output layer are adjusted and then the weights between the hidden layer and the input layer are adjusted. The choice of the learning rate is determined by trial-and-error. Rumelhart et al. (1986) described a process that adds a momentum term (μ) to the weight adjustment. The momentum term may be considered to increase the effective step size in shallow regions of the error surface (Hassoun, 1995) and can speed up the training process. Once an adjustment is carried out, it is saved and used to modify all subsequent weight adjustments. The modified adjustment of the delta weight, ΔW_{ji}, is given in Equation 1.4

$$\Delta W_{ji} = -\eta \frac{\partial E}{\partial W_{ji}} + \mu \Delta W_{ji} \tag{1.4}$$

The above process is repeated, and once the desired learning is achieved, the weights are fixed and the neural network can be used in practice. Some typical examples of the application of neural networks for prediction purposes in the field of biotechnology are given in Table 1.2.

1.2.3 Deep Learning

Deep learning process enables the computer to build a complex concept using many small concepts (related to various biological activities, e.g., transcription, translation, replication, etc.) to solve the associated problem in the biological world. The deep learning model is based on a feedforward deep network, which is also known as an

MLP. In order to resolve the difficulties of data expression profile in the deep learning process, the desired complicated map is broken into different nested series of simple maps and represented in different layers of a model.

1.2.3.1 Deep Learning Algorithm

Deep learning is a collection of new techniques including the recent class of ML applied in biology, health care, and drug discovery. After 2012, deep learning received more attention in computational biology (Rampasek and Goldenberg 2016). Recently, the deep learning-based methods became the state-of-the-art in prediction of various regulatory roles and expression pattern identification directly from DNA sequences (Park and Kellis 2015; Angermueller et al. 2016) (Table 1.3).

1.2.4 Support Vector Machines

SVM is a relatively new supervised ML statistical approach, which can be used for both the classification and regression challenges. SVM models the situation

TABLE 1.3
Application of Deep Learning in the Field of Biology

Type of Deep Learning	Study Focus	References
Deep learning	Prediction of sequence specific DNA- and RNA-binding proteins	Alipanahi et al. (2015)
Deep learning algorithm framework deepSEA	Prediction of eQTLs and disease associated variants from large scale chromatin profile data	Zhou and Troyanskaya (2015)
Deep learning	Prediction of single-cell DNA methylation	Angermueller et al. (2016)
Deep convolutional neural network (CNNs)	Prediction of regulatory code of genome wide SNPs	Kelley et al. (2016)
Deep convolutional neural network—deepLoc	Automated classification of protein subcellular localization	Kraus et al. (2017)
Deep learning—neural network	Protein subcellular localization classification from throughput microscopy images	Pärnamaa and Parts (2017)
Gene network (GNet)—LMM	Modeling and prediction of expression QTLs	Rakitsch and Stegle (2016)
Convolutional neural network based deep learning	Identification of Autism Spectrum Disorder (ASD) patient	Heinsfeld et al. (2018)
Deep PSL using stacked auto—encoder (SAE)	Prediction of human protein subcellular localization	Wei et al. (2017)
Convolutional neural network (CNN) based deep learning	Prediction of cancer using RNA sequence data sets	Xiao et al. (2018)
Deep transfer learning	Prediction of membrane protein & 3-D structure	Wang et al. (2017b)

TABLE 1.4
Application of SVM in the Field of Biological Science

Type of SVM	Study Focus	References
SVM Kernel	Prediction of DNA- and RNA-binding protein	Shao et al. (2009)
Hybrid fuzzy support vector regression	Quantitative prediction of peptide binding affinity	Uslan and Huseyin (2016)
SVM	Analysis of protein-protein interactions related to diabetes mellitus	Vyas et al. (2016)
SVM based gDNA-Prot	Prediction of DNA binding protein	Zhang et al. (2016)
SVM with nested cross-validation method	Protein–sugar binding site prediction	Banno et al. (2017)
SVM	Cancer subtype prediction based on integrated gene expression and protein network	Hung and Chiu (2017)
SVM using wavelet denoising	Prediction subcellular localization of gram-negative bacterial proteins	Yu et al. (2017)

by creating a feature space, which is a finite dimensional vector space. The fundamental principle of SVM is to separate the feature space into two classes by finding the hyper-plane differentiating the two classes very well. SVM methods are robust to study the interaction among common variants and rare variants in a small sample family. An efficient and effective method for SVM classifier using a coarse-grained parallel genetic algorithm (CGPGA) has been developed by Chen et al. (2016). CGPGA can be widely used to jointly select the feature subset and optimize the parameters for SVM in many practical applications of biological science (Table 1.4).

1.2.5 Clustering

Cluster analysis is a data explorative technique aiming to divide a multivariate dataset into clusters (groups) based on a set of measured variables in same group, which have similar subjects. Cluster analysis played an important tool in data analysis in a wide variety of fields including medicine, pharmaceutical science, biological science, statistics, and computer and information science. The classical k-means algorithm have efficiency in clustering large data sets and is used extensively in many applications (Milone et al. 2013; Wang et al. 2014). Several heuristic methods have been used to improve the accuracy and efficiency of the k-mean clustering algorithm for clustering the biological data (Wang et al. 2014) (Table 1.5).

1.2.6 Bayesian Networks

Bayesian networks (BNs) are probabilistic graphical distributions of a set of data with a possible mutual causal relationship also known as belief networks. This is one of the most popular supervised ML approaches based on the probabilistic classifier.

TABLE 1.5
Application of Clustering in the Field of Biological Science

Type of Clustering	Study Focus	References
Kohonen self organizing maps (SOP)	Clustering biological data: topology preservation in nonlinear dimension reduction	Milone et al. (2013)
Parallel clustering algorithm	Clustering large-scale biological data set	Wang et al. (2014)
Multiple kernel clustering (MKC) algorithms	Clustering with corrupted kernels data in biological applications	Li et al. (2017)
New hierarchical clustering algorithms	Prediction of coexpressed genes based on new cluster validation indices	Pagnuco et al. (2017)
Particle swarm clustering	Comparison and clustering of basal cell carcinoma (BCC), squamous cell carcinoma of the skin (SCC), and actinic keratosis (AK)	Poswar et al. (2017)

The network consists of nodes (each node in the graph represent a random variable) representing the probabilistic dependencies among the corresponding random variables (Figure 1.6). The conditional dependence between the nodes in the graph can be estimated by using other known statistical and computational methods.

Due to the complexity of learning a globally optimal network structure, vague empirical algorithms based on overlapping swarm (OS) intelligence (a modification of particle swarm optimization) have been developed to address the inference, structure-learning, and parameter estimation in BNs. The computational complexity of the

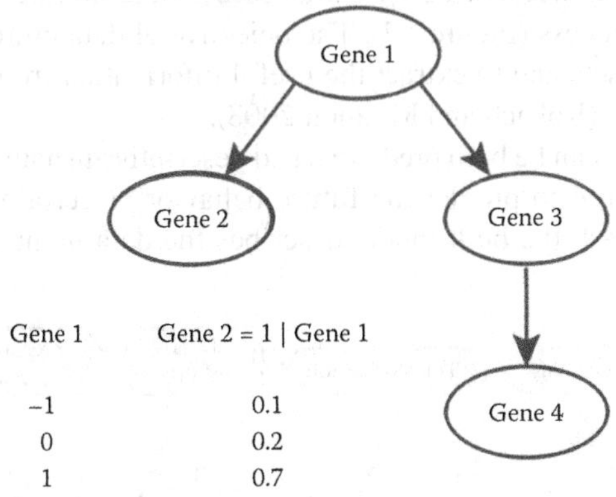

FIGURE 1.6 Baysian network example. (From Djebbari, A., Quackenbush, J., *BMC Systems Biology* 2, 2008: 57.). Here, each random variable corresponding to a gene showing three states corresponding to its transcriptional response: (i) −1 for under-expressed, (ii) 0 for unchanged, and (iii) +1 for overexpressed. The conditional probabilities for Gene 2 indicating the likelihood that Gene 2 is up-regulated given the transcriptional state of Gene 1.

TABLE 1.6
Application of BNs in the Field of Biological Science

Type of SVM	Study Focus	References
BNs-based Gaussian process regression (GPR) model	Process modeling and optimization	Yan et al. (2011)
Bayesian inference with support vector regression (SVR)	Soft sensor predictions in batch bioprocess	Yu (2012)
BNs-based Markov Chain Monto Carlo (MCMC) algorithm	Prediction of cancer related gene in acute myeloma leukemia cases	Wang et al. (2015)
Standard clustering techniques: PCA—BNs	Differential gene expression and gene ontology analysis of microarray data	Lowes et al. (2017)

Bayesian classifier learning structure can increase exponentially with the increase of descriptive variables in task. This can be overcome by learning of Bayesian classifier structure with the use of sophisticated algorithms such as genetic algorithm (Table 1.6).

1.2.7 Mining Methods

Data mining is the process of scrutinizing large database in different fields to identify specific patterns and to establish relationships among datasets. This is the computational process of analyzing big data to extract the pattern and get useful information (Gullo 2015). The widely used data mining techniques are induced learning based where a model is constructed explicitly or implicitly by generalization from the training of a data set. In the knowledge discovery database (KDD) process, data follows different phases like selection, preprocessing, transformation, and mining (Rokach and Maimon 2008; Gullo 2015). Data mining is a central step in the overall KDD process (Figure 1.7). The objective of data mining is to reduce the complexity of datasets and to extract the useful information from a large dataset to its maximum extent (Rokach and Maimon 2008).

Data mining task can be both predictive and descriptive in nature. In the predictive task, a model is build to predict the future behavior or certain features, whereas in the descriptive task the built model describes the data in an efficient and easily

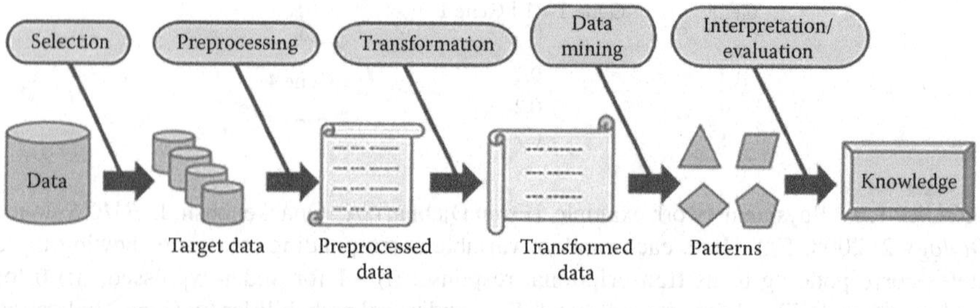

FIGURE 1.7 The KDD process. (From Gullo, F., *Physics Procedia* 62, 2015: 18–22.)

TABLE 1.7
Applications of Different Data Mining Methods in Biological Science

Data Mining Methods	Study Focus	References
ROSEFW-RF map reduce approach—random forest model	Large scale (big data) analysis of bioinformatics classification problem	Triguero et al. (2015)
DT and k-nearest neighbor algorithm	Improve the prediction accuracy of DNA variants identification in biological big data	Farid et al. (2016)
DAMIS (cloud technology) web based	Data mining classification, clustering and dimension reduction in biological datasets	Medvedev et al. (2017)
Graingenes—database for data mining	Inflow of data into meaningful metadata; curration of gene, QTLs, and genomic elements	Odell et al. (2017)
Classification, and clustering techniques	Discover valuable patterns as the process ontology	Khanbabaei et al. (2018)

understandable way. The descriptive tasks can perform the data characterization, data discrimination, association-rule discovery from data, and data clustering. The association analysis is used in computational biology for DNA microarray analysis, including monitoring of expression of thousands of genes or an entire genome (Table 1.7).

1.3 DATA MINING

Advancement of modern technology in medical science and biotechnology has acquired a great scale of biorelated data, among others. These data demand for thorough analysis in order to extract possible interesting pattern/information, which can be useful for various studies and investigations. Data mining is the computing process of extracting nonobvious and hidden information from a largely available data warehouse, which involves ML, statistical analysis, and database systems. As the name indicates, data mining is not to mine (extraction) the data itself but the discovery of knowledge from it. The apparent task here is to establish the association of huge amount of biorelated data and data mining methodologies in order to develop algorithms/models to solve complex biological problems. This section provides a general overview of some open challenges in data mining approaches motivating the development of new and effective tools, which can be used for the analysis of biorelated data sets.

Data mining is one of the important steps in the overall knowledge extraction/discovery process. The term knowledge discovery process (KDP) often gets confused with terminologies like knowledge discovery in databases, knowledge discovery, data mining, information extraction process, etc. However, researchers consider them as the synonyms and use them based on the convenient domain and its applications.

1.3.1 Knowledge Discovery Process

While executing the task of knowledge discovery for a particular decision-making problem, one has to perform some predefined steps. The number of steps usually varies based on models described by the developer, but mostly depends on the particular application. However, the present discussion is limited to some popular literature (Fayyad et al. 1996; Anand and Buchner 1998; Cabena et al. 1998; Cios et al. 2000) which has been employed successfully in several knowledge discovery tasks. All these models are often iterative in nature for certain steps initiated by a revision process, as per the requirement. In this direction Fayyad et al. (1996) proposed a primitive model consisting of nine-steps performed in succession. In this process, each step takes as input the information obtained as output from the successful completion of the preceding step. The graphical representation of the model proposed is illustrated in Figure 1.8.

In the step 1 (Figure 1.8), the KDD aims to understand the domain of application. In addition, the KDD tries to gather the prior information of the task and spot its prime objective(s). The second step performs the selection operation to figure out the most suitable data points from the available set. To some extent, variables/attributes of these points are also scrutinized in this step. In the third step, some preprocessing operations, such as cleaning and quality assessment of data points are carried out. This step also addresses the issue of missing data variables. The fourth step provides an appropriate representation of data points based on the requirement in hand. The transformation operations for data representation makes the feature reduction task more convenient and provides a feature subset, which cut down the computational burden incurred by the model. In the fifth step, the task is to make a suitable coordination of the goals of a problem at hand with the data mining technique. A complete analysis of the model is performed in the sixth step, and subsequently a hypothesis selection criterion is considered here. The most important component of the KDD process, called data mining operation, is performed in the seventh step for discovering interesting patterns of interest. The eighth step is to provide a proper

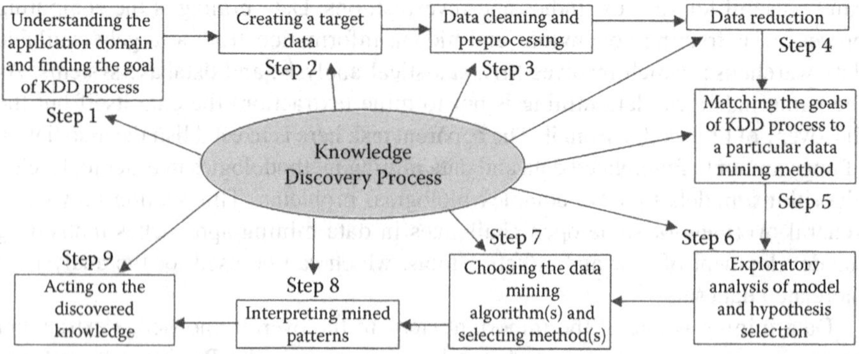

FIGURE 1.8 Schematic block diagram of the knowledge discovery process. (From Fayyad, U. et al. "*Advances in Knowledge Discovery and Data Mining.*" AAAI Press, Cambridge, 1996.)

reasoning of the discovered information from the preceding step. One may go for the visualization of extracted information for a deeper understanding and further processing. Once the discovered information is obtained in the form of knowledge, the ninth step acts on it and provides the directions for possible utilization of the information.

The above-mentioned KDD model is then revised to come up with reduced steps of operation, such as eight, six, and five steps. In this process, Anand and Buchner (1998) described an eight step-based model, Cios et al. (2000) presented a six step-based model, and Cabena et al. (1998) proposed a five step-based model. Although there are similarities in these models each of them has its advantages and disadvantages, and their suitability also depend on the application area and the purpose of their use.

1.3.2 Classification

Classification is one of the most important decision-making tools in data mining where the aim is to craft some rules in order to get a "nearly approximated answer" for a particular question. The answer may be a decision, predicting a behavior, the outcome of a process, or an information, which can act as an input to the next step of reasoning. The term "nearly approximated answer" indicates that the classification method normally finds a near optimal (based on the strength/efficiency of the rules) solution—out of many—for a particular problem.

At the beginning, the developer has to collect some prior information in the form of training samples, which should have a set of predefined meaningful attributes, including the most appropriate outcomes normally called class label/ground truth. The objective is to develop a hypothesis mapping the set of input attributes to an outcome. Once the classification model is developed, it is validated with an unseen data set, which comprises the same attributes as the training data set, but without class labels. In the operational step, the hypothesis or the algorithm analyzes the input variables and yields its prediction. The accuracy level of this prediction tells the goodness of the algorithm.

The principle of classification techniques is used for various activities. For example, financial forecasting in stock market might use a classification model to decide which stock is good to buy or sell. Medical practitioners might use this technique to get a solution for health issues. Although this approach is very convenient and easy to realize, it poses several challenges for many real-time problems. The model demands for a large number of authenticate training samples to estimate the model parameters accurately, and they should contain precise information with less ambiguity, uncertainties, and redundancy. Additionally, most of the real-time data do not have label information, which restrict their use for classification models. This particular challenge can be handled by clustering algorithms, as discussed in the following section.

1.3.3 Clustering

Clustering is the method of organizing a set of samples and identifying its subsets in such a way samples in a subset are similar to each other and dissimilar with the samples of other subsets. Clustering is an *unsupervised learning* phenomenon, where its goal is to discover a *structure* in the set of data points without class labels. In other words, the objective here is to partition the whole data set into nonoverlapping groups with the samples of similar behaviors. Let us understand this process with an example. Consider a deck of 52 numbers of *playing cards and the task is to group them using the clustering method. This particular problem can have many solutions because we can group the whole deck into many ways with similarly behaved cards. For example, a group of two (color-based), or four (shape-based), or thirteen (number-based) cards. Again, among each of these groups there may be several possible solutions. For example, in a two group-based solution, one may have groups with color and another with image. These possibilities lead to another task of finding the optimum one. These issues end up indicating that this task, although looking simple, is the most difficult one.* It can be proved that there exists no absolute "best" criterion, which would be independent of the final aim of clustering. Therefore, the developer—in concurrence with the user—has to provide a criterion of interest so that the result of the clustering will suit their needs. These criteria are broadly categorized into four types, such as connectivity-based (hierarchical clustering), centroid-based, distribution-based, and density-based clustering.

1.3.4 Regression

Regression analysis is a statistical technique for predictive modeling, which investigates and discovers the mapping between input and output variables. Input and output variables are termed as independent (or predictor) and dependent (or target variables), respectively. In a regression analysis, the number of dependent variables is normally less than the independent variables. This analysis is useful for naturally occurring variables rather than the manipulated variables obtained through an experiment. In addition, regression analysis is not suitable for investigating the causal relationships of variables. For that reason, it is logical to state that a variable "I predicts J," but not to state "I causes J." Regression analysis is mostly used for time series modeling, forecasting, and finding the appropriate relationship between the variables. With this method, one can handle the task of curve fitting, where the requirement is to estimate the occurrence of data points so that the predicted curve goes along with data points. Advantages of using regression analysis includes: discovering/indicating probable relationships between predictor and target variables, and estimating the level of influence of predictor variables on target variables. Similar to the classification, regression analysis also passes through two basic steps, such as training and testing. With the available training data points the regression model is developed, which is normally used to predict the dependent variables of the test data points through their independent variables.

1.3.5 Outlier Analysis

An outlier is an event/object/observation not conforming to an anticipated example or sample, and appearing to deviate noticeably from others in a data set, as if it were generated by a different mechanism. Most of the outlier analysis approaches remove the outlier with an assumption it is a noise. These types of samples do not hold on the conventional statistical explanation of an outlier as a rare object, and the outlier detection methods would unable to detect them. The existence of an outlier can have several reasons, for example, measurement or execution error, human error, natural deviations, or contamination with elements from outside the data set under consideration. Of late, outliers are considered as bad samples in a set of good samples, which are normally due to an incorrect assumption or the result of incorrectly coded experiments. Under such situations, the outliers should be discarded from the analysis. The objective of an outlier analysis method is to detect, rectify, or reduce the influence for reasonable analysis. However, in many cases it was observed that the consequences of the outlier removal result in the loss of important hidden information. These aspects thus alert us to handle the outliers in a strategic manner.

1.4 BIG DATA

In today's technological development, there are numerous transactions of data taking place every second, which are not in a specific format. This gives rise to a huge semistructured or unstructured datset. Big data is a very large, loosely structured dataset defining traditional storage. With the features available it is possible to store biology data repositories, data and back-ups about genes, proteins, and small molecules. The basic idea is to use the big data techniques in processing of unstructured data; performing biological analytics will solve the issues faced due to traditional data mining and processing (Bhosale and Gadekar 2014).

1.4.1 Why Big Data?

The necessity of big data lies in the requirement of different applications to process large volumes of data. The institutions need to employ data-driven decision-making to provide better results. Processing and integration of large volumes of data should make it a valid data, providing both panoramic and granular views to support the decision-making. It is accomplished with the support of big data using affordable computational and storage resources (Tekiner and Keane 2013). The challenges of handling large quantities of data when combining datasets across different archives, and cross-referencing datasets with patient records, treatments, and outcomes can be handled by analysis techniques using two primary methods of processing data, namely, cloud-based computing and big data (Lakshminarayanan et al. 2013).

1.4.2 Big Data Implementation Framework

There are wide varieties of big data implementation frameworks, and examples of famous open source frameworks are Hadoop, Flink, Apache Storm, and ApacheSamza (Matei et al. 2016). Hadoop is an open source implementation framework of Google Distributed computing. It uses a MapReduce pattern for processing data and can

TABLE 1.8
Functionalities of Different Components of Hadoop Framework Used for Big Data Implementation

Project	Functionalities of the Project	References
Ambari™	A web based tool to view the Pig, Hive and MapReduce applications	http://ambari.apache.org/
Avro™	Implementation for data serialization using Python, PHP, and most popular languages	http://avro.apache.org/
Cassandra™	A NoSQL database management system to handle large amount of data across many servers	http://cassandra.apache.org/
Chukwa™	A data collection system for managing large distributed systems	http://chukwa.apache.org/
HBase™	A structured data storage for large tables supported by a scalable distributed database	http://hbase.apache.org/
Hive™	To provides data summarization and ad hoc querying in a data warehouse infrastructure	http://hive.apache.org/
Mahout™	A Scalable ML and data mining library	http://mahout.apache.org/
Pig™	Apache Pig is an extension of Hadoop and is a high-level data processing language by maintaining scalability and reliability of Hadoop framework	http://pig.apache.org/
Spark™	Spark is a flexible in-memory framework that allows it to handle batch and real-time analytic and data processing workload with Hadoop data	http://spark.apache.org/
Tez™	Tez is a low level, distributed execution framework targeted towards data processing applications	http://tez.apache.org/
ZooKeeper™	ZooKeeper is a fault tolerant group of service across Hadoop cluster for configuration management and administration providing distributed synchronization	http://zookeeper.apache.org/

handle large volumes of data (http://hadoop.apache.org). It is associated with different components such as Ambari™, Sqoop, Avro™, Cassandra™, Chukwa™, HBase™, Hive™, Mahout™, Pig™, Spark™, Tez™, and ZooKeeper™. The most widespread components are Hadoop Distributed File System (HDFS) and MapReduce (Table 1.8).

1.4.3 MapReduce

Hadoop MapReduce is the processing component of Apache Hadoop, which processes data sets in a parallel, distributed system. The operations of MapReduce include: specify computation in terms of Map and Reduce function; parallel computation across large-scale clusters of machine; handling machine failures and performance issues; and ensuring efficient communication between the nodes. The key reason to perform mapping and reducing is to accelerate the execution of a specific process through splitting the process into a number of tasks, thereby enabling parallel work (Figure 1.9).

The map execution process consists of many phases for mapping, partitioning, shuffling, sorting, and reduction (Jiang et al. 2010). During the mapping phase, the

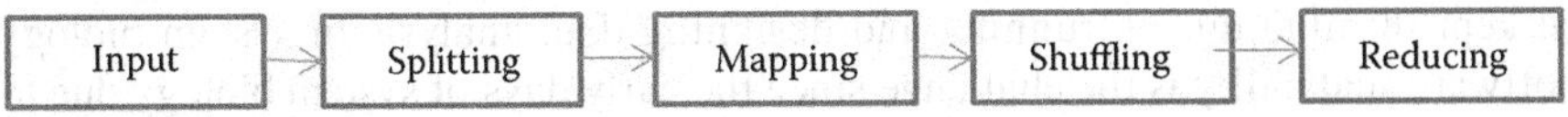

FIGURE 1.9 MapReduce operations.

assigned inputs are read from HDFS, followed by the processing of inputs into records (key/value pairs). The map function is applied to each record and the information is stored in MasterNode, at the end of the phase. In the partition phase, the reducer receiving each of the outputs is determined by each mapper. The number of partitions is equal to the number of reducers. The input data is fetched from all map tasks for the portion corresponding to the reduced task's bucket in the partition phase. In the sorting phase, the merge-sorts all map outputs into a single run. In the next phase, a user-defined reduction function is applied to the merged run and the output file is received in HDFS.

1.4.4 Big Data and Biology

Nowadays, biology is considered as "big-data science" with the recent advancement in high-throughput experimental technologies in last 20 years (Altaf-Ul-Amin et al. 2014). The hierarchy shown in Figure 1.10 summarizes the major types of molecules studied in system biology.

The significant advancement in next generation sequencing (NSG) and single-cell capture (SCC) technologies have increased interest in single-cell studies and thus enabled the generation of huge amounts (several Gigabytes) of sequencing data from a single-cell run (Yang et al. 2017). These samples of big data are generated by a number of large consortia including the Encyclopedia of DNA elements (ENCODE) project (www.encodeproject.org), the GTeX project (www.gtexportal.org), and LINCS program (www.lincsproject.org) (Dolinski and Troyanskaya 2015). Scalability and validation are the two fundamental issues, which must

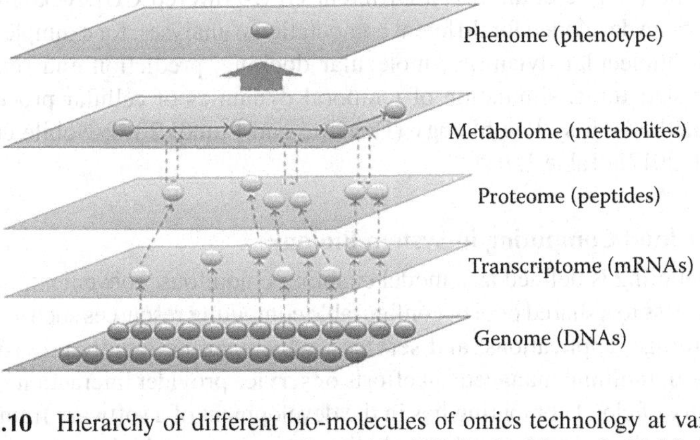

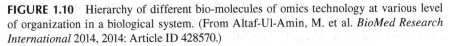

FIGURE 1.10 Hierarchy of different bio-molecules of omics technology at various level of organization in a biological system. (From Altaf-Ul-Amin, M. et al. *BioMed Research International* 2014, 2014: Article ID 428570.)

be kept in mind during running and designing data analysis in system biology. Software scalability is the challenge since the early days of system biology due to, for example, the high algorithmic complexity associated with multiple sequence alignment (Yang et al. 2017). Several efforts towards making a suitable computing environment for biological big data have been made with the development of new hardware technologies such as cluster computing, grid computing, graphical processing unit (GPU) technology, and cloud computing (Altaf-Ul-amin et al. 2014; Yang et al. 2017).

1.4.4.1 Cluster Computing in System Biology

Cluster computing is an early attempt to scaling the biological big data by networking individual computers into clusters through formation of a parallelized distributed-memory machine. For developing cluster-based software environments for big data analysis in system biology, communications protocols and software tools such as message passing interface (MPI) and parallel virtual machine (PVM) have been widely exploited (Table 1.9).

1.4.4.2 Grid Computing in System Biology

Grid computing forms a massively distributed high performance environment for handling big data in biology. Jacq et al. (2004) have reported their experience in the deployment of bioinformatic grid applications within the framework of the DataGrid project. The performed grid computing tests showed how resources (e.g., web portal, update, and distribution of biological databases) sharing improved the practices of bioinformatics (Table 1.9).

1.4.4.3 General Purpose Graphic Processing Units (GPGPUs) Computing in System Biology

GPGPUs are most popular among the scientific community as they allow more intensive investigations of biological systems with a simultaneous reduction of running time (Nobile et al. 2017). Different GPU-powered CUDA-based software tools have been developed for different computational analyses, for example, sequence alignment, molecular dynamics, molecular docking, prediction and searching of molecular structures, simulation of temporal dynamics of cellular processes, and analysis methods in system biology (Dematté and Prandi 2010; Nobile et al. 2017; Yang et al. 2017) (Table 1.9).

1.4.4.4 Cloud Computing in System Biology

Cloud computing is defined as a model enabling ubiquitous, convenient, on-demand network access to a shared pool of configurable computing resources such as networks, servers, storage, applications, and services, which can be rapidly provisioned and released with minimal management efforts or service provider interaction. The wider applicability of cloud computing lies in the development of a software framework for big data analysis in different fields including computational biology, bioinformatics, and system biology (O'Driscoll et al. 2013) (Table 1.9).

TABLE 1.9
Recent Applications of Different Computing Platform in Computational Biology

Type of Approaches	Study	References
	Cluster Computing	
MyGrid	An open source high-level middleware to support in silico experiments in biology on a Grid	Stevens et al. (2003)
MASON MPI based software using ClustalW algorithm	Performing multiple sequence alignment of a number of sequences	Ebedes and Datta (2004)
MC64-NW/SW for multicore64 Needleman–Wunsch/Smith–Waterman algorithm	Parallelizing/optimizing pairwise sequence alignment algorithm	Díaz et al. (2011)
MicroClAn framework	Evaluation of clustering algorithms for microarray data analysis	Bruno and Fiori (2013)
	Grid Computing	
GridBLAST	High-throughput implementation of BLAST based on Globus	Krishnan (2005)
Grid Technology-based European EGEE project infrastructure	Development of a high-performance pipeline for analyzing cDNA sequences	Trombetti et al. (2006)
Grid computing	Building phylogenetic or evolutionary trees for a set of DNA sequences	Joshi and Vadhiyar (2009)
	GUGUP Computing	
BarraCUDA	Sequence alignment based on Burrow-Wheeler Transform (BWT)	Klus et al. (2012)
GPU-REMuSic	Multiple sequence alignment based on Clustal with regular expression constraints	Lin and Lin (2014)
Molecular dynamics (MD) Program with myPresto/psygene algorithm	NonEwald scheme for long-range electrostatic interactions in biological macromolecules	Mashimo et al. (2013)
Nvidia GPUs—Hex spherical polar Fourier protein docking algorithm	Developing a fast protein docking tools for large scale protein–protein interactions (PPIs) networks	Ritchie and Venkatraman (2010)
	MEGADOCK	
(GPU—Katchalski-Katzir algorithm with Fast Fourier transform rigid docking scheme)	Application in protein–protein docking in large-scale interactome analysis	Ohue et al. (2014)
GPU-based CASSERT algorithm	Searching for similar 3D protein structures	Mrozek et al. (2014)
GPU—simulated annealing algorithm	Search of protein structure database for structural motif	Stivala et al. (2010)

(Continued)

TABLE 1.9 (*Continued*)
Recent Applications of Different Computing Platform in Computational Biology

Type of Approaches	Study	References
	Cloud Computing	
HadoopMapReduce programing model–cloud computing	Analysis of next-generation sequencing (NSG) data in timely and cost effective way	Shi et al. (2017)
Cloud computing with scheduling algorithm	Analysis of NSG data with reducing workflow execution time	Senturk et al. (2018)
CloudMap: A cloud-based pipeline	Mutant genome sequence analysis	Minevich et al. (2012)
Hadoop Blast (HBlast)—a parallelized BLAST algorithm	Proposal of a flexible method for virtual partitioning of databases and input query sequences	O'Driscoll et al. (2013)
Open source cloud environment with auto-scaling resources	Execution of bioinformatics and biomedical workflows	Krieger et al. (2017)

REFERENCES

Akbar, S., Hayat, M., Iqbal, M., Jan, M. A., "iACP-GAEnsC: Evolutionary genetic algorithm based ensemble classification of anticancer peptides by utilizing hybrid feature space." *Artificial Intelligence in Medicine* 79, 2017: 62–70.

Alipanahi, B., Delong, A., Weirauch, M. T., Frey, B. J., "Predicting the sequence specificities of DNA- and RNA-binding proteins by deep learning." *Nature Biotechnology* 33, 2015: 831–838.

Al-Shayji, K. A., Liu, Y. A., "Predictive modeling of large-scale commercial water desalination plants: Data-based neural network and model-based process simulation." *Industrial & Engineering Chemistry Research* 41, no. 25, 2002: 6460–6474.

Altaf-Ul-Amin, M., Afendi, F. M., Kiboi, S. K., Kanaya, S., "Systems biology in the context of big data and networks." *BioMed Research International* 2014, 2014: Article ID 428570.

Anand, S., Buchner, A., "*Decision Support Using Data Mining.*" Financial Times Pitman Publishers, London, 1998.

Angermueller, C., Lee, H., Reik, W., Stegle, O., "*Accurate prediction of single-cell DNA methylation states using deep learning.*" bioRxiv (2016) doi: 10.1101/055715.

Banno, M., Komiyama, Y., Caoa, W., Okua, Y., Ueki, K., Sumikoshi, K., Nakamura, S., Teradaa, T., Shimizu, K., "Development of a sugar-binding residue prediction system from protein sequences using support vector machine." *Computational Biology and Chemistry* 66, 2017: 36–43.

Behera, S. K., Meher, S. K., Park, H-S., "Artificial neural network model for predicting methane percentage in biogas recovered from a landfill upon injection of liquid organic waste." *Clean Technology and Environment Policy* 17, 2015: 443–453.

Behzad R-G., Rafiei, F., Niknafs, A. A., Behzad, Z., "Prediction of microRNA target genes using an efficient genetic algorithm based decision tree." *FEBS Open Bio* 5, 2015: 877–884.

Bhosale, H. S., Gadekar, D. P., "A review paper on big data and Hadoop." *International Journal of Scientific and Research Publications* 4, no. 10, 2014: ISSN 2250-3153.

Bonelli, M. G., Ferrini, M., Manni, A., "Artificial neural networks to evaluate organic and inorganic contamination in agricultural soils." *Chemosphere* 186, 2017: 124–131.

Bruno, G., Fiori, A., "MicroClAn: Microarray clustering analysis." *Journal of Parallel and Distributed Computing* 73, no. 3, 2013: 360–370.

Cabena, P., Hadjinian, P., Stadler, R., Verhees, J., Zanasi, A., "*Discovering Data Mining: From Concepts to Implementation*," Prentice-Hall Saddle River, New Jersey, 1998.

Chen, Z., Lin, T., Tang, N., Xia, X., "A parallel genetic algorithm based feature selection and parameter optimization for support vector machine." *Scientific Programming*, Article ID 2739621 2016: 1–10.

Cios, K., Teresinska, A., Konieczna, S., Potocka, J., Sharma, S., "Diagnosing myocardial perfusion from SPECT bull's-eye maps—a knowledge discovery approach." *IEEE Engineering in Medicine and Biology Magazine, special issue on Medical Data Mining and Knowledge Discovery* 19(4), 2000: 17–25.

D'heygere, T., Goethals, P. L. M., Niels, D. P., "Genetic algorithm for optimisation of predictive ecosystems models based on decision trees and neural networks." *Ecological Modelling* 195, 2006: 20–29.

DeLisle, R. K., Dixon, S. L., "Induction of Decision Trees via Evolutionary Programming." *Journal of Chemical Information and Computational Science* 44, no. 3, 2004: 862–870.

Demattè, L., Prandi, D., "GPU computing for systems biology." *Brief Bioinformatics* 11, no. 3, 2010: 323–333.

D'heygere, T., Goethals, P. L. M., De Pauw, N., "Use of genetic algorithm to select input variables in decision tree models for the prediction of benthic macro invertebrates." *Ecological Modelling* 160, 2003: 291–300.

Díaz, D., Esteban, F. J., Hernández, P., Caballero, J. A., Dorado, G., Gálvez, S., "Parallelizing and optimizing a bioinformatics pairwise sequence alignment algorithm for many-core architecture." *Parallel Computing* 37, no. 4–5, 2011: 244–259.

Djebbari, A., Quackenbush, J., "Seeded Bayesian Networks: Constructing genetic networks from microarray data." *BMC Systems Biology* 2, 2008: 57.

Dolinski, K., Troyanskaya, O. G., "Implications of big data for cell biology." *Molecular Biology of Cell* 26(14), 2015: 2575–2578.

Ebedes, J., Datta, A., "Multiple sequence alignment in parallel on a workstation cluster." *Bioinformatics* 20, no. 7, 2004: 1193–1195.

Farid, D. M., Al-Mamun, M. A., Manderick, B., Nowe, A., "An adaptive rule-based classifier for mining big biological data." *Expert Systems with Applications* 64, 2016: 305–316.

Fayyad, U., Piatesky-Shapiro, G., Smyth, P., Uthurusamy, R. (Eds.), "*Advances in Knowledge Discovery and Data Mining*." AAAI Press, Cambridge, 1996.

Foster, B., Collope, F., Ungar, L. H., "*Forecasting Using Neural Networks*." *AIChE Annual Meeting*, Chicago, USA, 1990.

Geurts, P., Irrthum, A., Wehenkel, L., "Supervised learning with decision tree-based methods in computational and systems biology." *Molecular Biosystems* 12, 2009: 1593–1605.

Grahovac, J., Jokić, A., Dodić, J., Vučurović, D., Dodić, S., "Modelling and prediction of bioethanol production from intermediates and byproduct of sugar beet processing using neural networks." *Renewable Energy* 85, 2016: 953–958.

Gullo, F., "From patterns in data to knowledge discovery: What data mining can do?" *Physics Procedia* 62, 2015: 18–22.

Hassoun, M. H., "*Fundamentals of Artificial Neural Networks*." MIT Press, Cambridge, 1995.

Haykin, S., "*Neural Networks: A Comprehensive Foundation*." Macmillan College Co., New York, 1994.

Heinsfeld, A. S., Franco, A. R., Craddock, R., Buchweitz, A., Meneguzzi, F., "Identification of autism spectrum disorder using deep learning and the ABIDE dataset." *NeuroImage: Clinical* 17, 2018: 16–23.

Hornik, K., Stinchcombe, M., White, H., "Multilayer feedforward networks are universal approximators." *Neural Network* 2, 1989: 359–366.

Hoskins, J. C., Himmelblau, D. M., "Artificial neural network models of knowledge representation in chemical engineering." *Computers & Chemical Engineering* 12, 1988: 881–890.

Hung, F-H., Chiu, H-W., "Cancer subtype prediction from a pathway-level perspective by using a support vector machine based on integrated gene expression and protein network." *Computer Methods and Programs in Biomedicine* 141, 2017: 27–34.

Jacq, N., Blanchet, C., Combet, C., Cornillot, E., Duret, L., Kurata, K., Nakamura, H., Silvestre, T., Breton, V., "Grid as a bioinformatic tool." *Parallel Computing* 30, no. 9–10, 2004: 1093–1107.

Jiang, D., Ooi, B. C., Shi, L., Wu, S., "The performance of MapReduce: An in-depth study." *Proceedings of the VLDB Endowment* 3, no. 1–2, 2010: 472–483.

Joshi, Y., Vadhiyar, S., "Analysis of DNA sequence transformations on grids." *Journal of Parallel and Distributed Computing* 69, no. 1, 2009: 80–90.

Kelley, D. R., Snoek, J., Rinn, J., "Basset: Learning the regulatory code of the accessible genome with deep convolutional neural networks." *Genome Research* 26, no. 7, 2016: 990–999.

Khanbabaei, M., Sobhani, F. M., Alborzi, M., Radfar, R., "Developing an integrated framework for using data mining techniques and ontology concepts for process improvement." *Journal of Systems and Software* 137, 2018: 78–95.

Klus, P., Lam, S., Lyberg, D., et al., "BarraCUDA—a fast short read sequence aligner using graphics processing units." *BMC Research Notes* 5, 2012: 27.

Kohonen, T., "*Self-Organization and Associative Memory.*" Springer-Verlag, Berlin, 1988.

Kraus, O. Z., Grys, B. T., Ba, J., Chong, Y., Frey, B. J., Boone, C., Andrews, B. J., "Automated analysis of high-content microscopy data with deep learning." *Molecular System Biology* 13, no. 4, 2017: 924.

Krieger, M. T., Torreno, O., Trelles, O., Kranzlmüller, D., "Building an open source cloud environment with auto-scaling resources for executing bioinformatics and biomedical workflows." *Future Generation Computer Systems* 67, 2017: 329–340.

Krishnan, A., "GridBLAST: A Globus-based high-throughput implementation of BLAST in a Grid computing framework." *Concurrency and Computations Practice and Experience* 17, no. 13, 2005: 1607–1623.

Lakshminarayanan, R., Kumar, B., Raju, M., "*Cloud computing benefits for educational institutions.*" arXiv preprint arXiv:1305.2616 (2013).

Li, T., Dou, Y., Liu, X., Zhao, Y., Lv, Q., "Multiple kernel clustering with corrupted kernels." *Neurocomputing* 267, 2017: 447–454.

Lin, C. Y., Lin, Y. S.,"Efficient parallel algorithm for multiple sequence alignments with regular expression constraints on graphics processing units." *International Journal of Computational Science and Engineering* 9, no. 1, 2014: 11–20.

López, M. E., Rene, E. R., Boger, Z., Veiga, M. C., Kennes, C., "Modelling the removal of volatile pollutants under transient conditions in a two-stage bioreactor using artificial neural networks." *Journal of Hazardous Materials* 324, 2017: 100–109.

Lowes, D. A., Galley, H. F., Moura, A. P. S., Webster, N. R., "Brief iso-flurane anaesthesia affects differential gene expression, gene ontology and gene networks in rat brain." *Behavioural Brain Research* 317, 2017: 453–460.

Mah, R. S., Chakravarthy, H. V., "Pattern recognition using artificial neural networks." *Computers & Chemical Engineering* 16, no. 4, 1992: 371–377.

Maier, H. R., Dandy, G. C., "Neural networks for the prediction and forecasting of water resources variables: A review of modelling issues and applications." *Environmental Modelling & Software* 15, 2000: 101–124.

Mashimo, T., Fukunishi, Y., Kamiya, N., Takano, Y., Fukuda, I., Nakamura, H., "Molecular dynamics simulations accelerated by GPU for biological macromolecules with a non-Ewald scheme for electrostatic interactions." *Journal of Chemical Theory and Computation* 9, no. 12, 2013: 5599–609.

Matei, Z., Xin, R. S., Wendell, P., Das, T., Armbrust, M., Dave, A., Meng, X. et al., "Apache Spark: A unified engine for big data processing." *Communications of the ACM* 59.11, 2016: 56–65, 2016.

Medvedev, V., Kurasova, O., Bernatavičienė, J., Treigys, P., Marcinkevičius, V., Dzemyda, G., "A new web-based solution for modelling data mining processes." *Simulation Modelling Practice and Theory* 76, 2017: 34–46.

Milone, D. H., Stegmayer, G., Kamenetzky, L., López, M., Carrari, F., "Clustering biological data with SOMs: On topology preservation in non-linear dimensional reduction." *Expert Systems with Applications* 40, no. 9, 2013: 3841–3845.

Minevich, G., Park, S. D., Blankenberg, D., Poole, J. R., Hobert, O., "CloudMap: A cloud-based pipeline for analysis of mutant genome sequences." *Genetics* 192, no. 4, 2012: 1249–1269.

Mrozek, D., Brozek, M., Małysiak-Mrozek, B., "Parallel implementation of 3D protein structure similarity searches using a GPU and the CUDA." *Journal Molecular Modeling* 20, no. 2, 2014: 1–17.

Nattkemper, T. W., "Multivariate image analysis in biomedicine." *Journal of Biomedical Informatics* 37, no. 5, 2004: 380–391.

Nguyen, T., Nahavandi, S., Creighton, D., Khosravi, A., "Mass spectrometry cancer data classification usingwavelets and geneticalgorithm." *FEBS Letters* 589, 2015: 3879–3886.

Nishiyama, K., "Decisiontree for the binding of dipeptides to the thermally fluctuating surface of cathepsin K." *Chemical Physics Letters* 647, 2016: 42–45.

Nobile, M. S., Cazzaniga, P., Tangherloni, A., Besozzi, D., "Graphics processing units in bioinformatics, computational biology and systems biology." *Brief Bioinformatics* 18, no. 1, 2017: 870–885.

Odell, S. G., Lazo, G. R., Woodhouse, M. R., Hane, D. L., Sen, T. Z., "The art of curation at a biological database: Principles and application." *Current Plant Biology* 2017, https://doi.org/10.1016/j.cpb.2017.11.001.

O'Driscoll, A., Daugelaite, J., Sleator, R. D., "Big data, Hadoop and cloud computing in genomics." *Journal of Biomedical Informatics* 46, no. 5, 2013: 774–781.

Ohue, M., Shimoda, T., Suzuki, S., Matsuzaki, Y., Ishida, T., Akiyama, Y., "MEGADOCK 4.0: An ultra–high-performance protein–protein docking software for heterogeneous supercomputers." *Bioinformatics* 30, no. 22, 2014: 3281–3283.

Pagnuco, I. A., Pastore, J. I., Abras, G., Brun, M., Ballarin, V. L., "Analysis of genetic association using hierarchical clustering and cluster validation indices." *Genomics* 109, no. 5–6, 2017: 438–445.

Park, Y., Kellis, M., "Deep learning for regulatory genomics." *Nature Biotechnology* 33, 2015: 825–826.

Pärnamaa, T., Parts, L., "Accurate classification of protein subcellular localization from high throughput microscopy images using deep learning." *G3 (Bethesda)* 7, no. 5, 2017: 1385–1392.

Poswar, F. O., Santos, L. I., Farias, L. C., Guimarães, T. A., Santos, S. H. S., Jones, K. M., Paula, A. M. B., et al., "An adaptation of particle swarm clustering applied in basal cell carcinoma, squamous cell carcinoma of the skin and actinic keratosis." *Meta Gene* 12, 2017: 72–77.

Raji, C. G., Vinod Chandra, S. S., "Graft survival prediction in liver transplantation using artificial neural network models." *Journal of Computational Science* 16, 2016: 72–78.

Rakitsch, B., Stegle, O., "Modelling local gene networks increases power to detect trans-acting genetic effects on gene expression." *Genome Biology* 17, 2016: 33.

Ramírez, M. C. V., Velho, H. F. de C., Ferreira, N. J., "Artificial neural network technique for rainfall forecasting applied to the São Paulo region." *Journal of Hydrology* 301, no. 1–4, 2005: 146–162.

Rampasek, L., Goldenberg, A., "TensorFlow: Biology's Gateway to Deep Learning?" *Cell Systems* 2, no. 1, 2016: 12–14.

Rene, E. R., "Removal of volatile organic compounds using compost biofiters." Ph.D Thesis, Submitted to the Dept Chem. Eng., IIT Madras, 2005.

Ritchie, D. W., Venkatraman, V., "Ultra-fast FFT protein docking on graphics processors." *Bioinformatics* 26, no. 19, 2010: 2398–2405.

Rokach, L., Maimon, O., "Data Mining with Decision Trees: Theory and Applications." In: Bunke, H. and Wang, P. S. P. (Eds.), *Series: Machine Perception and Artificial Intelligence*. Vol. 69. World Scientific Press, Singapore, 2008.

Rumelhart, D. E., Hinton, G. E., Williams, R. J., "Learning Internal Representations by Error Propagation." In: Rumelhart, D. E. and McClelland, J. L. (Eds.), *Parallel Distributed Processing*. MIT Press, Cambridge, Chapter 8, 318–362, 1986.

Senturk, I. F., Balakrishnan, P., Abu-Doleh, A., Kaya, K., Malluhi, Q., Çatalyürek, Ü. V., "A resource provisioning framework for bioinformatics applications in multi-cloud environments." *Future Generation Computer Systems* 78, no. 1, 2018: 379–391.

Shao, X., Tian, Y., Wu, L., Wang, Y., Jing, L., Deng, N., "Predicting DNA- and RNA-binding proteins from sequences with kernel methods." *Journal of Theoretical Biology* 258, no. 2, 2009: 289–293.

Sharma, R., Singh, K., Singhal, D., Ghosh, R., "Neural network applications for detecting process faults in packed towers." *Chemical Engineering and Processing* 43, no. 7, 2004: 841–847.

Shi, L., Wang, Z., Yu, W., Meng, X., "A case study of tuning MapReduce for efficient Bioinformatics in the cloud." *Parallel Computing* 61, 2017: 83–95.

Sözen, A., Arcaklioglu, E., Özalp, M., Çaglar, N., "Forecasting based on neural network approach of solar potential in Turkey." *Renewable Energy* 30, no. 7, 2005: 1075–1090.

Stevens, R. D., Robinson, A. J., Goble, C. A., "myGrid: Personalised bioinformatics on the information grid." *Bioinformatics* 19, 2003: i302–i304.

Stivala, A. D., Stuckey, P. J., Wirth, A. I., "Fast and accurate protein substructure searching with simulated annealing and GPUs." *BMC Bioinformatics* 11, 2010: 446.

Tarca, A. L., Carey, V. J., Chen, X. W., Romero, R., Drăghici, S., "ML and its application in biology." *PLoS Computational Biology* 3, no. 6, 2007: e116.

Tekiner, F., Keane, J. A., "Big data framework." *Proceeding of the 2013 IEEE International Conference on Systems, Man, and Cybernetics*, (2013): 1494–1499.

Triguero, I., Río, S., López, V., Bacardit, J., Benítez, J. M., Herrera, F., "ROSEFW-RF: The winner algorithm for the ECBDL'14 big data competition: An extremely imbalanced big data bioinformatics problem." *Knowledge-Based Systems* 87, 2015: 69–79.

Trombetti, G. A., Merelli, I., Milanesi, L., "High performance cDNA sequence analysis using grid technology." *Journal of Parallel and Distributed Computing* 66, no. 12, 2006: 1482–1488.

Uslan, V., Huseyin, S., "Quantitative prediction of peptide binding affinity by using hybrid fuzzy support vector regression." *Applied Soft Computing Journal* 43, 2016: 210–221.

Vyas, R., Bapat, S., Jain, E., Karthikeyan, M., Tambe, S., Kulkarni, B. D., "Building and analysis of protein-protein interactions related to diabetes mellitus using support vector machine, biomedical text mining and network analysis." *Computational Biology and Chemistry* 65, 2016: 37–44.

Wang, H., Huang, L., Jing, R., Yang, Y., Liu, K., Li, M., Wen, Z., "Identifying oncogenes as features for clinical cancer prognosis by Bayesian nonparametric variable selection algorithm." *Chemometrics and Intelligent Laboratory Systems* 146, 2015: 464–471.

Wang, H., Jing, X., Niu, B., "A discrete bacterial algorithm for features election in classification of micro array gene expression cancer data." *Knowledge-Based Systems* 126, 2017a: 8–19.

Wang, S., Li, Z., Yu, Y., Xu, J., "Folding membrane proteins by deep transfer learning." *Cell Systems* 5, no. 3, 2017b: 202–211.

Wang, M., Zhang, W., Ding, W., Dai, D., Zhang, H., Xie, H., et al., "Parallel clustering algorithm for large-scale biological data sets." *PLOS ONE* 9, no. 4, 2014: e91315.

Wei, L., Ding, Y., Su, R., Tang, J., Zou, Q., "Prediction of human protein subcellular localization using deep learning." *Journal of Parallel and Distributed Computing*, Article in Press (2017): https://doi.org/10.1016/j.jpdc.2017.08.009

Widrow, B., Sterns, S. D., "*Adaptive Signal Processing.*" Prentice-Hall, Englewood Cliffs, New Jersey, 1985.

Xiao, Y., Wu, J., Lin, Z., Zhao, X., "A deep learning-based multi-model ensemble method for cancer prediction." *Computer Methods and Programs in Biomedicine* 153, 2018: 1–9.

Yan, W., Hu, S., Yang, Y., Gao, F., Chen, T., "Bayesian migration of Gaussian process regression for rapid process modeling and optimization." *Chemical Engineering Journal* 166, no. 3, 2011: 1095–1103.

Yang, C., Huang, Q., Li, Z., Liu, K., Hu, F., "Big Data and cloud computing: Innovation opportunities and challenges." *International Journal of Digital Earth* 10, no. 1, 2017: 13–53.

Yu, J., "A Bayesian inference based two-stage support vector regression framework for soft sensor development in batch bioprocesses." *Computers & Chemical Engineering* 41, 2012: 134–144.

Yu, B., Li, S., Chen, C., Xu, J., Qiu, W., Wu, X., Chen, R., "Prediction subcellular localization of gram-negative bacterial proteins by support vector machine using wavelet denoising and chou's pseudo amino acid composition." *Chemometrics and Intelligent Laboratory Systems* 167, 2017: 102–112.

Zafar, M., Kumar, S., Kumar, S., "Optimization of napthalene biodegradation by a genetic algorithm based response surface methodology." *Brazilian Journal of Chemical Engineering* 27, no. 1, 2010: 89–99.

Zhang, Y-P., Wuyunqiqige, Zheng, W., Liu, S., Zhao, C., "gDNA-Prot: Predict DNA-binding proteins by employing support vector machine and a novel numerical characterization of protein sequence." *Journal of Theoretical Biology* 406, 2016: 8–16.

Zhang, Q., Xang, S. X., Mittal, G. S., Ti, S., "AE-automation and emerging technologies: Prediction of performance indices and optimal parameters of rough rice drying using neural networks." *Biosystems Engineering* 83, no. 3, 2002: 281–290.

Zheng, Z-Y., Guo, X-N., Zhu, K-X., Peng, W., Zhou, H-M., "Artificial neural network – Genetic algorithm to optimize wheat germ fermentation condition: Application to the production of two anti-tumor benzoquinones." *Food Chemistry* 227, 2017: 264–270.

Zhou, Y., Mao, H., Yi, Z., "Cell mitosis detection using deep neural networks." *Knowledge-Based Systems* 137, 2017: 19–28.

Zhou, J., Troyanskaya, O. G., "Predicting effects of noncoding variants with deep learning-based sequence model." *Nature Methods* 12, 2015: 931–934.

2 Mining and Analysis of Bioprocess Data

Prerna Grover, Aditya Shah, and Shampa Sen

CONTENTS

2.1 INTRODUCTION

It is a well-known fact that the demand and production of biopharmaceuticals has been increasing exponentially over the past few decades. Since the production of the earliest biopharmaceuticals like antibiotics began on a commercial scale, scientists and engineers have been trying to record and analyze data related to process parameters in order to optimize it. Optimization of these parameters is integral to the production process, as they affect the growth of the microorganisms employed by bioprocess industries, and the quality as well as quantity of the product formed. The most primitive form of data collection consisted of machine operators manually recording data like temperature, pH, and pressure and personally noting them down. Since then, several developments have been made and now the facilities manufacturing these products are equipped with sophisticated control systems, as well as systems for collecting and archiving the data. This archived data provides a wide scope for data mining, which in turn helps in enhancing the efficiency, consistency, and performance of the production process. This technique has offered a huge benefit to manufacturing units in the terms of the ease of characterizing and analyzing the process outcome, thereby improving the process economy [1].

In this chapter, the challenges associated with storage and analysis of data obtained from biotechnological processes has been discussed with special emphasis on various software designed and mathematical models proposed in bioprocess data mining. Applications of bioprocess data mining have been envisaged as an emerging discipline here.

2.2 CHARACTERISTICS AND TYPES OF BIOPROCESS DATA

Bioprocess data can be classified based on the frequency of measurement of parameters associated with the process carried out to produce the desired product of interest. Such data can be either monitored *online* or *off-line*. Online monitoring is vital for rapidly changing reaction parameters, while off-line monitoring is done at a slower pace [2].

Bioprocess data recording and handling is done electronically at every step, from material inputs to system outputs, control activities, system properties, and environmental constraints. Bioprocess data are heterogeneous in terms of time scale and data types with respect to the production cycle of the desired product. With respect to time scale, online parameters are calculated continuously while off-line parameters are measured within defined periods (as is required for system control actions). In addition to these measurements, some of the process parameters, such as concentration of product and quality related aspects are measured at a single time point only. Heterogeneity of process data with respect to data types can be reflected by the continuous or discrete nature of the process parameters. For example, parameters such as biomass concentration, pH, and temperature are regarded as continuous, whereas others such as feed input valve settings or gas sparging are considered to be discrete or binary, where an ON or OFF state is to be determined [3] (Table 2.1).

TABLE 2.1
Examples of Online and Off-line Bioprocess data

Online Monitored Data	Off-line Monitored Data
• pH	• Nutrient concentration
• Dissolved oxygen	• Biomass concentration
• Pressure	• Protein assays
• Temperature	• Cell types
• Flow rate	• Rheology
• Foam level	
• Stirrer speed	
• Broth weight	
• Gas composition	

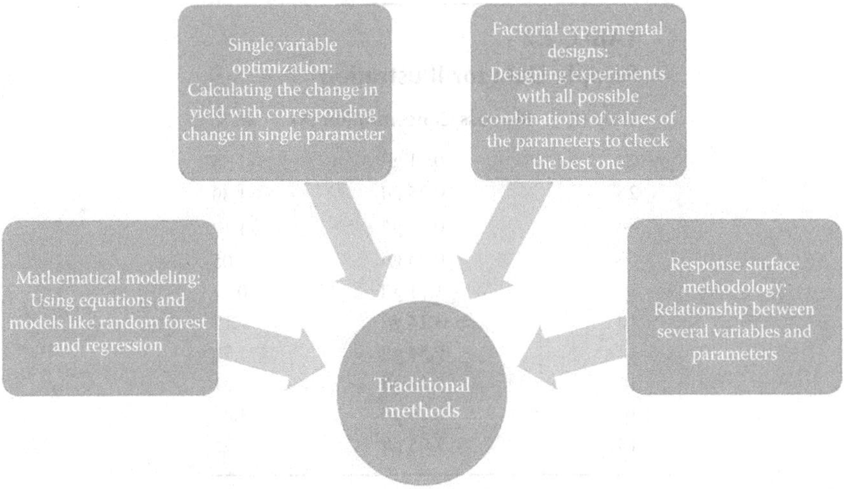

FIGURE 2.1 Traditional methods for bioprocess data analysis and optimization.

2.3 TRADITIONAL METHODS USED FOR BIOPROCESS OPTIMIZATION

Prior to the development of advanced data mining methods, four traditional methods have generally been used: mathematical modeling, single variable optimization, fractional factorial experimental designs, and response-surface methodology.

These methods have their own shortcomings like being too tedious, lacking automation, and the inability to perform advanced multivariable data analysis. Thus, with the advancements in computer science and in the field of data mining, several other methods like artificial neural networks are being experimented with and tried out for the purpose of bioprocess optimization (Figure 2.1).

2.4 ILLUSTRATION OF BIOPROCESS DATA ANALYSIS ON SMALL-SCALE MODEL

Before moving on to the actual industrial scale scenario of complex bioprocess data mining, let us first examine a small-scale example.

An aerobic fermentation of a particular species of bacteria in a simple batch mode is considered. The biomass concentration (x) in g/l for 10 different intervals of time is the input data. The data obtained is as follows (Table 2.2):

The data is analyzed to calculate the maximum specific growth rate (μ), which can then be used to calculate several other parameters. The relation between time (t) and biomass concentration (X), is given by the following equation [4]:

$$\ln X = \ln X_o + \mu t$$

TABLE 2.2
Sample Data for Illustration

Time (t)	Biomass Concentration (X)	ln X
0 s	0.21 g/l	−1.56
2 s	0.25 g/l	−1.39
4 s	0.31 g/l	−1.17
6 s	0.95 g/l	−0.05
8 s	1.71 g/l	0.54
10 s	3.24 g/l	1.17
12 s	5.54 g/l	1.71
14 s	6.12 g/l	1.81
16 s	6.24 g/l	1.83
18 s	7.05 g/l	1.95

Where X_o is the initial biomass concentration, and μ is the maximum specific growth rate.

Thus, by obtaining the equation of the line of best fit for the graph of ln X versus t, its slope may be estimated and thus, the value of μ.

Given below is a code in C++ to find out the equation of the line of best fit for any given data of time and biomass: the comment lines written in blue briefly explain the purpose of some particular lines of the code (Figure 2.2)

- Variable "n:" stores the number of data pairs;
- X[n]: an array consisting of n elements used to store the values of time intervals;
- Y[n]: an array consisting of n elements used to store the values of ln of Biomass concentrations (Figures 2.3 and 2.4).

When the code is compiled and executed, the following results are obtained in the output window (Figure 2.5).

From the output window, the equation of line of best fit is obtained:

$$Y = 0.226182x - 1.54164$$

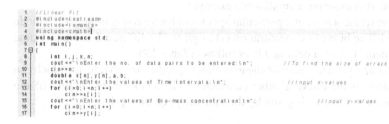

```
//Linear Fit
#include<iostream>
#include<iomanip>
#include<cmath>
using namespace std;
int main()
{
    int i,j,k,n;
    cout<<"\nEnter the no. of data pairs to be entered:\n";          //To find the size of arrays
    cin>>n;
    double x[n],y[n],a,b;
    cout<<"\nEnter the values of Time intervals:\n";                 //Input x-values
    for (i=0;i<n;i++)
        cin>>x[i];
    cout<<"\nEnter the values of Bio-mass concentration:\n";         //Input y-values
    for (i=0;i<n;i++)
        cin>>y[i];
```

FIGURE 2.2 First part of the program allotted to the input of data.

```
double xsum=0, x2sum=0, ysum=0, xysum=0;                    //variables for sums/sigma of xi, yi, xi^2, xiyi etc
for (i=0;i<n;i++)
{
    xsum=xsum+x[i];                                         //calculate sigma(xi)
    ysum=ysum+y[i];                                         //calculate sigma(yi)
    x2sum=x2sum+pow(x[i],2);                                //calculate sigma(x^2i)
    xysum=xysum+x[i]*y[i];                                  //calculate sigma(xi*yi)
}
a=(n*xysum-xsum*ysum)/(n*x2sum-xsum*xsum);                  //calculate slope
b=(x2sum*ysum-xsum*xysum)/(x2sum*n-xsum*xsum);              //calculate intercept
double y_fit[n];                                            //an array to store the new fitted values of y
for (i=0;i<n;i++)
    y_fit[i]=a*x[i]+b;                                      //to calculate y(fitted) at given x points
```

FIGURE 2.3 Portion of the program used to calculate the slope of the line and fitted values of the points.

```
cout<<"S.no"<<setw(5)<<"x"<<setw(19)<<"y(observed)"<<setw(19)<<"y(fitted)"<<endl;
cout<<"-----------------------------------------------------------------\n";
for (i=0;i<n;i++)                                           //print a table of x,y(obs.) and y(fit.)
    cout<<i+1<<"."<<setw(8)<<x[i]<<setw(15)<<y[i]<<setw(18)<<y_fit[i]<<endl;
cout<<"\nThe linear fit line is of the form:\n\n"<<a<<"x + "<<b<<endl;     //print the best fit line
return 0;
}
```

FIGURE 2.4 Display the results.

```
Enter the no. of data pairs to be entered:
10

Enter the values of Time intervals:
0
2
4
6
8
10
12
14
16
18

Enter the values of Bio-mass concentration:
-1.56
-1.39
-1.17
0.05
0.54
1.17
1.71
1.81
1.83
1.95
S.no    x         y(observed)          y(fitted)
-----------------------------------------------------------------
1.      0            -1.56              -1.54164
2.      2            -1.39              -1.08927
3.      4            -1.17             -0.636909
4.      6             0.05             -0.184545
5.      8             0.54              0.267818
6.      10            1.17              0.720182
7.      12            1.71               1.17255
8.      14            1.81               1.62491
9.      16            1.83               2.07727
10.     18             1.95               2.52964

The linear fit line is of the form:

0.226182x + -1.54164

--------------------------------
Process exited with return value 0
Press any key to continue . . .
```

FIGURE 2.5 Output window.

From the slope of the above equation, the value of the maximum specific growth rate (μ) is obtained as:

$$\mu = 0.226182$$

Thus, an important biological parameter was calculated from the basic data input using computational methods.

2.5 ANALYSIS OF BIOPROCESS DATA

The ultimate objective of bioprocess data mining is to maximize the efficiency of the process with respect to the qualitative aspects, such as product quality and product titer, as well as quantitative aspects such as product yield and efficiency of production. Analysis of the data collected from an array of process runs helps in achieving these goals.

Knowledge discovery with respect to bioprocess data helps in standardizing the process protocols and thereby helping the industries to enhance revenue generation. A typical knowledge discovery process consists of several iterative steps [5]. These steps include: *data preprocessing, feature selection*, and *data mining* (Figure 2.6).

Data preprocessing is done to convert this precursor data into feature. In the following step, the dimensions are reduced by selecting a set of data, which complies best with data mining. The data mining step involves automated or computed methods, like machine learning, to establish correlations between the parameters selected in the previous step, and for developing mathematical models to estimate the product yield.

2.5.1 Data Preprocessing

With the advent of technology in bioprocess production units, the scope of human errors and difficulties in characterization of data has been greatly reduced. This significantly reduces the need for carrying out the preprocessing techniques, such as data cleaning and denoising. However, the dynamic nature of data changing with time in culture processes, downstream processes, and fermentation poses a great deal of instigations, which need to be sought and sorted before data mining. Data preprocessing can be carried out through several approaches, two of the most common being: (1) Segmenting a temporal parameter profile into different time intervals by utilizing the triangular representation method. The evaluated first order and second order derivative of the profile for each interval deduce an increasing or decreasing trend [6,7]; and (ii) selection of temporal features with the help of wavelet decomposition [8,9]. Several other approaches including piecewise estimation of

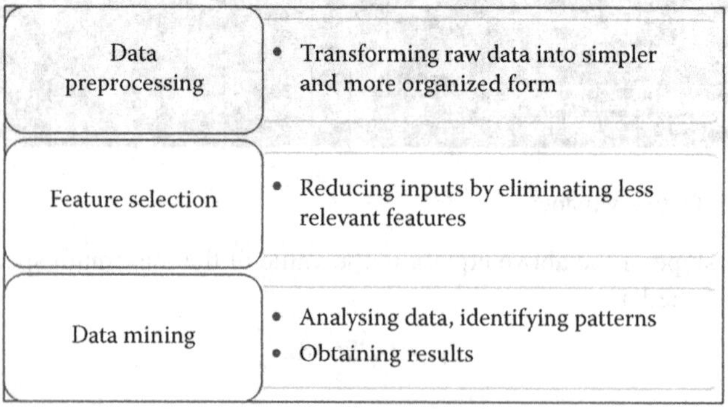

FIGURE 2.6 3 steps of data analysis.

features with the use of discrete Fourier transform methods and symbolic aggregate approximation (SAX) can be employed besides the most common ones [10,11].

Another problem arising in the preliminary stages of data analysis is the lack of alignment of the time series. This usually happens because of the presence of a lag phase or fluctuating growth rates [12,13]. This can lead to wrong interpretations and data modeling because the system avoids the possibility that the similar states do not represent the identical time points. This problem requires a preprocessing step to be rectified involving alignment of the time series of different processes [14,15]. This is done by a dynamic time wrapping strategy, which was initially developed for speech recognition [16] (Table 2.3).

2.5.2 Feature Selection

Feature selection is done in order to reduce the dimensions. In simpler terms, this stage consists of selecting features relevant to the process outcome from the set defined in data preprocessing. There are various approaches to carry out feature selection, the popular ones being wrapper and filter approaches. In the wrapper approach, the crucial parameters are defined by making a decision tree (DT), which predicts the low or high productivity of the process [8,17–19]. On the other hand, in the filter approach, the key features are selected on the basis of hypothesis testing [9,10]. All the approaches used for feature selection help in building mathematical models to predict the outcome of the process [20–22].

In addition to employing different approaches for feature selection, the sequence of events also needs to be taken care of because of the dynamic nature of process data. To prevent any redundancy, statistical methods are introduced to assess the importance of the selected features. There are chances different parameters or features from single runs, such as oxygen uptake rate and biomass growth, are often correlated. This relation may cause redundancy in the prediction of process outcomes. In order to avoid this, feature selection techniques are often used to choose from the entire feature set the features that are independent of each other. Many methods, such as nonnegative matrix factorization (NMF) and principle component analysis (PCA)

TABLE 2.3
Aspects of Data Preprocessing

Data Cleaning	Data Integration and Transformation	Data Reduction
• Fill in missing values • Smooth noisy data • Identify and remove outliers • Resolve inconsistencies	• Integrate data from multiple sources in a coherent format • Resolve data value conflicts arising due to variations in scale, units and representations	• Reduction in size of representation but retaining important information • Data compression • Discretization

are utilized to carry out this function [23]. These selected dimensions or features are further segregated into their respective groups, which belong to certain classes.

2.5.3 Data Mining

Data mining approaches can be broadly classified into *descriptive* or *predictive* approaches. Descriptive approaches look for motifs that help in carrying out data characterization, whereas predictive approaches depend upon previously observed parameters, which help in building mathematical models to predict the outcome of the process [23].

2.5.3.1 Descriptive Approaches

Descriptive methods can be carried out by either looking for patterns of interests and their temporal profile during multiple runs or data clustering.

In the first method stated, data mining is carried out by creating and executing algorithms to identify a data pattern from a large dataset. This approach assesses the pattern among features from multiple process runs. Any pattern occuring frequently is considered a pattern of interest as it renders significance to further statistical analysis [24,25]. The correlation among different features can be identified by tracking down these patterns, which further helps in framing association rules. For example, identification of patterns from multiple runs can help in determining the specific rates for particular metabolic processes.

In the second method of clustering, different runs can be clustered or grouped on the basis of similarity in the behavioral characteristics of certain features. With the help of some cluster visualization tools, such as Spot fire [26], the distinguishing features can also be determined. The element of utter importance in clustering approaches is the selection of method, which estimates the degree of similarity between two different runs on the basis of their time-dependent features. The sensitivity of this element is taken care of by carrying out the estimation of similarity in two steps. In the first step, the similarity between corresponding time dependent features of any two parallel runs is estimated [27], which is followed by the next step where the feature similarities obtained from the first step are collaborated to establish the total similarity between the process runs. The methods employed for establishing similarities between the runs, such as cosine similarity, Euclidean distance, Pearson's correlation coefficient, and information theory-based methods (e.g., mutual information) can be utilized for comparison of the same feature in multiple runs (to identify patterns), or to compare different features in a single run (to reduce the dimensions) [28].

2.5.3.2 Predictive Approaches

Predictive methods are based on assessing segregated runs and finding a valid relationship between the features and process outcomes. These established relationships, usually called models, are used to predict the process outcome and provide an easier way to describe how the predicted outcome can affect the process features. The methods employed in predictive approaches are support vector machines (SVM), DTs, regression and artificial neural networks (ANN). These methods are only helpful when the outcome has a definite discrete value. In the process runs where the outcome value falls in a continuous range, an outcome can be segregated to different classes [29].

Predictive approaches have found extensive application in the analysis of bioprocess data. For example, ANNs can be used to predict the outcomes of a fermentation process as a nonlinear function of experimental inputs [30–32]. DTs have also been used when the process outcomes can be segregated as a high productivity process or a low productivity process [33–35]. These methods are also used in association with optimization methods to help in maximizing the desired outcome [36]. Regression models based on partial least squares (PLS) have also been found to detect errors in the process, in addition to establishing a correlation between the inputs and outputs of the process [37–39].

Furthermore, the advancements in the application of predictive approaches for bioprocess data mining have been the introduction of the structural risk minimization (SRM) principle [40–43], which is extensively useful in determining the upper limit of the generalization error and kernel-based learning, mainly used in quantifying nonlinear and linear process data in conjunction with a sensitive model construction in which the features are sorted according to their significance [41]. These methods also form the foundation for SVM, which has found its application in analyzing certain data-rich concepts. [44]

2.6 SOFTWARES AND MATHEMATICAL MODELS USED

Software commonly used for data mining purposes can be either free open-source or proprietary data mining software. Free open-source data mining software offers free online access to data mining tools. These are mainly Carrot2, Chemicalize.org, ELKI, GATE, KNIME, Massive Online Analysis (MOA), MEPX, ML-Flex, MLPACK library, NLTK (Natural Language Toolkit), Open NN, Orange, R, scikit-learn, Torch, UIMA and Weka [45]. Proprietary data mining software levy subscriptions and monetary charges to its customers before handing any rights to access the data mining tools. The examples of such software packages are Angoss Knowledge STUDIO, Clarabridge, KXEN Modeler, LIONsolver, Megaputer Intelligence, Microsoft Analysis Services, Net Owl, Open Text Big Data Analytics, Oracle Data Mining, PSeven, Qlucore Omics Explorer, Rapid Miner, SAS Enterprise Miner, SPSS Modeler, STATISTICA Data Miner, Tanagra and Vertica [46–47].

The three most common approaches to model Bioprocess data are: multiple linear regression, regularized regression, and random forests.

2.6.1 Multiple Linear Regression

In this model, a combination of multiple predictor variables is used to model the response variable. The generalized equation for this model is:

$$\gamma = \beta_0 + \alpha_1\beta_1 + \alpha_2\beta_2 + \alpha_3\beta_3 + \cdots + \alpha_p\beta_p$$

where γ is the response variable, $\alpha_i(i = 1, 2, \ldots, p)$ are the predictor variables, and $\beta_i(i = 1, 2, \ldots, p)$ are their respective coefficients. β_0 represents the intercept. This equation can alternatively be represented in vector notation. It gives, $\gamma = H\theta$, where, H represents augmented predictor and θ represents parameter vector [48–49].

2.6.2 Regularized Regression

In this model, a penalty on the coefficient size is added to the error function. LASSO, standing for least absolute shrinkage and selection operator [50], is a widely accepted concept in which L_1 norm of coefficients is used to yield sparse solutions (models having multiple coefficients equal to zero).

From the generalized equation of multiple linear regression model, the L_1 norm can be depicted as:

$$\Phi = \|y - H\theta\|^2{}_2 + \lambda \|\theta\|_1$$

where $\|\theta\|_1$ represents L_1 norm and λ represents regularization parameter. The solutions for various different values of regularization parameters can be found by the use of numerous efficient existing algorithms [51].

2.6.3 Random Forests

Random forests rely upon DTs for modeling bioprocess data. DTs can fall under the category of either *classification tree* or *regression tree.* A common term used to represent both these types together is *Classification and Regression Tree* (CART). DTs show hierarchy in their structures, in which decisions can be made either on the basis of process outcome (in regression tree) or on the basis of class (in classification tree). The concept behind using DTs as a mathematical model is that the foresaid method compares values of the features with a threshold selected by the user. This returns a set of if-then rules arranged in a hierarchy. These hierarchical rules are then represented in the form of a tree, in which the comparison of each input feature value is equivalent to a node. Ultimately, the output value is depicted by the leaves [52–53].

2.7 CASE STUDIES

2.7.1 Artificial Neural Networks for Analyzing Wine Production

In wine production, the quality of wine is determined by its taste and smell and is an essential process outcome to be considered apart from the yield. In a study by Sophocles Vlassides and his team at the University of California, in 2000, they analyzed historical data obtained from wines produced in 1996. Three parameters were analyzed: grape maturity at harvest (2 levels), fermentation temperature (3 levels), and skin contact time prior to inoculation (3 levels). Data from 31 batches was analyzed. First, a sensory analysis was done with the help of 15 professional wine tasters who rated the wine on the basis of sweetness, sourness, and viscosity on a scale of 1 to 9. An ANN was then developed to obtain three outputs: fermentation kinetics, chemical characteristics, and sensory characteristics of the wine. The results showed that the ANN could predict the outcome fairly accurately in accordance with the results given by the wine tasters [54].

2.7.2 Multivariate Data Analysis for Recombinant Immunoglobulin Production

In a study published in 2012 by Huong Le et al. of the University of Minnesota, data obtained from 134 process parameters was analyzed from 243 runs of a production process. The study was based on Genentech's Vacaville manufacturing facility, which produces a recombinant immunoglobulin. For the analysis, kernel-based support vector regression and partial least square regression methods were used to predict the yield. Another important parameter analyzed was lactate consumption. The results accurately predicted the output and established the relationship between lactate consumption and yield [55].

2.8 CONCLUSION

In contrast to traditional fermentation and bioprocess plants, modern plants ensure control systems are highly sophisticated to maintain the consistency of the product and the sensitivity of the process. However, fluctuations are inevitable in a process plant. Keen understanding of the factors causing these fluctuations can significantly help in enhancing the performance and controlling the outcome of the process. With the expansion of the bioprocess industry, the archived process data is also expanding, which has an enormous scope in identifying the patterns and relationships between the process features and outcomes. The objective of bioprocess data mining is mainly to find out ways for process innovation or process advancement. Since in a bioreactor the process outcome ultimately depends on the physiology of the cells, interpretations by the field experts can greatly help in relating the patterns discovered to the outcomes. This can further be used in modulating the cellular physiology according to different process parameters.

The opportunities lying in bioprocess data mining and the benefits gained are enormous. This is the consequence of advancements in the tools of data mining, and insights into knowledge discovery of bioprocess functions, which have been present since the past decade. These advanced tools holding wide applications for exploring bioprocess data are considered to be potentially highly and invariably rewarding in the future.

REFERENCES

1. Walsh, G. "Biopharmaceutical benchmarks 2006." *Nature Biotechnology* 24, no. 7, 2006: 769–776.
2. Teixeira, A. P., R. Oliveira, P. M. Alves, and M. J. T. Carrondo. "Advances in on-line monitoring and control of mammalian cell cultures: Supporting the PAT initiative." *Biotechnology Advances* 27, no. 6, 2009: 726–732.
3. Kirchman, D., H. Ducklow, and R. Mitchell. "Estimates of bacterial growth from changes in uptake rates and biomass." *Applied and Environmental Microbiology* 44, no. 6, 1982: 1296–1307.

4. Ashoori, A., B. Moshiri, A. Khaki-Sedigh, and M. R. Bakhtiari. "Optimal control of a nonlinear fed-batch fermentation process using model predictive approach." *Journal of Process Control* 19, no. 7, 2009: 1162–1173.
5. Piatetsky-Shapiro, G. *Advances in Knowledge Discovery and Data Mining*. Edited by Usama M. F., Padhraic S., and Ramasamy U. Vol. 21. Menlo Park: AAAI press, 1996.
6. Cheung, J. T.-Y. and G. Stephanopoulos. "Representation of process trends—Part II. The problem of scale and qualitative scaling." *Computers & Chemical Engineering* 14, no. 4–5, 1990: 511–539.
7. Cheung, J. T.-Y. and G. Stephanopoulos. "Representation of process trends—Part I. A formal representation framework." *Computers & Chemical Engineering* 14, no. 4–5, 1990: 495–510.
8. Bakshi, B. R. and G. Stephanopoulos. "Representation of process trends—IV. Induction of real-time patterns from operating data for diagnosis and supervisory control." *Computers & Chemical Engineering* 18, no. 4, 1994: 303–332.
9. Bakshi, B. R. and G. Stephanopoulos. "Representation of process trends—III. Multiscale extraction of trends from process data." *Computers & Chemical Engineering* 18, no. 4, 1994: 267–302.
10. Moult, J. "Rigorous performance evaluation in protein structure modelling and implications for computational biology." *Philosophical Transactions of the Royal Society of London B: Biological Sciences* 361, no. 1467, 2006: 453–458.
11. Tompa, M., N. Li, T. L. Bailey, G. M. Church, B. De Moor, E. Eskin, and A. V. Favorov et al. "Assessing computational tools for the discovery of transcription factor binding sites." *Nature Biotechnology* 23, no. 1, 2005: 137–144.
12. Huang, J., H. Nanami, A. Kanda, H. Shimizu, and S. Shioya. "Classification of fermentation performance by multivariate analysis based on mean hypothesis testing." *Journal of Bioscience and Bioengineering* 94, no. 3, 2002: 251–257.
13. Kamimura, R. T., S. Bicciato, H. Shimizu, J. Alford, and G. Stephanopoulos. "Mining of biological data I: Identifying discriminating features via mean hypothesis testing." *Metabolic Engineering* 2, no. 3, 2000: 218–227.
14. Sakoe, H. and S. Chiba. "Dynamic programming algorithm optimization for spoken word recognition." *IEEE Transactions on Acoustics, Speech, and Signal Processing* 26, no. 1, 1978: 43–49.
15. Keogh, E., K. Chakrabarti, M. Pazzani, and S. Mehrotra. "Locally adaptive dimensionality reduction for indexing large time series databases." *ACM Sigmod Record* 30, no. 2, 2001: 151–162.
16. Keogh, E. and C. A. Ratanamahatana. "Exact indexing of dynamic time warping." *Knowledge and Information Systems* 7, no. 3, 2005: 358–386.
17. Buck, K. K. S., V. Subramanian, and D. E. Block. "Identification of critical batch operating parameters in fed-batch recombinant *E. coli* fermentations using decision tree analysis." *Biotechnology Progress* 18, no. 6, 2002: 1366–1376.
18. Coleman, M. C., K. K. S. Buck, and D. E. Block. "An integrated approach to optimization of Escherichia coli fermentations using historical data." *Biotechnology and Bioengineering* 84, no. 3, 2003: 274–285.
19. Stephanopoulos, G., G. Locher, M. J. Duff, R. Kamimura, and G. Stephanopoulos. "Fermentation database mining by pattern recognition." *Biotechnology and Bioengineering* 53, no. 5, 1997: 443–452.
20. Tai, Y. C. and T. P. Speed. "A multivariate empirical Bayes statistic for replicated microarray time course data." *The Annals of Statistics* 34, no. 5, 2006: 2387–2412.
21. Storey, J. D., W. Xiao, J. T. Leek, R. G. Tompkins, and R. W. Davis. "Significance analysis of time course microarray experiments." *Proceedings of the National Academy of Sciences of the United States of America* 102, no. 36, 2005: 12837–12842.

22. Bar-Joseph, Z., G. Gerber, I. Simon, D. K. Gifford, and T. S. Jaakkola. "Comparing the continuous representation of time-series expression profiles to identify differentially expressed genes." *Proceedings of the National Academy of Sciences* 100, no. 18, 2003: 10146–10151.
23. Mercier, S. M., B. Diepenbroek, R. H. Wijffels, and M. Streefland. "Multivariate PAT solutions for biopharmaceutical cultivation: Current progress and limitations." *Trends in Biotechnology* 32, no. 6, 2014: 329–336.
24. Kamimura, R. T., S. Bicciato, H. Shimizu, J. Alford, and G. Stephanopoulos. "Mining of biological data II: Assessing data structure and class homogeneity by cluster analysis." *Metabolic Engineering* 2, no. 3, 2000: 228–238.
25. Agrawal, R. and R. Srikant. "Fast algorithms for mining association rules." In *Proc. 20th int. conf. very large data bases, VLDB*, Vol. 1215, pp. 487–499. 1994.
26. Seno, M. and G. Karypis. "Lpminer: An algorithm for finding frequent itemsets using length-decreasing support constraint." In *Data Mining, 2001. ICDM 2001, Proceedings IEEE International Conference on*, pp. 505–512. IEEE, 2001. Ahlberg, C., 1996 Spotfire: An information exploration environment. SIGMOD Rec. 25, 25–29.
27. D'haeseleer, P. "How does gene expression clustering work?" *Nature Biotechnology* 23, no. 12, 2005: 1499–1501.
28. Lapointe, J., C. Li, J. P. Higgins, M. Van De Rijn, E. Bair, K. Montgomery, and M. Ferrari et al. "Gene expression profiling identifies clinically relevant subtypes of prostate cancer." *Proceedings of the National Academy of Sciences of the United States of America* 101, no. 3, 2004: 811–816.
29. Papa, J. P., A. X. Falcao, and C. T. N. Suzuki. "Supervised pattern classification based on optimum-path forest." *International Journal of Imaging Systems and Technology* 19, no. 2, 2009: 120–131.
30. Read, E. K., R. B. Shah, B. S. Riley, J. T. Park, K. A. Brorson, and A. S. Rathore. "Process analytical technology (PAT) for biopharmaceutical products: Part II. Concepts and applications." *Biotechnology and Bioengineering* 105, no. 2, 2010: 285–295.
31. Glassey, J., G. A. Montague, A. C. Ward, and B. V. Kara. "Enhanced supervision of recombinant E. coli fermentation via artificial neural networks." *Process Biochemistry* 29, no. 5, 1994: 387–398.
32. Glassey, J., G. A. Montague, A. C. Ward, and B. V. Kara. "Artificial neural network based experimental design procedure for enhancing fermentation development." *Biotechnology and Bioengineering* 44, no. 4, 1994: 397–405.
33. Bachinger, T., U. Riese, R. K. Eriksson, and C.-F. Mandenius. "Electronic nose for estimation of product concentration in mammalian cell cultivation." *Bioprocess and Biosystems Engineering* 23, no. 6, 2000: 637–642.
34. Vlassides, S., J. G. Ferrier, and D. E. Block. "Using historical data for bioprocess optimization: Modeling wine characteristics using artificial neural networks and archived process information." *Biotechnology and Bioengineering* 73, no. 1, 2001: 55–68.
35. Coleman, M. C. and D. E. Block. "Retrospective optimization of time-dependent fermentation control strategies using time-independent historical data." *Biotechnology and Bioengineering* 95, no. 3, 2006: 412–423.
36. Kirdar, A. O., J. S. Conner, J. Baclaski, and A. S. Rathore. "Application of multivariate analysis toward biotech processes: Case study of a cell-culture unit operation." *Biotechnology Progress* 23, no. 1, 2007: 61–67.
37. Vapnik, V. N. and V. Vapnik. *Statistical learning theory.* Vol. 1. New York: Wiley, 1998.
38. Vapnik, V. *The nature of statistical learning theory.* New York: Springer science & business media, 2013.
39. Yang, L., R. Jin, and J. Ye. "Online learning by ellipsoid method." In *Proceedings of the 26th Annual International Conference on Machine Learning*, pp. 1153–1160. ACM, 2009.

40. Weinberger, K. Q. and L. K. Saul. "Distance metric learning for large margin nearest neighbor classification." *Journal of Machine Learning Research* 10, 2009: 207–244.
41. Lanckriet, G. R. G., T. De Bie, N. Cristianini, M. I. Jordan, and W. S. Noble. "A statistical framework for genomic data fusion." *Bioinformatics* 20, no. 16, 2004: 2626–2635.
42. Lanckriet, G. R. G., N. Cristianini, P. Bartlett, L. El Ghaoui, and M. I. Jordan. "Learning the kernel matrix with semidefinite programming." *Journal of Machine learning Research* 5, 2004: 27–72.
43. Brown, M. P. S., W. N. Grundy, D. Lin, N. Cristianini, C. W. Sugnet, T. S. Furey, M. Ares, and D. Haussler. "Knowledge-based analysis of microarray gene expression data by using support vector machines." *Proceedings of the National Academy of Sciences* 97, no. 1, 2000: 262–267.
44. Charaniya, S., S. Mehra, W. Lian, K. P. Jayapal, G. Karypis, and W.-S. Hu. "Transcriptome dynamics-based operon prediction and verification in Streptomyces coelicolor." *Nucleic Acids Research* 35, no. 21, 2007: 7222–7236.
45. Tong, S. and D. Koller. "Support vector machine active learning with applications to text classification." *Journal of Machine Learning Research* 2, 2001: 45–66.
46. Tong, S. and E. Chang. "Support vector machine active learning for image retrieval." In *Proceedings of the ninth ACM international conference on Multimedia*, pp. 107–118. ACM, 2001.
47. Tibshirani, R. "Regression shrinkage and selection via the lasso." *Journal of the Royal Statistical Society. Series B (Methodological)* 58, 1996: 267–288.
48. Andersen, P. K. and L. T. Skovgaard. "Multiple regression, the linear predictor." In *Regression with Linear Predictors*, pp. 231–302. Springer, New York, 2010.
49. Miller, A. *Subset selection in regression*. CRC Press, 2002.
50. Mercier, S. M., B. Diepenbroek, M. C. F. Dalm, R. H. Wijffels, and M. Streefland. "Multivariate data analysis as a PAT tool for early bioprocess development data." *Journal of Biotechnology* 167, no. 3, 2013: 262–270.
51. Goldrick, S., R. Turner, M. Kuiper, K. Lee, R. Pradhan, and S. Farid. "*Application of multivariate data analysis in the monitoring and control of mammalian cell processes*," Paper Presented at: Cell Culture Engineering XV, Palm Springs, CA, USA, May 8–13, 2016.
52. Mercier, S. M., B. Diepenbroek, R. H. Wijffels, and M. Streefland. "Multivariate PAT solutions for biopharmaceutical cultivation: Current progress and limitations." *Trends in Biotechnology* 32, no. 6, 2014: 329–336.
53. Thibault, J., V. Van Breusegem, and A. Chéruy. "On-line prediction of fermentation variables using neural networks." *Biotechnology and Bioengineering* 36, no. 10, 1990: 1041–1048.
54. Syu, M.-J. and G. T. Tsao. "Neural network modeling of batch cell growth pattern." *Biotechnology and Bioengineering* 42, no. 3, 1993: 376–380.
55. Yang, X.-M. "Optimization of a cultivation process for recombinant protein production by Escherichia coli." *Journal of Biotechnology* 23, no. 3, 1992: 271–289.

3 Data Mining in Nutrigenomics

Avipsha Sarkar, Shreyasi Kundu, Shakti Singh, and Shampa Sen

CONTENTS

3.1 INTRODUCTION

The concept of "personalized medicine" has evolved based on the predicted pharmacogenetic responses of individuals. These response predictions were greatly enhanced by the Human Genome Project and the resultant detection of single nucleotide polymorphisms (SNPs) within human societies. Nutritional genomics provide genetic uprising, which incorporates (Subbiah 2007):

1. *Nutrigenomics*: To understand the changes occurring in the proteome and metabolome of an individual after the dietary compounds interact with the genome.
2. *Nutrigenetics*: Involves an individual's gene-based reactions to various dietary compounds—thus, it helps in the development of a variety of nutraceuticals, well suited to the genetic makeup of the person.

Nutrigenomics explores the effects of nutrients on the genome, proteome and metabolome, and nutrigenetics. Figure 3.1 below depicts a scheme to estimate the risk of a particular disease and the health condition of an individual by utilizing several nutrigenomic techniques incorporating dietary, genetic, and metabolic knowledge. On the basis of more than one interrelated dimension, a "systems" biomarker has been designed by the amalgamation of various aforesaid data.

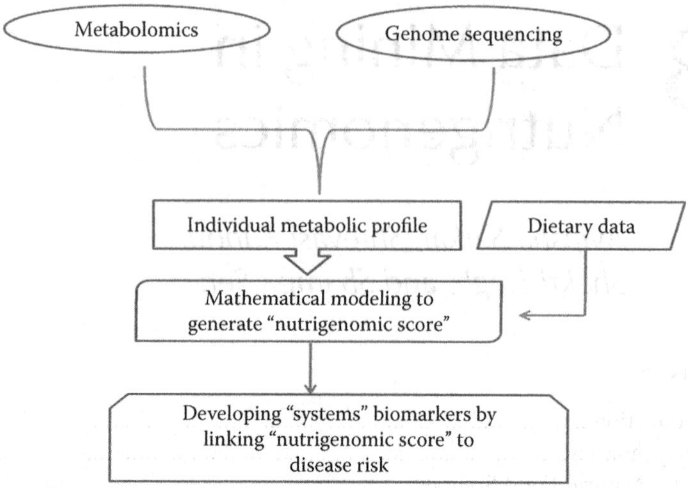

FIGURE 3.1 Nutrigenomic biomarker of health.

Nutraceuticals are the products derived from food sources having extra health benefits in addition to their basic nutritional value. They can be categorized as: dietary supplements, functional foods, medicinal foods, and pharmaceuticals. Nutrigenomics has a wide range of applications and can be used to understand the consequence of nutritional compounds on genetic constitution, function, and expression profile along with the transcriptome of an individual. Intake of nutraceutical substitutes in the form of tablets or pill capsules are also known to prevent multifactorial diseases like cancer, CVD, type II diabetes mellitus, and some monogenic disorders, for example, phenylketonuria, galactosemia, lactose intolerance, etc. Nutrigenetics and nutrigenomics are two different fields, which have two completely different approaches for understanding the genetic interaction with the dietary compounds. However, their common final aspiration is to personalize diet, understand genetic polymorphisms, and to offer potent gateways and hence augment human health (Mutch et al. 2005). Nutritional interventions like, antioxidants, vitamins (e.g., vitamin A, E, D and C, etc.), flavonoids, omega-3 fatty acids, etc., aim to prevent the pathogenesis of diabetes mellitus, metabolic syndromes, and their complications. Plant-derived food products show also positive effects on the reduction of chronic diseases due to the presence of phytochemicals. Various food materials like green tea, vitamin E, soy, vitamin D, lycopene, and selenium are taken from natural sources and are being used to alleviate human health (Cencic and Chingwaru 2010). So, inborn error metabolism can be rectified artificially by giving nutraceuticals, or personalized diet supports, which have future aspects of personalized nutrigenome medications. Data mining is the procedure of finding patterns in data to calculate a result, or to predict future outputs. Using data mining tools and techniques, huge amounts of data can be handled in a short period of time. Cluster detection, memory-based reasoning, market basket analysis, genetic algorithms, link analysis, decision trees, and neural nets are some of the powerful techniques used for data mining and evaluation purposes. Data mining is playing a major role in nutrigenomics analysis, as it helps to study the present status of nutraceuticals in the market, as well as control the market value of those medicated products and the customers' responses to them.

3.2 NUTRIGENOMICS

Nutrigenomics deal with the study of the effects of nutrients on the expression of an individual's genetic makeup. This, in turn, provides information for the biochemical functions of the body. Better nutritional advice can be provided to research groups and subgroups, or the general population by utilizing nutrigenomics (Fenech 2015). Revolutionary changes in genetic research have occurred due to the novel techniques used to study and analyze the complete genome. This has also helped in understanding the exact mechanism of diseases. Genomic research has changed into statistical evaluation of large masses of data, otherwise known as big data (Palotie and Ripatti 2017). A personalized diet or nutrition can be important for its role in a healthy lifestyle, as well as in chronic and acute diseases. Due to the polymorphism in the genetic blueprint of our body, food-processing mechanisms are different for each individual. So, nutrigenomics is a powerful area of genomic study by which a relation between dietary intake and its effect can be analyzed, making disease prevention possible. "Omics" define different disciplines like transcriptomics, proteomics, and metabolomics, studying the way these genes–proteins–metabolites interact, which is useful for personalized nutrition—nutrigenomics (Affolter et al. 2009). Analytical softwares like genomics analysis tools (CGA™ Tools) can be used to analyze human genome.

The recent interest of nutrigenomics study can be used to evaluate the genetic variations of different individuals in the laboratory. By analyzing sophisticated analytical technologies of metadata we can analyze the systemic chemical process, and can identify and quantify cellular metabolites. For example, some nutrigenomic research tools, available in the field of nutrigenomics, are the synthesis of mRNA (transcriptomics), polymerase chain reaction (PCR), geneticp (SNPs), and protein synthesis (proteomics); additionally, microarray data in the field of nutritional genomics (nutrigenomics) are available. The nutrigenomics provide an approach to integrating the various available technologies for systems biology. *OMICS* provide reports on methods and research, including bioinformatics, computational biology, metadata, data standards, data sharing, databases, biomedical informatics, biobanks, and the methods along with the statistical and algorithmic developments of cloud computing (Dove et al. 2012). Understanding the function of metabolic stress and its link to various metabolic syndromes is a key objective nutrigenomics aspires to. It can be used for preventing disease by assembling and studying metabolic stress, diabetes, and phenotypes combining inflammation and insulin stress (Afman and Müller 2006). Improvements can be made in the field of nutrition exploration with novel tools like genome editing and stem cell-based techniques (Mathers 2017).

3.2.1 Natural Food Sources

In daily life we consume different food and food products form a variety of sources that has potential nutritional value. The following substances provide nutrition

TABLE 3.1
Nutraceuticals Contain Natural Food Source

Components	Sources	Health Benifits
Carotinoids • Beta–carotene • Lycopene	Carrots, various fruits Tomatoes and tomato products	Neutralize free radicals, boost the cellular antioxidant defense
Fatty acids	Tree nuts	Reduce coronary heart disease
Phenols Caffeic acid, ferullic acid	Apples, pears, citrus fruits etc.	Enhance detoxification property and boost cellular antioxidant defense
Flavonoids	Onions, apples, tea, broccoli	Neutralize free radicals, boost cellular antioxidant defense
Phytoestrogens Isoflavones (daizein, genistein)	Soybeans and soy-based foods	Maintain bone health, healthy brain and immune responses
Probiotics Biofidobacteria, Lactobacilli	Yogurt, dairy sources	Improve gastrointestinal health and immunity
Dietary fiber Insoluble fiber	Wheat bran	Maintenance of healthy digestive tract

to our body: carbohydrates, proteins, fats, vitamins, minerals, fiber, water, etc. A healthy human being contains roughly 1% carbohydrates, 16% proteins, 16% fats, 6% minerals, 62% water, and small amounts of vitamins. Food contains different nutritious components and after consuming it complex molecules are broken down as smaller absorbable molecules, with the help of different kinds of enzymes and juices, to restore energy to our body. This energy and nutrition production mechanism is different in all individuals due to the presence of different levels of polymorphism. Dietary gene interactions are highly complex and their regulations are difficult to understand. These interactions are controlled by environmental factors, the individual's genotype, and their diet (Mead 2007). Table 3.1 represents varieties of nutraceuticals, currently available in markets (Taylor 2004), which contain natural food components of benefit to human health.

3.2.2 Nutraceuticals

Nutraceuticals is the combination of nutrients and pharmaceuticals. Nutraceuticals provide health benefits and work as a modern medicine, which help in maintaining health and act against various diseases. Nutraceuticals contain various bioactive compounds like carotenoids, phytosterols, phytoestrogens, phenols, saponins, and cyclic phenolics, which are proven to be essential in disease prevention (Pradeep and Mallikarjuna 2012). Varieties of nutraceuticals available in markets are:

- *Dietary supplements*: Contain nutrients derived from food products available in liquid and capsule form.

- *Functional foods*: Contain natural dietary supplements, or sometimes dietary supplements, which are added to obtain a functional food; for example, vitamin D added to milk.
- *Medical food*: This is specially designed for the prevention of specific diseases, or some medicinal plants able to cure certain diseases.
- *Pharmaceuticals*: This refers to the medical applications of genetically modified animals and crops.

Nutrients, dietary supplements, and herbals are major contents of nutraceuticals (Kumari et al. 2015). After understanding an individual's nutritional needs by analyzing the data according to his/her genetic materials, nutritious components can be given artificially in the form of tablets or pills, such as vitamin, calcium tablets, etc. Potential databases are now available for human gene sequencing and play a significant role in nutrigenomics. So, by detecting an individual's nutrigenomic factors, medical practitioners can prescribe nutraceuticals to prevent diseases arising due to deficiencies in food nutritional value. Personalized nutrition will be the future—designing and prescribing a diet for an individual based on their genome and their genetic variations (Pavlidis et al. 2015). Nutrigenomic data can also predict digestive enzyme production rates with the help of transcriptomics data analysis of mass protein production, as well as the proper protein folding analysis based on an individual's genomic information.

3.2.3 In Health Care

Gene function is known to play a major role in our dynamic health states and DNA has an observable trait. On a molecular level nutrients transmit signals, which can be translated into changes in gene, protein, and metabolite expression. The phenotypic appearance of monogenic disorders, like lactose intolerance or phenylketonuria, is highly influenced by an individual's diet. Therefore nutrigenomics can enhance prevention since it helps in identifying haplotype arrangements of particular mutations altering an affected person's response to diet. In case of multifactorial disorders like obesity, cancer, or cardiovascular disorders (CVD), nutrigenomics can be utilized to alter the onset—as well as development—of the disease since reports suggest they are prone to dietary intrusion (Gorduza et al. 2008). Nutrigenomics could be applied to everyday life, as the future of nutrition science offers new tools for dieticians to design and prescribe diets for each person, based on their genome and their genetic variations. The two most known genes *FTO* rs9939609 and the *MC4R* rs17782313 polymorphisms were found to be associated with type 2 diabetes which can be altered by diet. These were studied while investigating the effects of the Mediterranean diet on two polymorphisms (rs9939609 and rs17782313, respectively), where no association to diabetes type 2 was found (Ortega-Azorín et al. 2012). Genetic variation may affect appetite and calorific intake, as well as insulin signaling, inflammation, adipogenesis (formation of fat cells), and lipid metabolism. This means individual variations seen in body weight and composition are likely influenced by genetic makeup, as well as diet and activity patterns. Dietary factors can work to stabilize the genome when genetic abnormalities have occurred (Mead 2007).

3.3 DATA MINING

3.3.1 Nutrigenomics Database versus Data Mining

Mısırlı et al. in 2016 have stated that the development of a new and conventional biological system (after combining the simpler devices and parts) has become one of the most important endeavors in synthetic biology. But the procedure is not very successful since there is a shortage of completely categorized and distinct devices. The information should be made available in a compact format through novel computational advancements, making data mining more accessible. Scientists have introduced SyBiOnt, which helps in gathering information on various biological elements and their relationships with each other. SyBiOntKB is developed using SyBiOnt and also offer the inclusion of already existing biological standards and ontologies (Mısırlı et al. 2016).

The field of bioinformatics has been actively trying to overcome the key challenge of biological data amalgamation. The conventional methods for data incorporation are:

- *Data warehousing*: Where data from several databases are combined together into a single database (Goble and Stevens 2008, Balakrishnan et al. 2012, Contrino et al. 2011).
- *Federated data integration*: Where the user can make parallel queries in multiple databases, but the results will appear in an integrated format (Belleau et al. 2008, Cheung et al. 2007, Lenzerini 2002, Stein 2003, Antezana et al. 2009).

Traditional statistical techniques cannot completely analyze "big data" sets stepwise and investigate their various patterns and relationships. However, this can be done by data mining, which is also an integral tool in nutritional epidemiology. Present day health care produces huge amounts of computerized data associated with patient care, keeping all the records for various regulatory needs (Raghupathi 2016). A new handy and convenient tool has been developed offering new ways to evaluate the patients' food choices and nutritional behavior (Ngo et al. 2009). Haslam and James (2005) performed data mining in the field of nutrition among 1140 children with respect to their obesity status. They investigated the dietary patterns possibly related to obesity among them. The methodical examination of the relationship between the metabolites 3D structures and their biological activities was performed using the "KNApSAcK Metabolite Activity" database (Afendi et al. 2013, Ohtana et al. 2014).

3.3.2 Data Mining Approaches

Nutrigenomics research is gaining momentum day by day, thereby augmenting the amount of microarray data associated to this field. This calls for an immediate and well-organized infrastructure for data storage to enhance the workflow. Saito et al. (2005) developed a web-based database assimilating all the microarray publications along with expression data related to nutrigenomics. The data present in this database are linked to other databases like PubMed and other microarray databases (Saito et al. 2005).

Scientists developed a database, known as nutritional phenotype database (dbNP), to connect nutrition studies with all the information required for the study including the diet taken, genetic variation and other phenotypic information. The databases aid in describing the two axes of nutrition, namely the effects of food ingestion and exposure. The most important constituents of these axes are metabolomics, genetics, proteomics, food composition, functional assays, and intake of food, which should be adapted during research in the field. The complete removal of restrictions, with respect to information retrieval in dbNP, requires the inclusion of good quality data, disseminated networking, setting up of rigorous protocols, and annotations definite to nutrition (Ommen et al. 2010).

Scientists have developed a database known as the Soybean Genomics and Microarray Database, which aids in obtaining information about soybean genomics (Alkharouf and Matthews 2004). This database is available as an open source system and consists of all the gene expression reports, and the relationship of soybean with their cyst nematode.

The database developed by Lee et al. in 2011 comprises information about interactions between genes and the environment. It is known as GxE and aids in mining all the data related to particular interactions significant to cardiovascular disorders, nutrition, type 2 diabetes, and blood lipids (Lee et al. 2011). GxE shows the importance of nutrition-associated diseases in the field of medicine.

Woolf et al. (2011) developed vProtein, a database for the identification of essential amino acid supplements derived from plant-based nutrition. Determination of the quantity of each food necessary is the main aim of this database (Woolf et al. 2011). The database is based on the information already presented by the U.S. Department of Agriculture about various foods.

Shen et al. in 2005 have developed a specified plant microarray database for visualizing and analyzing the data statistically. It is known as BarleyBase, and is significant for both nutrigenomics as well as plant genomics.

De Santis et al. in 2015 aimed at characterizing the nutrigenomics profile of *Salmo salar*, that is, the scientists introduced nutritional stress using soybean meal derived from plant:

1. *GSEA (gene-set enrichment analysis)*: The software R function *gage* was used to analyze the sequences uniquely annotated. This software was taken from the package GAGE, that is, the generally applicable gene-set enrichment (Luo et al. 2009). The sole purpose of the analysis was the identification of the changes in mechanism, as were already suggested by the changes in coordinated expression in the sample gene sets.
2. *Hierarchical clustering*: The resemblance linking the samples were evaluated and analyzed by the R package *pvclust* (Suzuki and Shimodaira 2006) after the entire data was analyzed by hierarchical clustering.
3. *Differential expression summary*: R package *ggplot2* was used to plot all the genes having differential expression (Wickham 2016).

3.4 FUTURE PROSPECTS

One of the new fields of "omics" science research includes nutrigenomics. Present day medicine is highly dependent on nutrigenomics. In the future, the concept of nutrigenomics could be quite essential in controlling diseases related to nutrition (Wiwanitkit 2012), as data mining in nutrigenomics has a huge potential in the field of medicine. The scientific community has been focusing on the correlation of data mining and nutrigenomics to enhance the potential of this field. The utilization of genetic information for personalized dietary guidance can be applied in the field of medicine; however, it might give rise to diverse ethical and social concerns. The revelation of an individual's genetic constituents might not be desirable. Mathematical models can be designed by developing novel tools (statistical, bioinformatics), which can assimilate enormous data sets. These data sets should consider both diet and genetic outcomes in big or/and different population. Predictive tools like "decision trees" can be used for the prediction of target diseases affecting a person due to his/her diet. This can then be used by medical professionals, or nutritionists, to prevent a particular disease by altering a patient's diet. New tools should be developed to make these predictions more accurate.

REFERENCES

Afendi, F. M., N. Ono, Y. Nakamura, K. Nakamura, L. K. Darusman, N. Kibinge, and A. H. Morita et al. "Data mining methods for omics and knowledge of crude medicinal plants toward big data biology." *Computational and Structural Biotechnology Journal* 4, no. 5, 2013: 1–14.

Affolter, M., F. Raymond, and M. Kussmann. "Omics in nutrition and health research." *Nutrigenomics and Proteomics in Health and Disease: Food Factors and Gene Interactions* 2009: 11–29.

Afman, L. and M. Müller. "Nutrigenomics: From molecular nutrition to prevention of disease." *Journal of the American Dietetic Association* 106, no. 4, 2006: 569–576.

Alkharouf, N. W. and B. F. Matthews. "SGMD: The soybean genomics and microarray database." *Nucleic Acids Research* 32, no. suppl_1, 2004: D398–D400.

Antezana, E., M. Kuiper, and V. Mironov. "Biological knowledge management: The emerging role of the semantic web technologies." *Briefings in Bioinformatics* 10, no. 4, 2009: 392–407.

Balakrishnan, R., J. Park, K. Karra, B. C. Hitz, G. Binkley, E. L. Hong, J. Sullivan, G. Micklem, and J. Michael Cherry. "YeastMine—an integrated data warehouse for Saccharomyces cerevisiae data as a multipurpose tool-kit." *Database* 2012, 2012.

Belleau, F., M.-A. Nolin, N. Tourigny, P. Rigault, and J. Morissette. "Bio2RDF: Towards a mashup to build bioinformatics knowledge systems." *Journal of Biomedical Informatics* 41, no. 5, 2008: 706–716.

Cencic, A. and W. Chingwaru. "The role of functional foods, nutraceuticals, and food supplements in intestinal health." *Nutrients* 2, no. 6, 2010: 611–625.

Cheung, K.-H., A. K. Smith, K. Y. L. Yip, C. J. O. Baker, and M. B. Gerstein. "Semantic Web approach to database integration in the life sciences." *Semantic Web* 2007: 11–30.

Contrino, S., R. N. Smith, D. Butano, A. Carr, F. Hu, R. Lyne, and K. Rutherford et al. "modMine: Flexible access to modENCODE data." *Nucleic Acids Research* 40, no. D1, 2011: D1082–D1088.

De Santis, C., K. L. Bartie, R. E. Olsen, J. B. Taggart, and D. R. Tocher. "Nutrigenomic profiling of transcriptional processes affected in liver and distal intestine in response to a soybean meal-induced nutritional stress in Atlantic salmon (Salmo salar)." *Comparative Biochemistry and Physiology Part D: Genomics and Proteomics* 15, 2015: 1–11.

Dove, E. S., V. Özdemir, and Y. Joly. "Harnessing omics sciences, population databases, and open innovation models for theranostics—guided drug discovery and development." *Drug Development Research* 73, no. 7, 2012: 439–446.

Fenech, M. "Nutrigenomics and nutrigenetics: The new paradigm for optimising health and preventing disease." *Journal of Nutritional Science and Vitaminology* 61, no. Supplement, 2015: S209–S209.

Goble, C. and R. Stevens. "State of the nation in data integration for bioinformatics." *Journal of Biomedical Informatics* 41, no. 5, 2008: 687–693.

Gorduza, E. V., L. L. Indrei, and V. M. Gorduza. "Nutrigenomics in postgenomic era." *Revista medico-chirurgicala a Societatii de Medici si Naturalisti din Iasi* 112, no. 1, 2008: 152–164.

Haslam, D. W. and W. P. James. "Obesity." *Lancet* 366, 2005: 1197–1209. CrossRef| PubMed| Web of Science® Times Cited 834.

Kumari, M., J. Shashi, and S. Jagdeep. "Nutraceutical–medicine of future." *Journal of Global Biosciences* 4, no. 7, 2015: 2790–2794.

Lee, Y.-C., C.-Q. Lai, J. M. Ordovas, and L. D. Parnell. "A database of gene-environment interactions pertaining to blood lipid traits, cardiovascular disease and type 2 diabetes." *Journal of Data Mining In Genomics & Proteomics* 2, no. 1, 2011: 106.

Lenzerini, M. "Data integration: A theoretical perspective." In *Proceedings of the twenty-first ACM SIGMOD-SIGACT-SIGART symposium on Principles of database systems*, pp. 233–246. ACM, 2002.

Luo, W., M. S. Friedman, K. Shedden, K. D. Hankenson, and P. J. Woolf. "GAGE: Generally applicable gene set enrichment for pathway analysis." *BMC Bioinformatics* 10, no. 1, 2009: 161.

Mathers, J. C. Nutrigenomics in the modern era. *Proceedings of the Nutrition Society* 76, no. 3, 2017: 265–275.

Mead, M. N. "Nutrigenomics: The genome–food interface." *Environmental Health Perspectives* 115, no. 12, 2007: A582.

Mısırlı, G., J. Hallinan, M. Pocock, P. Lord, J. A. McLaughlin, H. Sauro, and A. Wipat. "Data integration and mining for synthetic biology design." *ACS Synthetic Biology* 5, no. 10, 2016: 1086–1097.

Mutch, D. M., W. Wahli, and G. Williamson. "Nutrigenomics and nutrigenetics: The emerging faces of nutrition." *The FASEB Journal* 19, no. 12, 2005: 1602–1616.

Ngo, J., A. Engelen, M. Molag, J. Roesle, P. García-Segovia, and L. Serra-Majem. "A review of the use of information and communication technologies for dietary assessment." *British Journal of Nutrition* 101, no. S2, 2009: S102–S112.

Ohtana, Y., A. A. Abdullah, Md. Altaf-Ul-Amin, M. Huang, N. Ono, T. Sato, and T. Sugiura et al. "Clustering of 3D-structure similarity based network of secondary metabolites reveals their relationships with biological activities." *Molecular Informatics* 33, no. 11–12, 2014: 790–801.

Ommen, B., J. Bouwman, L. O. Dragsted, C. A. Drevon, R. Elliott, P. Groot, and J. Kaput et al. "Challenges of molecular nutrition research 6: The nutritional phenotype database to store, share and evaluate nutritional systems biology studies." *Genes & Nutrition* 5, no. 3, 2010: 189.

Ortega-Azorín, C., D. Corella, E. Ros, E. Gómez-Gracia, E. M. Asensio, F. Arós, and G. Sáez-Tormo et al. "Associations of the FTO rs9939609 and the MC4R rs17782313 polymorphisms with type 2 diabetes are modulated by diet, being higher when adherence to the Mediterranean diet pattern is low." *Cardiovascular Diabetology* 11, no. 1, 2012: 137.

Palotie A. and Ripatti S. "Finland establishing the internet of genomics and health data." *Duodecim* 133, no. 8, 2017: 771–5.

Pavlidis, C., G. P. Patrinos, and T. Katsila. "Nutrigenomics: A controversy." *Applied & Translational Genomics* 4, 2015: 50–53.

Pradeep K. S. and R. Mallikarjuna. Phytochemicals in vegetables and their health benefits. *Asian J. Agric. Rural Dev* 2, 2012: 177–183.

Raghupathi, W. "Data mining in healthcare." *Healthcare Informatics: Improving Efficiency through Technology, Analytics, and Management* 2016: 353–372.

Saito, K., S. Arai, and H. Kato. "A nutrigenomics database–integrated repository for publications and associated microarray data in nutrigenomics research." *British Journal of Nutrition* 94, no. 4, 2005: 493–495.

Shen, L., J. Gong, R. A. Caldo, D. Nettleton, D. Cook, R. P. Wise, and J. A. Dickerson. "BarleyBase—an expression profiling database for plant genomics." *Nucleic Acids Research* 33, no. suppl_1, 2005: D614–D618.

Stein, L. D. "Integrating biological databases." *Nature Reviews Genetics* 4, no. 5, 2003: 337–345.

Subbiah, MTR. "Nutrigenetics and nutraceuticals: The next wave riding on personalized medicine." *Translational Research* 149, no. 2, 2007: 55–61.

Suzuki, R. and H. Shimodaira. "pvclust: Hierarchical clustering with P-values via multiscale bootstrap resampling." *Bioinformatics* 22, no. 12, 2006: 1540–2.

Taylor, C. L. "Regulatory frameworks for functional foods and dietary supplements." *Nutrition Reviews* 62, no. 2, 2004: 55–59.

Wickham, H. *ggplot2: elegant graphics for data analysis*. Springer, 2016.

Wiwanitkit, V. "Database and tools for nutrigenomics: A brief summary." *Journal of Medical Nutrition and Nutraceuticals* 1, no. 2, 2012: 87.

Woolf, P. J., L. L. Fu, and A. Basu. "vProtein: Identifying optimal amino acid complements from plant-based foods." *PLOS One* 6, no. 4, 2011: e18836.

4 Machine Learning in Metabolic Engineering

Sayak Mitra

CONTENTS

4.1 METABOLIC ENGINEERING: AN INTRODUCTION

With an ever-increasing demand for bioproducts in today's society, the main aim of biotechnological industries is to optimize their resources for maximum production at low cost. Most bioprocess industries employ microorganisms for production of primary metabolites (such as commercially important enzymes, ethanol, citric acid, nutritional amino acids, and many more), secondary metabolites (antibiotics, antioxidants, alkaloids, and many more), or biomass constituents (single cell protein, lipids, and others). However, living systems naturally produce a substance just in amounts required for their survival, after which, the cell employs certain mechanisms to stop the wastage of any more cellular resources in the production of the metabolite. The amount of metabolite thus obtained is obviously not sufficient for its industrial-scale production. To tackle this problem, the field of biochemical engineering has evolved, of which metabolic engineering is an inherent component.

4.1.1 Why Engineer Metabolism?

From the perspective of a biochemical engineer, the cell growing inside a bioreactor can be considered to be a factory, which takes in substrates and gives us products. This factory has its own production processes, which can be manipulated and optimized for enhanced production. These production processes are carried out using the metabolic pathways of the cell, in which the substrates get oxidized to produce energy and metabolites important for the growth and survival of the cell.

In any industry, one needs to be completely familiar with the process flow taking place, from the source of raw materials to the last product purification step being carried out. This needs extensive knowledge of the complete layout of the industry. Such information cannot be simply qualitative, but also needs to be quantitative so that the process can be optimized, thus preventing any wastage of resources. This is very similar to the cellular factory, which has a source of raw materials (uptake of substrates), a processing unit (the cellular metabolism), and a finished product dispatch unit (secretion of product), as depicted in Figure 4.1. In essence, metabolic engineering is the quantitative analysis and optimization of this factory. This entails a detailed knowledge of all the branch points and connecting nodes of the cell's metabolic network, and their influence on the flow and speed of the production process. In analogy to the various parameters affecting the production process in an industry, the metabolism of the cell is also affected by various parameters, which are expressed quantitatively by a set of control coefficients, as will be discussed later. Imbued with all this knowledge and the skills of genetically manipulating a cell, the biochemical engineer is able to reorient the metabolism of a cell to meet our specific needs.

4.1.2 Mathematical Models in Metabolic Rengineering

The metabolic pathway associated with the concerned production process is decided based upon a literature survey or from online metabolic databases, like KEGG (Gombert and Nielsen 2000; Kanehisa et al. 2017). A cell feeds on a substrate and utilizes it via metabolic pathways to give a product (secreted out of the cell) or biomass constituents (polymeric constituents present inside the cell, such as DNA, RNA, protein, and others). A metabolic pathway is, in essence, a linear/branched combination of chemical reactions. The individual compounds formed, and subsequently utilized via these reactions, are called intermediates. A general mass balance equation of all the aforementioned cellular components is given by (Stephanopoulos, Aristidou, and Nielsen 1998):

$$\sum \mathit{Substrates} = \sum \mathit{Products} + \sum \mathit{Biomass\ constituents} + \sum \mathit{Intermediates} \tag{4.1}$$

Each component of a chemical reaction has a stoichiometric coefficient. For substrates, this coefficient is −ve, as substrates are always consumed or depleted, while that for products is always +ve, as products are always formed. Biomass constituents and intermediates may have −ve or +ve stoichiometric coefficients, depending on whether they are being used up or produced in a particular reaction. To explain this more clearly, let us take only the pathway metabolite into consideration of the equation, corresponding to metabolic intermediates. If there are "i" intermediates (I) being involved in "j" reactions, the last term on the RHS of the above equation is expanded as:

$$\begin{aligned} &g_{11}I_1 + g_{12}I_2 + g_{13}I_3 + \cdots + g_{1M}I_M \\ &\quad + g_{21}I_1 + g_{22}I_2 + g_{23}I_3 + \cdots + g_{2M}I_M \\ &\quad + \cdots + g_{j1}I_1 + g_{j2}I_2 + g_{j3}I_3 + \cdots + g_{jM}I_M \end{aligned}$$

(a)

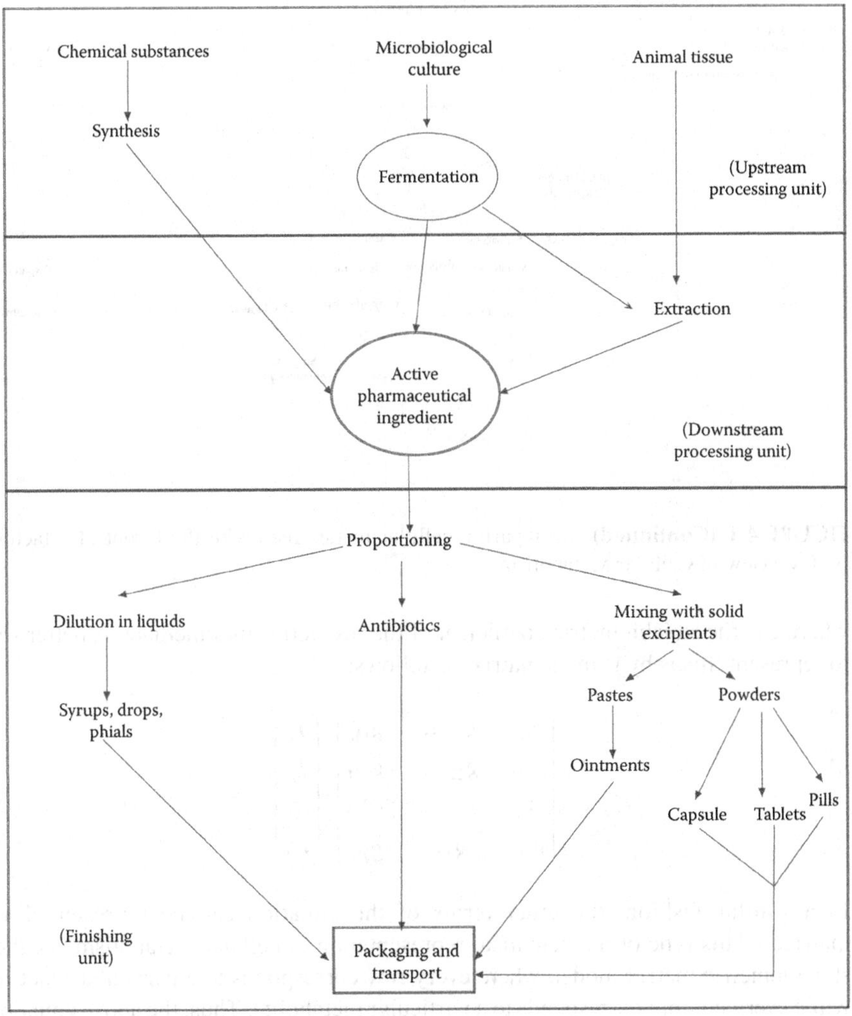

FIGURE 4.1 Comparing cellular metabolism with the layout of a factory. (a) General Layout of a Pharmaceutical Industry. *(Continued)*

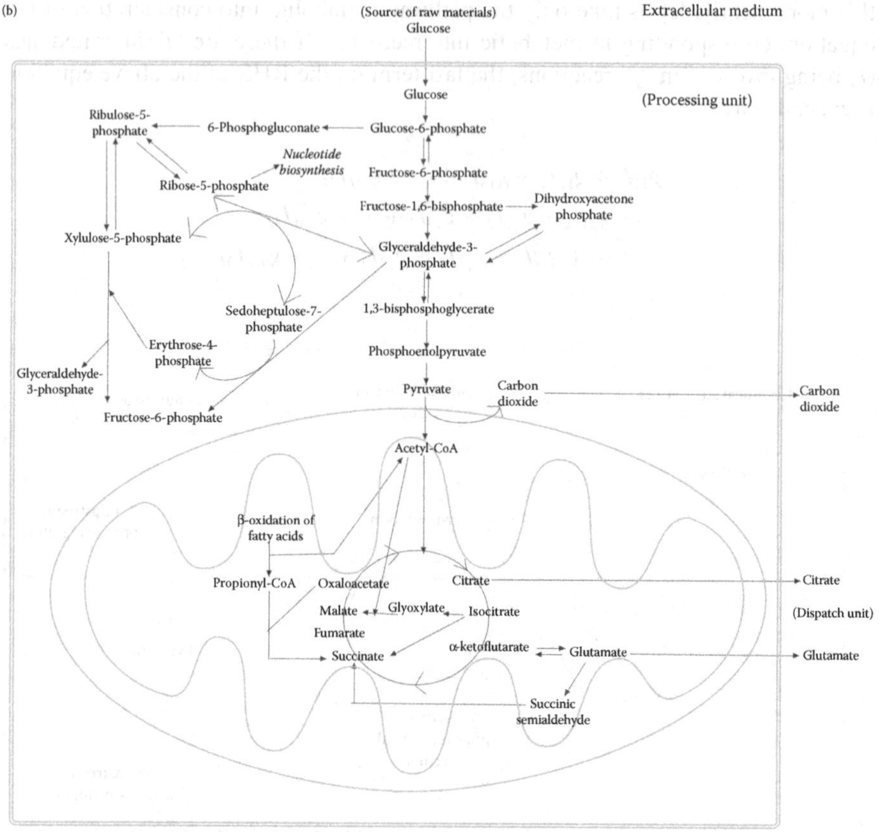

FIGURE 4.1 (Continued) Comparing cellular metabolism with the layout of a factory. (b) Overview of Cellular Metabolism.

where g is the stoichiometric coefficient of the respective intermediate. A better way to represent this is by using a matrix, as follows:

$$\begin{pmatrix} g_{11} & g_{12}\cdots & g_{1M} \\ g_{21} & g_{22}\cdots & g_{2M} \\ \vdots & \vdots & \vdots \\ g_{j1} & g_{j2} & g_{jM} \end{pmatrix} \cdot \begin{pmatrix} I_1 \\ I_2 \\ \vdots \\ I_M \end{pmatrix}$$

In a similar fashion, the other terms of the equation can be represented via matrices. This type of mathematical representation of cellular metabolism is called stoichiometric matrix model, where every row corresponds to a particular reaction, while every column corresponds to a particular metabolite. Thus, the above equation, in matrix notation is (Stephanopoulos, Aristidou, and Nielsen 1998):

$$AS + BP + \Gamma X_{macro} + GX_{met} = 0 \tag{4.2}$$

where A, B, Γ, and G represent the matrices of stoichiometric coefficients.

However, to alter the metabolism of cell, we also need to know the rates at which the metabolic intermediates are formed and/or consumed. This can be represented by rate matrix models as follows:

$$r_{met} = G^T \nu \tag{4.3}$$

where "r_{met}" is the net rate of formation or consumption of the intermediates, while "ν" is the rate matrix containing all the rates of the individual reactions of the metabolic pathway. The transpose of the stoichiometric coefficient matrix (G^T) is to be multiplied with the rate matrix to obtain the matrix r_{met}. Hence, the net rate of product formation can be expressed as:

$$r_{met} = \begin{pmatrix} g_{11} & g_{21} \cdots & g_{j1} \\ g_{12} & g_{22} \cdots & g_{j2} \\ \vdots & \vdots & \vdots \\ g_{1M} & g_{2M} & g_{jM} \end{pmatrix} \cdot \begin{pmatrix} \nu_1 \\ \nu_2 \\ \vdots \\ \nu_3 \end{pmatrix} \tag{4.4}$$

The stoichiometric matrix models and the rate matrix models are the first steps towards engineering cellular metabolism. It is on the basis of these models that we perform metabolic flux analysis and metabolic control analysis.

4.1.3 Metabolic Flux Analysis

Flux is the rate at which a metabolite flows into a particular metabolic pathway, which is influenced by the rate of the competing reactions (Stephanopoulos 1999). The rate matrix models discussed above are used to quantify the intracellular fluxes of the metabolic network. In any metabolic pathway, we need to identify fluxes, which we can measure experimentally, such as rate of secretion of certain intermediates into the medium or rate of formation of biomass constituents. We can thus segregate the rate matrix obtained above into two sections—one containing experimentally measured fluxes ($G_m^T \nu_m$) and the other containing the fluxes needing to be calculated ($G_c^T \nu_c$). If we assume the system to be at steady state, the net rate of formation or consumption of the intermediates becomes 0, reducing Equation 4.3 to:

$$G_m^T \nu_m + G_c^T \nu_c = 0 \tag{4.5}$$

$$\nu_c = -(G_c^T)^{-1} G_m^T \nu_m \tag{4.6}$$

Under certain conditions, however, this method of flux calculation fails, as reviewed by Weichert (Wiechert 2001). In these cases, one or more carbon atoms of a particular metabolite are replaced by the ^{13}C isotope. By using analytical instruments, such as NMR and GC-MS, we can track the rate at which these heavy carbon atoms flow through the metabolic pathway, allowing us to determine the metabolic flux.

After performing metabolic flux analysis, we obtain a flux map containing the flux split ratios of all the metabolites (Manish, Venkatesh, and Banerjee 2007; Sarma et al. 2017). We are also able to identify where the control points of the concerned pathway lie, and if they are stubborn or amenable to manipulation. However, for a quantitative estimation of the extent to which these control points can be manipulated and the subsequent effects on the corresponding fluxes, we perform a metabolic control analysis.

4.1.4 Metabolic Control Analysis

Every reaction occurring in the cell is catalyzed by enzymes, and metabolic control analysis is a measure of how much influence an enzyme has over the rate of the corresponding reaction. This is best expressed by the flux control coefficient (C^J), which describes the change in the flux of a metabolite (J) by an infinitesimal change in the enzyme activity (E) (Kacser and Burns 1973; Heinrich and Rapoport 1974b, 1974a):

$$C^J = \frac{E}{J}\frac{dJ}{dE} \tag{4.7}$$

Enzyme activity is a system parameter, which we can change at will, while flux is a system variable, which is completely dependent on the parameter and that we cannot influence directly.

Enzyme activity is generally manipulated by certain effectors (inhibitors or activators) that bind to the enzyme, thus influencing its job as a catalyst. The response coefficient describes the influence exerted by such an effector (e_i) on the flux of the metabolite (J_k):

$$R_{X_i}^{J_k} = \frac{e_i}{J_k}\frac{dJ_k}{de_i} \tag{4.8}$$

Another coefficient—the elasticity coefficient—expresses the relationship between metabolite concentration (X_j) and rate of the reaction (υ_i), both of which are system variables:

$$\varepsilon_{X_j}^{i} = \frac{X_j}{\nu_i}\frac{\partial \nu_i}{\partial X_j} \tag{4.9}$$

Going back to the analogy of a metabolic pathway with a factory layout, we get to know what changes need to be made at which points in the factory layout and by how much, after performing metabolic control analysis. Hereafter, the biochemical engineer performs genetic engineering or other techniques to effectively change the factory processes.

To manually analyze the mathematical models of any metabolic pathway is a daunting task due to the huge number of variables and parameters involved. Various software packages have been developed to facilitate metabolic engineering by automating the analysis of these mathematical models, after which we can optimize the cell by experimental means. These software packages are evolving continuously, and they have been discussed in the next section.

4.2 SOFTWARES USED IN METABOLIC ENGINEERING

Software packages and tools are an integral part of metabolic engineering for modeling metabolism and analysis of the mathematical models discussed above. Manual planning of how to design optimized cells is limited by the huge number of genes and corresponding metabolic reactions. In this section, we shall discuss in brief the various software packages and tools commonly used for metabolic engineering. These software packages typically simulate the metabolic pathways in a quantitative perspective by means of constraints-based flux analysis, which employs various optimization techniques to maximize/minimize one or more system variables based on alteration effects of one or more system parameters (objective function). For instance, it is possible to maximize biomass formation rates by choosing the right constraints reflecting the physicochemical conditions affecting the metabolism of the cells (Kim, Kim, and Lee 2008; Park, Kim, and Lee 2009). Many algorithms have been developed, which can be used to identify mutable targets based on flux distribution of the metabolic pathway, such as minimization of metabolic adjustment (MoMA), regulatory on/off regularization (ROOM), OptKnock, OptStrain, and many others (Segre, Vitkup, and Church 2002; Shlomi, Berkman, and Ruppin 2005; Boghigian et al. 2010). A few of these software packages have been discussed below.

4.2.1 MatLab

Developed for the calculation and manipulation of matrices, MatLab undoubtedly finds many applications in implementation and analysis of mathematical models of cellular metabolism.

Solving these mathematical models, as described in the previous section, is in essence solving optimization problems to evaluate the consequences of engineering the strain on its metabolic capabilities. This requires the optimization tool embedded in MatLab, known as "optimtool." Exploiting MatLab as a platform to run a variety of programs, many packages have been developed for in silico analysis of metabolic fluxes. One such package is the PFA toolbox (Morales et al. 2016), which runs by means of MatLab functions, and solves every step of a typical metabolic flux analysis problem, including the proposed "*interval* MFA" and "*possibilistic* MFA" (provide interval estimates rather than point wise solutions when multiple flux values are possible). PFA toolbox uses YALMIP as the optimizing tool, and also has an inbuilt graphic user interface for *possibilistic* perspectives of parameters and constraints. Cipher for evolutionary design (CiED) is another example, which is compiled via MatLab, and is used to evolve microbial strains in silico using an optimization function selecting beneficial mutations combined with a genome-scale metabolic model (Chemler et al. 2010). The most famous example of the same is, however, FiatFlux, which shall be discussed in the following section.

4.2.2 FiatFlux

FiatFlux (Zamboni, Fischer, and Sauer 2005) consists of two modules. In the first module, called the RATIO module, metabolic flux ratios are calculated by analyzing ^{13}C-labelling experiments obtained from mass spectrometry (MS). These split ratios of in vivo fluxes are represented as probabilistic equations. The software is preconfigured to derive metabolic flux ratios from [1-^{13}C] and [U-^{13}C] labeled glucose-labeling experiments conducted in yeast, *E. coli*, *Bacillus subtilis*, and others. In the second module, known as the NETTO module, the aforementioned equations, along with biomass requirements, are imposed as constraints in mass balance calculations using the experimentally measured extracellular fluxes, thus yielding the net carbon fluxes within the metabolic pathway. The major advantage of this software is that it is user-friendly, and anyone can reorient the software to meet their specific needs, such as the use of a different ^{13}C-labeled substrate or a different organism.

4.2.3 OptFlux

OptFlux (Rocha et al. 2010) is an open-source, user-friendly and modular software very commonly used for metabolic design and engineering. Implemented using the programming language Java, OptFlux is fully compatible with SBML and *Cell Designer.* Apart from containing provisions for phenotype simulation and introduction of experimentally measured fluxes, OptFlux also acts as a platform for implementing the strain optimization algorithms OptKnock and OptStrain algorithms. Both these algorithms employ a bi-level optimization strategy to select the superior mutant strain. They also use mixed integrated linear programming (MILP), which has the ability to optimize one objective (such as biomass production) within a second competing objective (such as metabolite production) (Boghigian et al. 2010). However, for OptKnock and OptStrain to determine the optimum number of mutations suitable for strain design, genomic space for knockout candidates needs to be exhaustively searched, which is computationally expensive. To overcome this limitation, the OptGene algorithm was developed, which combines evolutionary algorithms (EA) and simulated annealing (SA) (Boghigian et al. 2010).

4.2.4 Other Softwares

13CFLUX-2 (Weitzel et al. 2013) is used to plan and analyze ^{13}C-labeling experiments to determine in vivo metabolic fluxes. It uses a modified XML language, called FluxML. However, 13CFLUX-2 requires that several steps be carried out manually. ^{13}C-labeling-based metabolic flux analysis can be automated and standardized using Flux-P (Ebert et al. 2012), which is based on the Bio-jETI workflow framework. Isotopomer Network Compartmental Analysis (INCA) (Young 2014) is the first software able to perform ^{13}C metabolic flux analysis under both steady-state and nonstationary (transient labeling experiments) conditions, thus expanding the scope of metabolic engineering to photosynthetic organisms and mammalian cultures.

Although the software packages discussed above are very commonly used to plan in silico how to manipulate the cells, they are required to analyze data from huge datasets (such as the pathway databases or protein databases), which becomes computationally expensive. Moreover, a lot of experimental work, which is used as a constraint in flux balance analysis, is involved before one can use these software packages. Performing these experiments for every strain separately is time-consuming, and calls for the use of highly sophisticated and expensive equipment. This problem of redundancy is best solved by the use of machine learning techniques, via which the machine is trained to predict results for repeated rounds of engineering cellular metabolism.

4.3 MACHINE LEARNING TECHNIQUES

Machine learning emphasizes the development of programs that accesses data and learns from experiences. The process of learning starts with the observation of data directly through instructions and experiences, followed by analysis of the data to check for possible patterns and finally perform decision-making. The popularly used machine learning algorithms are:

- *Supervised Machine Learning*: A data mining task where a function is inferred from labeled training data. This training data consists of a set of paired training examples, which has an input object or vector and the output object in the form of a supervisory signal. Based on the analysis of training data, the inferred function is generated helping to map new examples on the data set. The popular supervised machine learning algorithms are Logistic Regression, Decision Trees, Support Vector Machines (SVM), K—Nearest Neighbor (KNN), Naïve Bayes, Random Forest, Linear Regression, and Polynomial Regression algorithm.
- *Unsupervised Machine Learning*: In case of unsupervised learning, the objective is to find hidden information or structure from unlabeled training data. The popular unsupervised learning algorithms are K—Means Clustering, Hierarchical Clustering, and Hidden Markov model.
- *Semisupervised Machine Learning*: A variant of supervised learning which basically uses larger number of unlabeled data to train smaller number of labeled data. It is a combination of supervised and unsupervised machine learning technique.
- *Reinforced Machine Learning*: A machine learning technique inspired by behavioristic psychology focusing on the mechanisms of software agents taking actions in an environment in order to maximize the concept of cumulative award.

At the core of engineering metabolism lies the most important step of characterizing metabolism. Metabolomics is the field of study concerned with discovery of the complete set of metabolites in a cell through various analytical techniques such as GC-MS. This is accompanied by determining how these metabolites interact with each other, or, in other words, characterizing the metabolic pathways. To store the enormous amounts of data discovered through these methods, various databases have been created, such as KEGG (Kanehisa et al. 2017), EcoCyc (Karp 2002a), MetaCyc (Karp 2002b), and many more.

Research in the field of biology is an iterative cycle between mathematically modeling natural phenomenon, implementing the model via wet lab experimentations, and subsequently improving the model to better explain or manipulate the biological process. This cycle maps different levels of biological organization. For instance, modeling individual metabolic reactions or manipulating individual enzyme molecules ultimately has a consequence on the organism's metabolism at large (Kell 2006). Machine learning techniques and concepts aim to automate this iterative cycle for an enhanced knowledge discovery. Machine learning, in essence, involves teaching predictive models to the machine so that it can predict the outcome for any input. These predictive models are based on the observations of a system, thus allowing the experimental nature of biology to be a good candidate for implementing machine learning techniques. Such predictive models should be interpretable, easy to use, and flexible (Geurts, Irrthum, and Wehenkel 2009).

Given the annotated genome of a cell, it is possible to predict the metabolic pathways present in the cell. This has been carried out on *Helicobacter pylori*, by developing the PathoLogic algorithm, which uses MetaCyc or EcoCyc databases as a template for predicting the pathways (Paley and Karp 2002). This algorithm searches for pathways in the aforementioned databases containing at least one enzyme, which is common with the organism whose metabolism is being predicted. Once a number of pathways from the databases are chosen, the algorithm starts eliminating them one by one after every iteration using a set of manually described criteria. The last pathway to remain is predicted to be present in the concerned organism. However, as the size of the MetaCyc database increased, the algorithm started giving many false positive predictions as well. To correct this issue, in a later work (Dale, Popescu, and Karp 2010), a data-driven, transparent, and tunable pathway prediction algorithm was developed based on various machine learning techniques. A "gold-standard" dataset, containing various datasets from a wide variety of organisms, was constructed by taking information from many databases to train the machine learning algorithms. Various features were defined and computed for all the pathways in the gold standard dataset, which was subsequently divided into training and test sets. The training data were used for feature selection and parameter estimation for the predictor types of the data set, and the test data were used for the evaluation of the predictors. The PathoLogic algorithm takes as input the newly sequenced and annotated genome, and predicts the metabolic pathways using the best-evaluated predictor. A similar approach was followed when mining high-dimensional metabolic data for patterns to screen metabolic-based diseases using machine learning (Baumgartner, Böhm, and Baumgartner 2005). Based on a database of metabolic data curated from cases of phenylalanine hydroxylase deficiency, medium-chain acyl-CoA dehydrogenase deficiency, and 3-methylcrotonyl CoA carboxylase deficiency, various classification models were constructed and their parameters estimated. The validated model can be used for an automated prediction and diagnosis of metabolic disorders when certain markers of the patient's metabolism are given as input.

Machine learning has also been applied in the construction of metabolic networks for the calculation of parameters in stoichiometric and kinetic models. The approach has significant application in feature analysis of models relevant to bioreactors. SVF, RF, and several machine leaning algorithms have been used in the analysis of points in NMR spectra and MS spectrograms. Some of the metabolomics software also uses machine-learning methods, which provide mechanisms for metabolomics analysis from data preprocessing and metabolic network analysis. The popularly used tools are mummychog, Cytoscape, Galaxy, FingerID, SIRIUS, Metaboanalyst, KNIME, Weka, Orange, and TensorFlow (Shannon et al. 2003; Li et al. 2013; Guitton et al. 2017).

Some of the commonly used machine learning techniques used in engineering of cellular metabolism, or for analysis of metabolomics data, have been discussed below.

4.3.1 Decision Tree Learning

The most commonly used nonparametric method, decision tree learning showcase excellent interpretability, efficiency, and accuracy (Geurts, Irrthum, and Wehenkel 2009). A decision tree is, in fact, a simple flowchart, where each node asks a "Yes/No" question. For each node, there are two branches representing either a "Yes" or a "No." The questions asked at each node can be described as criteria of some sort to decide whether the process proceeds to the next step or if it should be stopped at that step (Baumgartner, Böhm, and Baumgartner 2005; Dale, Popescu, and Karp 2010). The ranking of nodes is decided by the capacity of a particular feature to shortlist the number of predictions, until a particular solution is reached (Boccard, Veuthey, and Rudaz 2010). A decision tree has basically a flowchart-like structure where the internal nodes present the test for an attribute and the branches are basically the resultant outcomes of the tests conducted and the leaf nodes are the class labels. The topmost node is the root node. The popularly used decision tree learning algorithms are ID3 (Iterative Dichotomiser), C4.5, Classification and Regression Tree (CART) algorithm, CHAID and MARS (Breiman et al. 1984; Quinlan 1996). A detailed explanation of decision trees, how the results can be represented in tabular format, and their applications in biology has been provided by Geurts, Irrthum, and Wehenkel (2009). A cell can employ many metabolic pathways to synthesize a particular compound. Carbonell et al. (2011) proposed a novel retrosynthetic approach to determine which metabolic pathway to target to enhance production of the compound of interest using decision trees. The decision tree thus developed aimed to rank pathways on the basis of thermodynamic feasibility of the individual reactions of each pathway, performance and homogeneity of each enzyme involved, toxicity of the final product, and nominal flux of each pathway. Decision trees have also been used to predict the stability of a protein upon point mutations (change of a particular amino acid) with 82% accuracy (Huang, Gromiha, and Ho 2007).

Diabetes is a metabolic disorder, which has affected millions of people across the globe. The accurate prediction and detection of diabetes has always been a challenge. The study by AVN et al. (2017) mentioned the use of the Decision Tree C5.0 algorithm for the diagnosis and prediction of metabolic disorder-related diabetic diseases. The C5.0 algorithm in machine learning is used for the design of the decision tree with emphasis on information gain and entropy for the creation of the tree. Each stage is divided for absolute attributes, and the topmost attribute is predicted having maximum information gain.

Another study by Karimi-Alavijeh, Jalili, and Sadeghi (2016) has implemented the support vector machine (SVM), decision tree, and also the hybrid of the two techniques, for the prediction of metabolic syndrome from Isfahan Cohort Study participants. Participants without the metabolic syndrome were selected according to ATPIII criteria and were considered for the study. The variables included in the study were gender, age, weight, and physical aspects such as body mass, waist circumference, and twenty other variables. The sensitivity, specificity, and accuracy criteria were considered for validation purposes. The results revealed the SVM method to be more accurate and efficient than the decision tree-based approach based on sensitivity and specificity criteria.

4.3.2 Association Rule Learning

Association rule learning is used to determine multifeature relationships between two disjoint datasets. If the two datasets are A and B, the relation between them is represented as $A \rightarrow B$, which is the basis of association rule learning. This association is usually a many-to-one relation. For example, every gene sequence annotated as belonging to a mitochondrion belongs to a eukaryotic cell, and this can be represented as "Gene: Mitochondrial → Origin: Eukaryota." Boudellioua et al. (2016) developed a system to envisage metabolic pathways in prokaryotes by obtaining data from UniProtKB/Swiss-Prot database, a protein sequence database. The association algorithm used was Apriori, which could determine relations between various attributes of UniProtKB, such as InterPro signatures and organism taxonomy. Association rule mining is also the basis of revealing microbial genotype-phenotype associations by using algorithms such as NETCAR (Tamura and D'haeseleer 2008) and CPAR (MacDonald and Beiko 2010). The former was used to mine out clusters of orthologous (COGs) of proteins related to six microbial phenotypes, via association rule learning, from a dataset of 11,969 COG profiles (corresponding to 155 prokaryotic organisms). COGs of proteins provide a link between the organism's genotype and phenotype profiles. The latter also works on a similar basis by using evolutionary ancestry of the organism as a pattern in phenotype prediction. In essence, CPAR generalizes over the maximum number of samples in a class, and then focuses on the samples not covered by the generalization. CPAR has been reported to provide a slightly higher accuracy coupled with a faster running time over NETCAR.

4.3.3 Clustering

Cluster analysis is a multivariate statistical method involving sorting data into groups/clusters, such that data of a certain cluster share a specific property not present in other clusters. Data obtained from metabolomics experiments can be grouped into clusters based on several features. The parameters usually chosen as the basis of segregation are intensities of the metabolites and experimental conditions. The clustering of metabolomics data can be broadly divided into two categories (Figure 4.2)—one-mode clustering and two-mode clustering (Hageman et al. 2008). In *one-mode clustering*, only a single parameter is changed at a time, and it is further subdivided into two methods based on the parameter being changed. In *metabolite-based clustering*, a biological system is subjected to a particular experimental condition, and the numerous metabolites of the cell are grouped into clusters based on their relative intensities under

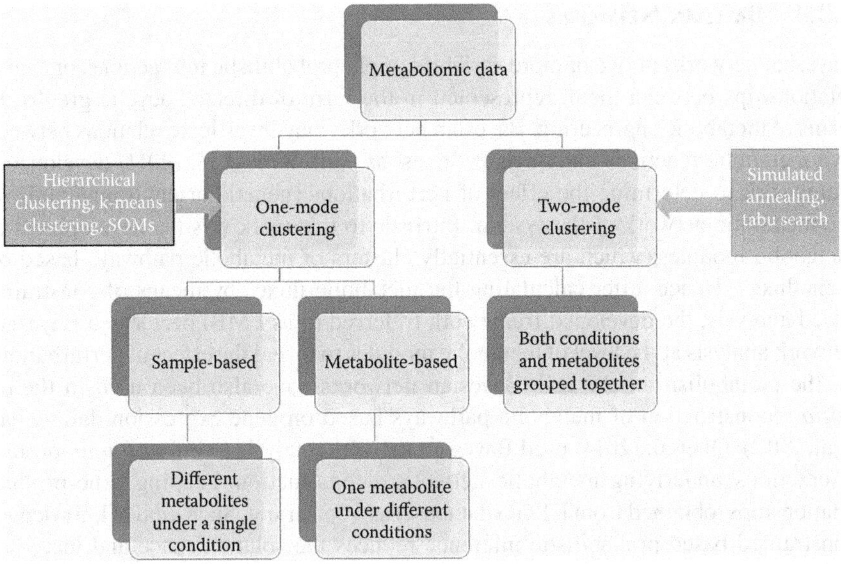

FIGURE 4.2 Types of clustering methods for metabolomics data.

that experimental condition. It is assumed that metabolites of a particular cluster are obtained from the same biosynthetic pathway, or are subject to the same control loops. This method of clustering allows identification of biologically important metabolic markers, which can be used to identify species or even diseases (Meinicke et al. 2008). In *sample-based clustering*, on the other hand, the intensities of a single metabolite under different experimental conditions are used as the basis for grouping the metabolites. This method of clustering reveals information on how the metabolism of a cell can be correlated with its phenotypic features (Fiehn 2002). *Two-mode clustering* involves changing both the parameters simultaneously and clustering metabolites and experimental conditions together accordingly. The algorithms of hierarchical clustering, k-means clustering, and one-dimensional self-organizing maps (SOM) fall under the one-mode clustering approach (Beckonert et al. 2003; Meinicke et al. 2008), while the two-mode clustering relies on global optimization approaches, including simulated annealing (SA) and tabu search (TS) algorithms (Mechelen, Bock, and De Boeck 2004; Prelić et al. 2006). Hageman et al. (2008) developed a genetic algorithm to perform two-mode clustering of metabolomics data, thus determining which groups of metabolites respond to which experimental conditions. More details on genetic algorithms have been discussed in the following sections. Cluster analysis has been applied during the metabolic profiling of seven potato genotypes (Roessner 2001). This work revealed that although the phenotype was the same (increased accumulation of starch), the genotypes formed prominent clusters, implying that different metabolic conditions of a system can have the same phenotypic result (Trethewey 2004). However, cluster analysis has certain disadvantages, such as difficulty in interpretation (as the experimental conditions inevitably overlap), propagation of error (formed due to similar reasons), and poor reproducibility of results, as reviewed by Andreopoulos et al. (2009).

4.3.4 Bayesian Networks

Bayesian networks of two or more variables imply probabilistic interactions, or causal relationships between them, represented in the form of directed acyclic graphs. In terms of metabolic engineering, Bayesian networks may investigate relations between two metabolic reactions and/or their fluxes, as well. Kim et al. (2011) developed a framework to determine the effect of perturbations (genetic or environmental) on the metabolic network of the system. Intrinsic to this work was the development of metabolic modules, which are essentially clusters of metabolic pathways based on their fluxes. Hence, after calculating the metabolic fluxes by means of constraint-based analysis, the developed framework (referred to as FMB) performed Bayesian network analysis at the level of metabolic modules to reveal the effect of perturbations on the metabolism of the cell. Bayesian networks have also been used in the *ab initio* reconstruction of metabolic pathways based on gene expression data (Zhao et al. 2012). Qi et al. (2014) used Bayesian network analysis to infer protein–protein interactions underlying metabolic network reconstruction, keeping gene-product relationships (obtained from KEGG database) as a constraint. Such kind of knowledge-constrained based probabilistic inference reduces the solution space and increases the accuracy of prediction. In a similar work (Zhu et al. 2012), a Bayesian network analysis was used to reconstruct the metabolic network of a dividing yeast population by "stitching together" six types of data—metabolite concentration profiles, RNA expression patterns, variations in DNA, DNA-protein binding, protein–metabolite interaction, and protein–protein interaction. The network thus constructed has also been used to derive causal relationships of the metabolites with the genome, thus allowing the discovery of novel genes that could potentially regulate the cellular metabolism. In order to determine relevant biological markers for the identification of *Bacillus* spores, a combination of Bayesian networks and genetic algorithms was used to reduce the data dimensionality, that is, choose only a few of the data generated by mass spectroscopy experiments (Correa and Goodacre 2011).

4.3.5 Genetic Algorithms (GA)

Genetic algorithms are effective multivariate analysis approaches for analyzing metabolomics datasets (Zou and Tolstikov 2009). Inspired by the process of natural selection, which promises maximum survival of the fittest species under adverse conditions, GA is used to fish out certain critical parameters or the smallest subsets out of the datasets which can effectively describe the datasets (Yeo et al. 2015). GA starts with organizing the given dataset into smaller subsets, each of which is tested for its prediction ability. The subsets obtained in the next step of GA of higher classification accuracy are created by mechanisms inspired by reproduction, selection, mutation, crossover, and migration. This process is repeated for subsequent generations, until a subset with a predetermined accuracy level is reached (Trevino and Falciani 2006). In this way, the entire population of data is "evolved," or optimized, towards a global optimum. For example, GA were developed to select the best parameters suitable for preprocessing of LC-MS-based metabolomics data (Yeo et al. 2015). Boghigian et al. (2010) performed metabolic engineering on a strain of *E. coli* by means of

GAs. Suitable gene targets were identified by evolving the metabolic network with thermodynamically-favorable pathways. This was achieved by developing an optimization framework based on GAs and thermodynamically weighted elementary modes analysis (which, in essence, is metabolic flux analysis). Using this framework, they were able to optimize biomass and product formation simultaneously.

The term "metabolic engineering" was initially coined in the 1990s. At that time, metabolic engineering was a more theoretical branch of biochemical engineering. Since then, this field of research has diversified to a great extent, becoming more concrete rather than an abstract concept. Moreover, it has become a multidisciplinary area of research, which further facilitates its use in various biotechnological industries. However, metabolic engineering has still not matured as an independent field of research due to various reasons, as discussed in the next section.

4.4 FUTURE ASPECTS

Metabolic engineering is rapidly becoming an important parcel of upstream processing in various bioprocess industries. At present, however, research in the field of metabolic engineering is hampered by the huge amount of time taken to perform various experiments, and the need for a sophisticated and expensive experimental setup, and the attached costs. Machine learning methodologies are able to help in this regard, as they allow the engineer to predict outcome of strain optimization procedures, thus eliminating the need for redundant "trial-and-error" experiments for every strain being employed by the industry. Once an optimized strain is designed by in silico means, it can be validated in vitro by experimental means, and the same algorithm can be reused for optimizing another taxonomically similar strain, both of which share similar metabolic features. In this way, the machine is trained to virtually engineer the cellular metabolism for every varied input (wild-type strain). However, as of now, machine-learning techniques are mostly limited to analysis of metabolomics data. Hence, there is a long way to go before computer science can effectively play an integral part in biological sciences for the ease of research and production processes in multiple industries. This gap can be bridged by making more and researchers and industry stakeholders aware of the huge potential of metabolic engineering approaches in the reduction of costs, while enhancing productivity of bioprocess industries. Moreover, a common platform needs to be developed for effective communication between biochemical engineers and computer science engineers.

REFERENCES

Andreopoulos, B., A. An, X. Wang, and M. Schroeder. 2009. "A Roadmap of Clustering Algorithms: Finding a Match for a Biomedical Application." *Briefings in Bioinformatics* 10 (3): 297–314. doi:10.1093/bib/bbn058.

Avn, U., R. Gullapudi, M. Gonuguntla, and N. Rao MR. 2017. "Machine Learning Techniques For Predicting The Wellbeing In Diabetes." *International Journal of Pure and Applied Mathematics* 116 (6): 185–189.

Baumgartner, C., C. Böhm, and D. Baumgartner. 2005. "Modelling of Classification Rules on Metabolic Patterns Including Machine Learning and Expert Knowledge." *Journal of Biomedical Informatics* 38 (2): 89–98. doi:10.1016/j.jbi.2004.08.009.

Beckonert, O., M. E. Bollard, T. M.D. Ebbels, H. C. Keun, H. Antti, E. Holmes, J. C. Lindon, and J. K. Nicholson. 2003. "NMR-Based Metabonomic Toxicity Classification: Hierarchical Cluster Analysis and K-Nearest-Neighbour Approaches." *Analytica Chimica Acta* 490 (1–2): 3–15. doi:10.1016/S0003-2670(03)00060-6.

Boccard, J., J. L. Veuthey, and S. Rudaz. 2010. "Knowledge Discovery in Metabolomics: An Overview of MS Data Handling." *Journal of Separation Science* 33 (3): 290–304. doi:10.1002/jssc.200900609.

Boghigian, B. A., H. Shi, K. Lee, and B. A. Pfeifer. 2010. "Utilizing Elementary Mode Analysis, Pathway Thermodynamics, and a Genetic Algorithm for Metabolic Flux Determination and Optimal Metabolic Network Design." *BMC Systems Biology* 4: 49. doi:10.1186/1752-0509-4-49.

Boudellioua, I., R. Saidi, R. Hoehndorf, M. J. Martin, and V. Solovyev. 2016. "Prediction of Metabolic Pathway Involvement in Prokaryotic Uniprotkb Data by Association Rule Mining." *PLOS ONE* 11 (7): 1–16. doi:10.1371/journal.pone.0158896.

Breiman, L., J. H. Friedman, R. A. Olshen, and C. J. Stone. 1984. *Classification and Regression Trees*. Wadsworth, Monterey, California.

Carbonell, P., A. G. Planson, D. Fichera, and J. L. Faulon. 2011. "A Retrosynthetic Biology Approach to Metabolic Pathway Design for Therapeutic Production." *BMC Systems Biology* 5: 122. doi:10.1186/1752-0509-5-122.

Chemler, J. A., Z. L. Fowler, K. P. McHugh, and M. A. G. Koffas. 2010. "Improving NADPH Availability for Natural Product Biosynthesis in Escherichia Coli by Metabolic Engineering." *Metabolic Engineering* 12 (2): 96–104. Elsevier. doi:10.1016/j.ymben.2009.07.003.

Correa, E. and R. Goodacre. 2011. "A Genetic Algorithm-Bayesian Network Approach for the Analysis of Metabolomics and Spectroscopic Data: Application to the Rapid Identification of Bacillus Spores and Classification of Bacillus Species." *BMC Bioinformatics* 12 (1): 33. BioMed Central Ltd. doi:10.1186/1471-2105-12-33.

Dale, J. M., L. Popescu, and P. D. Karp. 2010. "Machine Learning Methods for Metabolic Pathway Prediction." *BMC Bioinforma* 11: 15. doi:10.1186/1471-2105-11-15.

Ebert, B. E., A.-L. Lamprecht, B. Steffen, and L. M. Blank. 2012. "Flux-P: Automating Metabolic Flux Analysis." *Metabolites* 2 (4): 872–90. doi:10.3390/metabo2040872.

Fiehn, O. 2002. "Metabolomics - The Link between Genotypes and Phenotypes." *Plant Molecular Biology* 48 (1–2): 155–71. doi:10.1023/A:1013713905833.

Geurts, P., A. Irrthum, and L. Wehenkel. 2009. "Supervised Learning with Decision Tree-Based Methods in Computational and Systems Biology." *Molecular BioSystems* 5 (12): 1593. doi:10.1039/b907946g.

Gombert, A. K. and J. Nielsen. 2000. "Mathematical Modelling of Metabolism." *Current Opinion in Biotechnology* 11 (2): 180–86. doi:10.1016/S0958-1669(00)00079-3.

Guitton, Y., M. Tremblay-Franco, G. L. Corguillé, J. F. Martin, M. Pétéra, P. Roger-Mele, and A. Delabrière et al. 2017. "Create, Run, Share, Publish, and Reference Your LC–MS, FIA–MS, GC–MS, and NMR Data Analysis Workflows with the Workflow4Metabolomics 3.0 Galaxy Online Infrastructure for Metabolomics." *International Journal of Biochemistry and Cell Biology* 93: 89–101. Elsevier Ltd. doi:10.1016/j.biocel.2017.07.002.

Hageman, J. A., R. A. van den Berg, J. A. Westerhuis, M. J. van der Werf, and A. K. Smilde. 2008. "Genetic Algorithm Based Two-Mode Clustering of Metabolomics Data." *Metabolomics* 4 (2): 141–49. doi:10.1007/s11306-008-0105-7.

Heinrich, R. and T. A. Rapoport. 1974a. "A Linear Steady-State Treatment of Enzymatic Chains." *European Journal of Biochemistry* 42: 97–105.

Heinrich, R. and T. A. Rapoport. 1974b. "A Linear Steady-State Treatment of Enzymatic Chains. General Properties, Control and Effector Strength." *European Journal of Biochemistry* 42 (1): 89–95.

Huang, L.-T., M. Michael Gromiha, and S.-Y. Ho. 2007. "iPTREE-STAB: Interpretable Decision Tree Based Method for Predicting Protein Stability Changes upon Mutations." *Bioinformatics (Oxford, England)* 23 (10): 1292–93. doi:10.1093/bioinformatics/btm100.

Kacser, H. and J. Burns. 1973. "The Control of Flux." *Symposia of the Society for Experimental Biology* 27: 65–104.

Kanehisa, M., M. Furumichi, M. Tanabe, Y. Sato, and K. Morishima. 2017. "KEGG: New Perspectives on Genomes, Pathways, Diseases and Drugs." *Nucleic Acids Research* 45 (D1): D353–61. doi:10.1093/nar/gkw1092.

Karimi-Alavijeh, F., S. Jalili, and M. Sadeghi. 2016. "Predicting Metabolic Syndrome Using Decision Tree and Support Vector Machine Methods." *ARYA Atherosclerosis* 12 (3): 146–152.

Karp, P. D. 2002a. "The EcoCyc Database." *Nucleic Acids Research* 30 (1): 56–58. doi:10.1093/nar/30.1.56.

Karp, P. D. 2002b. "The MetaCyc Database." *Nucleic Acids Research* 30 (1): 59–61. doi:10.1093/nar/30.1.59.

Kell, D. B. 2006. "Metabolomics, Modelling and Machine Learning in Systems Biology - Towards an Understanding of the Languages of Cells: Delivered on 3 July 2005 at the 30th FEBS Congress and 9th IUBMB Conference in Budapest." *FEBS Journal* 273 (5): 873–94. doi:10.1111/j.1742-4658.2006.05136.x.

Kim, H. U., T. Y. Kim, and S. Y. Lee. 2008. "Metabolic Flux Analysis and Metabolic Engineering of Microorganisms." *Molecular BioSystems* 4 (2): 113–20. doi:10.1039/B712395G.

Kim, H., T. Kim, and S. Lee. 2011. "Framework for Network Modularization and Bayesian Network Analysis to Investigate the Perturbed Metabolic Network." *BMC Systems Biology* 5 (Suppl 2): S14. BioMed Central Ltd. doi:10.1186/1752-0509-5-S2-S14.

Li, S., Y. Park, S. Duraisingham, F. H. Strobel, N. Khan, Q. A. Soltow, D. P. Jones, and B. Pulendran. 2013. "Predicting Network Activity from High Throughput Metabolomics." *PLoS Computational Biology* 9 (7). doi:10.1371/journal.pcbi.1003123.

MacDonald, N. J. and R. G. Beiko. 2010. "Efficient Learning of Microbial Genotype-Phenotype Association Rules." *Bioinformatics* 26 (15): 1834–40. doi:10.1093/bioinformatics/btq305.

Manish, S., K. V. Venkatesh, and R. Banerjee. 2007. "Metabolic Flux Analysis of Biological Hydrogen Production by Escherichia Coli." *International Journal of Hydrogen Energy* 32 (16): 3820–30. doi:10.1016/j.ijhydene.2007.03.033.

Mechelen, I. V., H. H. Bock, and P. De Boeck. 2004. "Two-Mode Clustering Methods: A Structured Overview." *Statistical Methods in Medical Research* 13 (5): 363–94. doi:10.1191/0962280204sm373ra.

Meinicke, P., T. Lingner, A. Kaever, K. Feussner, C. Göbel, I. Feussner, P. Karlovsky, and B. Morgenstern. 2008. "Metabolite-Based Clustering and Visualization of Mass Spectrometry Data Using One-Dimensional Self-Organizing Maps." *Algorithms for Molecular Biology* 3 (1): 1–18. doi:10.1186/1748-7188-3-9.

Morales, Y., G. Bosque, J. Vehí, J. Picó, and F. Llaneras. 2016. "PFA Toolbox: A MATLAB Tool for Metabolic Flux Analysis." *BMC Systems Biology* 10 (1): 1–10. BMC Systems Biology. doi:10.1186/s12918-016-0284-1.

Paley, S. M. and P. D. Karp. 2002. "Evaluation of Computational Metabolic Pathways Predictions for Helicobacter Pylori." *Bioinformatics* 18 (5): 715–24.

Park, J. M., T. Y. Kim, and S. Y. Lee. 2009. "Constraints-Based Genome-Scale Metabolic Simulation for Systems Metabolic Engineering." *Biotechnology Advances* 27 (6): 979–88. Elsevier B.V. doi:10.1016/j.biotechadv.2009.05.019.

Prelić, A., S. Bleuler, P. Zimmermann, A. Wille, P. Bühlmann, W. Gruissem, L. Hennig, L. Thiele, and E. Zitzler. 2006. "A Systematic Comparison and Evaluation of Biclustering Methods for Gene Expression Data." *Bioinformatics* 22 (9): 1122–29. doi:10.1093/bioinformatics/btl060.

Qi, Q., J. Li, and J. Cheng. 2014. "Reconstruction of Metabolic Pathways by Combining Probabilistic Graphical Model-Based and Knowledge-Based Methods." *BMC Proceedings* 8 (Suppl 6): 1–10. doi:10.1186/1753-6561-8-S6-S5.

Quinlan, J. R. 1996. "Improved Use of Continuous Attributes in C4.5." *Journal of Artificial Intelligence Research* 4: 77–90. doi:10.1613/jair.279.

Rocha, I., P. Maia, P. Evangelista, P. Vilaça, S. Soares, J. P. Pinto, J. Nielsen, K. R. Patil, E. C. Ferreira, and M. Rocha. 2010. "OptFlux: An Open-Source Software Platform for in Silico Metabolic Engineering." *BMC Systems Biology* 4. doi:10.1186/1752-0509-4-45.

Roessner, U. 2001. "Metabolic Profiling Allows Comprehensive Phenotyping of Genetically or Environmentally Modified Plant Systems." *The Plant Cell Online* 13 (1): 11–29. doi:10.1105/tpc.13.1.11.

Sarma, S., A. Anand, V. K. Dubey, and V. S. Moholkar. 2017. "Metabolic Flux Network Analysis of Hydrogen Production from Crude Glycerol by Clostridium Pasteurianum." *Bioresource Technology* 242: 169–77. Elsevier Ltd. doi:10.1016/j.biortech.2017.03.168.

Segre, D., D. Vitkup, and G. M Church. 2002. "Analysis of Optimality in Natural and Perturbed Metabolic Networks." *Proceedings of the National Academy of Sciences USA* 99 (23), 15112–15117.

Shannon, P., A. Markiel, O. Ozier, N. S. Baliga, J. T. Wang, D. Ramage, N. Amin, B. Schwikowski, and T. Ideker. 2003. "Cytoscape: A Software Environment for Integrated Models of Biomolecular Interaction Networks." *Genome Research* 13: 2498–2504. doi:10.1101/gr.1239303.metabolite.

Shlomi, T., O. Berkman, and E. Ruppin. 2005. "Regulatory On/off Minimization of Metabolic Flux Changes after Genetic Perturbations." *Proceedings of the National Academy of Sciences USA* 102 (21): 7695–7700.

Stephanopoulos, G. 1999. "Metabolic Fluxes and Metabolic Engineering." *Metabolic Engineering* 1 (1): 1–11. doi:10.1006/mben.1998.0101.

Stephanopoulos, G. N., A. A. Aristidou, and J. Nielsen. 1998. *Metabolic Engineering: Principles and Methodologies*. Metabolic Engineering. Vol. 54. California, USA doi:10.1016/B978-0-12-666260-3.50019-4.

Tamura, M. and P. D'haeseleer. 2008. "Microbial Genotype-Phenotype Mapping by Class Association Rule Mining." *Bioinformatics* 24 (13): 1523–29. doi:10.1093/bioinformatics/btn210.

Trethewey, R. N. 2004. "Metabolite Profiling as an Aid to Metabolic Engineering in Plants." *Current Opinion in Plant Biology* 7 (2): 196–201. doi:10.1016/j.pbi.2003.12.003.

Trevino, V. and F. Falciani. 2006. "GALGO: An R Package for Multivariate Variable Selection Using Genetic Algorithms." *Bioinformatics* 22 (9): 1154–56. doi:10.1093/bioinformatics/btl074.

Weitzel, M., K. Nöh, T. Dalman, S. Niedenführ, B. Stute, and W. Wiechert. 2013. "13CFLUX2 - High-Performance Software Suite for 13C-Metabolic Flux Analysis." *Bioinformatics* 29 (1): 143–45. doi:10.1093/bioinformatics/bts646.

Wiechert, W. 2001. "13C Metabolic Flux Analysis." *Metabolic Engineering* 3 (3): 195–206. doi:10.1006/mben.2001.0187.

Yeo, H. C., B. K.-s. Chung, W. Chong, J. X. Chin, K. S. Ang, M. Lakshmanan, Y. S. Ho, and D.-Y. Lee. 2015. "A Genetic Algorithm-Based Approach for Pre-Processing Metabolomics and Lipidomics LC – MS Data." *Metabolomics* 12: 5. Springer US. doi:10.1007/s11306-015-0884-6.

Young, J. D. 2014. "INCA: A Computational Platform for Isotopically Non-Stationary Metabolic Flux Analysis." *Bioinformatics* 30 (9): 1333–35. doi:10.1093/bioinformatics/btu015.

Zamboni, N., E. Fischer, and U. Sauer. 2005. "FiatFlux—a Software for Metabolic Flux Analysis from 13C-Glucose Experiments." *BMC Bioinformatics* 6: 209. doi:10.1186/1471-2105-6-209.

Zhao, Y., M. H. Chen, B. Pei, D. Rowe, D. G. Shin, W. Xie, F. Yu, and L. Kuo. 2012. "A Bayesian Approach to Pathway Analysis by Integrating Gene-Gene Functional Directions and Microarray Data." *Statistics in Biosciences* 4 (1): 105–31. doi:10.1007/s12561-011-9046-1.

Zhu, J., P. Sova, Q. Xu, K. M. Dombek, E. Y. Xu, H. Vu, Z. Tu, R. B. Brem, R. E. Bumgarner, and E. E. Schadt. 2012. "Stitching Together Multiple Data Dimensions Reveals Interacting Metabolomic and Transcriptomic Networks That Modulate Cell Regulation." *PLoS Biology* 10 (4): e1001301. doi:10.1371/journal.pbio.1001301.

Zou, W. and V. V. Tolstikov. 2009. "Pattern Recognition and Pathway Analysis with Genetic Algorithms in Mass Spectrometry Based Metabolomics." *Algorithms* 2 (2): 638–66. doi:10.3390/a2020638.

5 Big Data and Transcriptomics

Sudharsana Sundarrajan, Sajitha Lulu, and Mohanapriya Arumugam

CONTENTS

5.1 BACKGROUND

DNA encodes the information required to determine the properties and functions of the cells in living organisms. Using this blueprint, the cells dynamically access and translate specific instructions via gene expression by switching on and off a specific set of genes. The data encoded in the particular genes are transcribed into

RNA molecules, which are further translated into proteins or directly utilized to fine tune the gene expression. The set of RNAs transcribed at certain time and condition reveals the current state of the cell. Transcriptome analysis characterizes the transcriptional activity by focusing on the subset of target genes (and transcripts) or profile thousands of genes in parallel to provide a global picture of cellular function. Thus, gene expression analysis provides a snapshot of actively expressed genes and transcripts under various conditions and time.

The term transcriptome includes all kinds of complementary RNA synthesized during transcription of the genome at a point of time. The microRNA (miRNA) and messenger RNA (mRNA) are of greatest interest in transcriptome research. The expression levels of mRNA reflects the gene activity of the cell, while the difference in the miRNA expression levels, which is a major posttranslational regulator of the gene activity, are thought to be associated with several cellular functions. Global analysis of gene expression is a major advancement in biomedical research. The analysis of RNA expression has three fundamental applications: disease subclassification, identification of key genes, and elucidation of biological pathways.[1]

The current chapter discusses the current state-of-the-art technologies in transcriptome research. In addition, the pros and cons of the technologies will be discussed along with their potential application in the health care sector.

5.2 METHODS IN TRANSCRIPTOMICS

Studies on individual transcripts started in early 1970s. In 1980s, low-throughput sequencing methods, namely Sanger sequencing, were used to sequence random transcripts called expressed sequence tags (ESTs). With the advent of high-throughput methods, such as sequencing by synthesis, Sanger sequencing was taken over. Few other methods, which were laborious and captured small subsection of the transcriptome, such as northern blot and reverse transcriptase quantitative PCR (RT-qPCR) were also available before the evolution of current transcriptomics techniques. Serial analysis of gene expression (SAGE) was one of the early transcriptome techniques, which worked based on the Sanger sequencing method by concatenating random transcript fragments. Later on, during early 2000s, the contemporary microarray and RNA-Seq techniques were developed. In late 2000s, a variety of microarrays were produced, which were utilized to measure the expression of known genes of various model organisms. The advancement in scientific technologies allowed the measurement of more genes in a single array. The improvement in the fluorescence detection also increased the measurement accuracy and sensitivity of low abundance transcripts. RNA-Seq refers to the sequencing of cDNAs, and the number of counts of each transcript determines their abundance. In 2004, massively parallel signature sequencing (MPSS) utilized complex hybridization series to sequence 16–20 bp sequences. After 2008, RNA-Seq gained popularity and allowed the measurement of 10^9 transcript sequences.

5.3 MICROARRAY

The microarray technique involves hybridization of mRNA molecule to the DNA template of its origin. Thousands of spotted samples, known as probes (with known

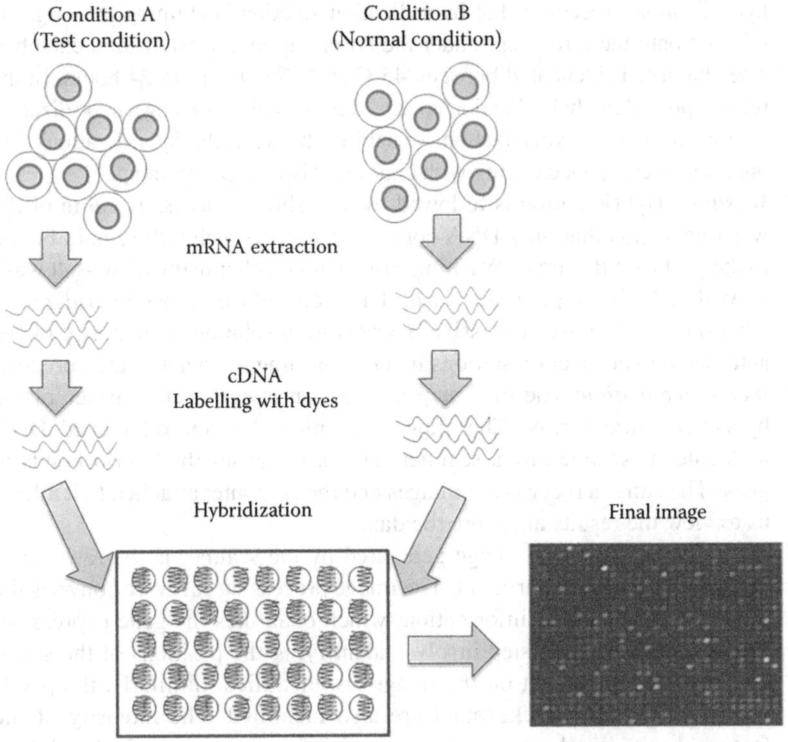

FIGURE 5.1 Schematic of the experimental steps to investigate differential gene expression.

identity), are immobilized on a solid support (microscopic glass slides, silicon chips, or nylon membrane). The spots can be cDNA, DNA, or oligonucleotides, which are used to determine the complementary binding of the unknown sequences leading to analysis of numerous gene expressions. An orderly arrangement of the probes on the array is essential since the spots on the array are used for the identification of the genes.[2]

The experimental design includes the following four steps (Figure 5.1):

1. *Sample preparation and labeling*: The first step is the extraction of RNA from the tissues of interest using either column or using solvents such as phenol-chloroform.

   Labeling step depends on the technology being used. Affymetrix uses biotin-labeled complementary RNA for hybridizing, while in the past radioactive labels were used. Many laboratories now use fluorescent labels, with two dyes Cy3 (excited by a green laser) and Cy5 (excited by a red laser). In the majority of the experiments the two samples are hybridized to the arrays and each sample is labeled with one dye, which allows measuring the gene expression in both the samples simultaneously.
2. *Hybridization*: In this step the probes on the solid support and the labeled target (RNA) form heteroduplexes through Watson–Crick base pairing. Hybridization can be achieved either manually or using robotic technology. In the manual

hybridization procedure, the hybridization solution containing the target is injected onto the array kept under the cover slip in a hybridization chamber. The chamber is incubated between 45°C and 65°C for 12 to 24 hours. In the robotic procedure, hybridization is carried out robotically using a hybridization station. In robotic hybridization, variability between the hybridizations and operators were reduced along with controlled hybridization setup.
3. *Washing*: Hybridization is followed by a washing process. The aim of the washing step is that only DNA complementary to each will remain bound to the probe on the array. Washing can be done either using a low-salt wash or with a high temperature wash. The step reduces cross hybridization reactions and removes excessive hybridization solution. A majority of the automated hybridization stations include washing as an automated process.
4. *Image acquisition*: The final step is to get an image of the surface of the hybridized microarray. The slide containing the heteroduplexed DNA molecules is scanned by a scanner. The laser causes the hybrid bonds to glow. The camera records the images and the computer attached to it allows us to view the results and store the data.
5. *Image processing*: The image generated by the scanner is the raw image generated after the experiment. The image processing software converts the images into numerical information, which quantifies the gene expression. The feature extraction step involves identifying the positions of the spots. The pixels in each spot on the image are identified. Similarly, the pixels contributing to the background are also identified. The intensity of the feature, background, and quality control information are calculated. The feature extraction software computes the numerical measurements of gene expression from the image.

5.3.1 Data Analysis

Microarray studies can be divided into three classes based on their aims, which includes class comparison, class discovery, and class prediction. The groups involved in the study can represent different biological states, which may be disease states, treatment groups, or histological subtypes. The steps involved in the conversion of complex microarray data into a easily understandable biological data are summarized in Figure 5.2. The major steps include identification of significantly regulated genes, identification of global patterns of gene expression, and determination of the biological significance of the mined genes and group of genes. The experiments with adequate numbers of replicates will produce lists of differentially expressed genes, which are identified by appropriate statistical tests. Similarly unsupervised clustering methods can identify global patterns among the differentially expressed genes. The biological significance of the list of differentially regulated genes is analyzed individually using gene annotations.[3]

5.3.1.1 Identification of Differentially Expressed Genes

Differentially expressed genes are those whose expression levels differ significantly under different conditions. The fold change determines the difference between the

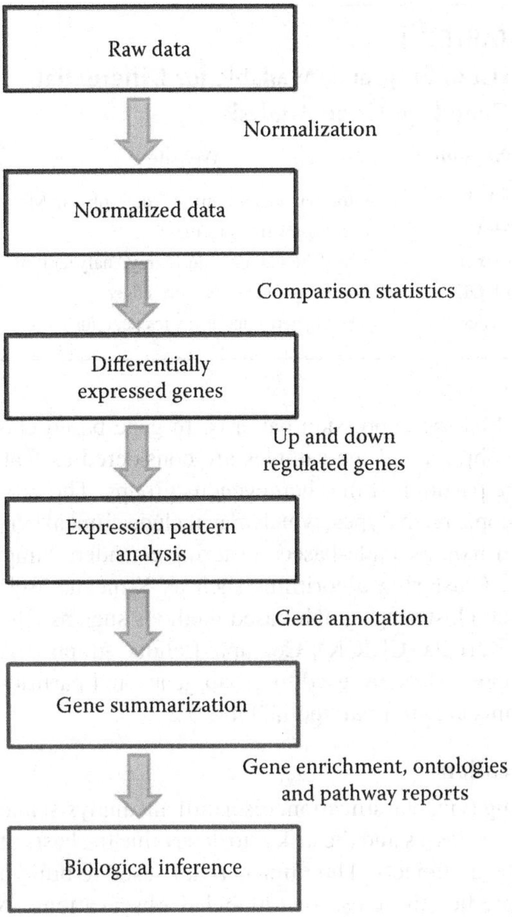

FIGURE 5.2 Steps involved in microarray data analysis.

groups and their biological significance. Various univariate statistical analyses such as t-test, two-sample t-tests, F-statistic, and modified t-test (SAM) are available to determine the relative expression of genes from normalized microarray data. For multiple classes analysis of variance (ANOVA) is used.[4] Various software packages, such as Bioconductor packages, implemented in R are available to identify the differentially expressed genes. A list of programs available for differential gene expression analysis is summarized in Table 5.1.

5.3.1.2 Clustering Gene Expression Data

Clustering is the process of grouping objects into a set of disjoint classes, namely clusters. The objects within a class are highly similar with each other and the objects in other classes are more dissimilar. Clustering is the first step of gene expression matrix analysis. One of the major characteristics of gene expression data is that it can be clustered to both genes and samples. The coexpressed genes can be grouped

TABLE 5.1
List of Programs Available for Differential Gene Expression Analysis

Program	Website
SAM	http://www-stat.stanford.edu/~tibs/SAM/
MeV	http://www.tm4.org/mev/
iArray	http://zhoulab.usc.edu/iArrayAnalyzer.htm
EDGE	http://www.genomine.org/edge/
Cyber-T	http://cybert.microarray.ics.uci.edu/

into clusters based on the expression patterns. In gene-based clustering, the genes are considered as objects and the samples are considered as features. In addition, the samples can be partitioned into homogenous groups. The groups correspond to particular macroscopic phenotypes, which may include clinical syndromes or disease types. On the other hand, sample-based clustering considers samples as objects and genes as features. Clustering algorithms such as K-means, Self-organizing map (SOM), hierarchical clustering, graph-based methods such as Cluster Identification via Connectivity Kernels (CLICK), Corrupted clique graph (CAST), and model-based clustering approaches are used to group genes and partition samples.[5] Other clustering algorithms are summarized in Table 5.2.

5.3.1.3 Classification

Supervised learning (viz. classification, discriminant analysis, and class prediction) considers predefined classes and the task is to determine the basis of the classification from the set of labeled objects. The information is used to build a classifier, which can be used to predict the class of unlabeled observations. Normalized gene expression data is used as the input for building the classification rule. Various algorithms such as k Nearest Neighbours (kNN), Artificial Neural network (ANN), random forest, Support vector machines (SVM), and weighted voting algorithms were used.[6] The machine-learning application tools used for classification are summarized in Table 5.3.

TABLE 5.2
List of Programs Available for Cluster Analysis

Program	Website
Cluster and Treeview	http://rana.lbl.gov/EisenSoftware.htm
dChip	http://biosun1.harvard.edu/complab/dchip/
MeV	http://www.tm4.org/mev/
MAGIC Tools	http://www.bio.davidson.edu/projects/magic/magic.html
CAGED	http://www.genomethods.org/caged

TABLE 5.3
List of Programs for Classification

Program	Website
Weka	http://www.cs.waikato.ac.nz/ml/weka/
SAS	http://www.sas.com/technologies/analytics/datamining/miner/
IBM/SPSS	http://www.spss.com/software/modeling/modeler-pro/
SVMlight	http://svmlight.joachims.org/
LIBSVM	http://www.csie.ntu.edu.tw/~cjlin/libsvmtools/

5.3.1.4 Biological Significance

Once the differential genes are mined, the biological significance of the list of genes can be determined. Various databases and tools offer broad biological themes for the list of differential genes. This tools aids in the identification of particular genes of interest from a list of potential targets. A list of commonly used gene annotation resources is summarized in Table 5.4.

TABLE 5.4
List of Tools Available for Functional Annotation

Program	Website
oPOSSUM	http://www.cisreg.ca/oPOSSUM/
MATCH	http://www.gene-regulation.com/pub/programs.html#match
ConTra	http://bioit.dmbr.ugent.be/ConTra/index.php
Whole genome rVISTA	http://genome.lbl.gov/vista/index.shtml
TFSCAN	http://mobyle.pasteur.fr/cgi-bin/portal.py?form=tfscan
TFSEARCH	http://www.cbrc.jp/research/db/TFSEARCH.html
TransFind	http://transfind.sys-bio.net/
GenMAPP2	www.genmapp.org
ArrayXPath	http://www.snubi.org/software/ArrayXPath/
GO-cluster	http://www.mpibpc.mpg.de/go-cluster/
GO-view	http://db.math.macalester.edu/goproject
Onto-express	http://vortex.cs.wayne.edu/Projects.html
Gominer	http://discover.nci.nih.gov/gominer/
PathExpress	http://bioinfoserver.rsbs.anu.edu.au/utils/PathExpress/
Ingenuity pathway analysis	http://www.ingenuity.com/
Pathway studio	http://www.ariadnegenomics.com/products/pathway-studio/
KOBAS	http://kobas.cbi.pku.edu.cn
GSEA	http://www.broadinstitute.org/gsea/
DAVID	https://david.ncifcrf.gov/
Gene ontology consortium	http://geneontology.org
OBBO	http://www.obofoundry.org, http://obofoundry.github.io
Reactome	http://www.reactome.org/
KEGG pathway database	http://www.genome.jp/kegg/pathway.html

5.3.2 Quality Control

Reasons for quality variation of the experiments is due to the contaminations on the chip, nonuniform dispersal of fluids, or problems during experiment handling. The combined experimental and bioinformatics procedures require rigorous quality control measures to be monitored at each and every step to get accurate results. Collections of standard set of protocols for microarray experiments are available at Brown Lab, Stanford and The Institute of Genomics Research (TIGR).

5.3.2.1 Monitoring RNA and Array Quality

Before running the microarray experiments, the quality of the RNA extracted from the experimental samples should be checked. Agarose gel electrophoresis with ethidium bromide can be used to evaluate the quality of the samples.

Similarly, the quality of the array should be checked before proceeding with the experiment. The spots on the array are checked for the uniformity of the oligonucleotide products coated on them. The slides are scanned to verify the spiked products, which are later eliminated by a washing procedure before proceeding to the hybridization step.

1. *Signal saturation metric*: The fluorescent signal saturation is one of the major concerns affecting the spot intensity measurement. The microarray quantification software packages calculate the signal saturation. A feature is identified as saturated if 50% of the pixels in the spot are saturated. The best way is to eliminate the frequently saturated gene measurements.
2. *Signal intensity*: Signal intensity is the average raw signal intensity from both channels independently. The value is highly dependent on the labeling technique. Low signal intensity values denote problems during labeling, poor DNA quality, and inappropriate amounts of the sample being used.
3. *Signal intensity ratio*: The value is the ratio between the red and green channel intensities, which is calculated from the processed red and green signals. The value should be close to 1. The deviation from the value indicates problem with the normalization procedure.
4. *Signal-to-noise ratio*: The signal-to-noise ratio is calculated by dividing signal intensity by the background noise. The ratio denotes how well the spots can be detected above the background intensity. The signal-to-noise ranges between 100 and 30, and less than 30 indicates poor reliability.

Other quality control measures include investigating the negative controls, nonuniform features, and background noise.

5.3.3 Tools and Databases

Various tools and databases are available, which are used to store and analyze microarray data. Some of the tools and databases used for the analysis are summarized in the following Table 5.5.

TABLE 5.5
List of Databases and Tools Available for Microarray Data Analysis

Gene expression database	ArrayExpress, Gene Expression Omnibus (GEO), Genevestigator, L2L Microarray Database (L2L MDB), 4DXpress, Stanford Tissue Microarray Database, Prostate Expression Database, Oncomine
General programing tool	Biobase, tkWidgets, BASE, Chipster, EzArray, GeneX, MARS, CarmaWeb, TM4, ArrayTrack, dChio, EMMA2, FGDP, Gecko, GEPAT, MiMiR, EMAAS, SBEAMS, WebArray, 2HAPI, 4DXpress, ArrayPipe, ArrayPlex, ArrayQuest, Asterias, Bioconductor, Biosphere, CARPET, CIBEX, Celsius, ChIPOTle, CoCo, Expression Profiller, EzArray, GECKO, GEISHA, GEPAS, GEPAT, GeneCruiser, GenePattern, GenePublisher, GeneX, Genevestigator, Genopolis, J-Express, KMD, M-CHIPS, MAGMA, MIDAW, NTAP, PEPR, RACE, RAD, SBEAMS, TAMEE, Webarray, arrayMagic, dChip, lumi

5.3.4 Applications in Health Care

Microarrays are efficiently used in clinical diagnosis, which ranges from identification disease-relevant genes to diagnosis of the diseases using its biomarkers. The microarrays are employed for genotyping and determination of disease-causing genes, mutation analysis, screening of single nucleotide polymorphisms (SNP), detection of chromosomal abnormalities, and global determination of posttranslational modifications (Figure 5.2).

The global analysis of the gene expression aids in the identification of biomarkers involved in the disease. The identification of the biomarkers can be achieved by comparing the differential gene expression of the genes from healthy and diseased samples. By analyzing the genes associated with the different stages of the disease, or condition from the clinical samples, we can identify genes expressed in the particular stage of the disease.

Rus et al. used cDNA microarray enclosing samples from 21 different SLE patients and 12 controls to identify candidate molecular markers of the disease. The expression pattern of 375 potential genes was analyzed using the peripheral blood mononuclear cells (PBMC). 50 genes were observed to possess 2.5 fold differences in their expression level. The study found that the majority of the genes previously not associated with SLE belonged to various families such as the IL-1 cytokine family, TNF receptor family, IL-8, and its receptor family. The same group has investigated the gene expression profiling in PBMC from 12 and 14 patients with active and inactive disease, respectively. 14 upregulated and 15 downregulated genes were found to be distinct between the active and inactive state of the disease. Most of the genes belonged to the protease family, adhesion molecules, and neurotrophic factors.[7]

Adib et al. compared the differential gene expression profiles between 4 normal ovarian samples, and 6 pairs of primary and omental serous adenocarcinoma. 421 genes and 118 genes were found to have more than two fold and three fold over-expression respectively in primary ovarian cancer compared to normal tissue. The genes were found to be involved in epithelia and cell–cell contacts. The study also identified a distinct biomarker, MGB2, which is significantly overexpressed in primary and metastatic ovarian cancer compared to normal ovarian tissue. The investigation also identified other potential biomarkers such as HPN, IFI-15K, KLK6, CP, SLP1, and HE4, which have previously proven to be associated with other types of cancer.[8]

Sundarrajan et al. analyzed the gene expression profiles of lesional and nonlesional samples of psoriatic transcriptome. The results indicated around 244 genes were upregulated and 183 genes were downregulated. The genes were found to be involved in intracellular cascades of the immune response, cytoskeleton and cell adhesion, regulation of cell proliferation, and apoptosis.[9] The study also identified a few potential biomarkers not previously studied in psoriasis. In another study, Sundarrajan et al. used supervised analysis utilizing a random forest (RF) algorithm to distinguish the psoriatic samples from the normal samples. The differential gene expression profiles were first clustered based on the coexpression profile using weighted gene coexpression network analysis (WGCNA) to identify the top 50 genes implicated in the disease, which could be used as a predictor for the next supervising analysis. The genes were found to be involved in functions such as cell cycle, cell proliferation, cytoskeleton rearrangement, keratinocyte proliferation, angiogenesis, and immune response, which are impaired in the disease condition. The RF classifier was trained using the expression profiles from two different microarray datasets. The classifier was tested against three different datasets where it was found to have a specificity value ranging between 0.96 and 1.[10]

Golub et al. studied 38 bone marrow samples from acute leukemic patients. The acute leukemia patients were grouped into acute myeloid leukemia (AML) and acute lymphoblastic leukemia (ALL) categories. The investigators used supervised analysis for class prediction using 50 genes, which were differentially expressed between 11 and 27 AML and ALL samples, respectively. The predictor of 50 genes was utilized to test set of 34 new leukemic samples. 29 of 34 samples were correctly classified. The genes used for the prediction were involved in cell cycle, cell adhesion, transcription, and oncogenes, which were found to be involved in cancer pathogenesis. The second part of the same study used class discovery on the initial 38 leukemic samples to determine the efficiency of the global gene expression analysis in distinguishing AML and ALL. They used SOM to distinguish between the two groups. 24 of the 25 ALL samples clustered together in one group and 10 out of 13 AML samples were clustered in a second class. The results of the class discovery studies indicated that it is possible to discover the diagnostic classes of the cancer when morphological tests were unavailable whereas the biological and clinical information were available.[11]

Various applications of microarray technology have been utilized in the area of cardiomyopathy. Some studies involved comparisons of failing and nonfailing heart, before and after left ventricular assist device (LVAD) implantation, and dilated and hypertrophic cardiomyopathy. The microarray analysis of failing and nonfailing hearts provides therapeutic insights. The gene expression profile comparison identified that the genes involved in fatty acid metabolism were up-regulated and genes involved in glucose metabolism were down-regulated in failing hearts. These results provide an insight into drugs prescribed to shift fatty acid to glucose metabolism. In a supervised analysis of the gene expression of the dataset of the patients with heart failure, the investigators observed changes in the pattern of gene expression at different clinical stages.[12–13] In the study conducted by Steenman et al, on failing and non failing hearts, a distinct cluster of patients emerged during unsupervised analysis, who require the highest medical urgency while awaiting cardiac transplantation.[14]

Microarray technology has been utilized to investigate the differential gene expression of pathogens. An oligonucleotide microarray targeting the 16S rRNA gene was developed to detect a panel of 40 predominant human intestinal bacterial pathogens in human fecal samples. A ViroChip system containing 1600 unique 70-mer long oligonucleotide probes, covering 140 viral genomes, was used to detect severe acute respiratory syndrome viruses. The oligonucleotide microarrays were developed to analyze and identify drug-resistant *M. tuberculosis* strains, and the results were comparable to those of the standard antimicrobial susceptibility tests.[15]

5.4 DEEP SEQUENCING (REF-SEQ)

RNA sequencing (RNA-Seq) is a recently developed transcriptome profiling technique utilizing deep-sequencing technologies. It allows entire transcriptomes to be surveyed at single-base resolution and concurrently profile the gene expression on the genome scale. RNA-Seq determines previously undetected changes occurring in a disease states, in response to therapeutics, under different environmental conditions and across a broad range of other studies. RNA-seq can detect both known and novel features in a single assay, gene fusions, transcript isoforms, single nucleotide variants, and others. The major strength of RNA-Seq lies in the identification of novel features of the transcriptome along with its ability to capture the whole transcriptome, which includes both known and unknown regions.[16]

RNA-Seq utilizes deep-sequencing technologies. The RNA of interest is isolated from the tissue of interest. The RNA is converted into a library of cDNA fragments, which are attached to the adaptors to one or both ends. Each molecule (with or without amplification) is sequenced to obtain short sequences from one end (single-end sequencing), or both ends (pair-end sequencing). Each read has a typical size between 30 and 400 bp, depending on the sequencing technology. The reads are aligned either to the reference transcripts or the reference genome, or assembled from a *de novo* genomic sequence (Figure 5.3). The results can be used to analyze both genomic and posttranslational mutations and mRNA expression levels of the genes. The sequencing platforms include Illumina IG, Applied Biosystems SOLiD, Roche 454 Life Science, or Helicos Biosciences tSMS. There are three main sequencing technologies including pyro-sequencing developed by 454 Life Sciences, reversible dye-terminator-based technology by Illumuna, and SOLiD technology by Applied Biosystems.[17]

5.4.1 Data Processing

5.4.1.1 Quality Assessment

Since RNA-Seq involves numerous processes, which range from sample preparation to sequencing, it is not a straightforward method to identify and quantify the RNA species from the reads sequenced. The quality assessment requires to filter data, remove the low-quality sequences, contaminators, adaptors, and overrepresented

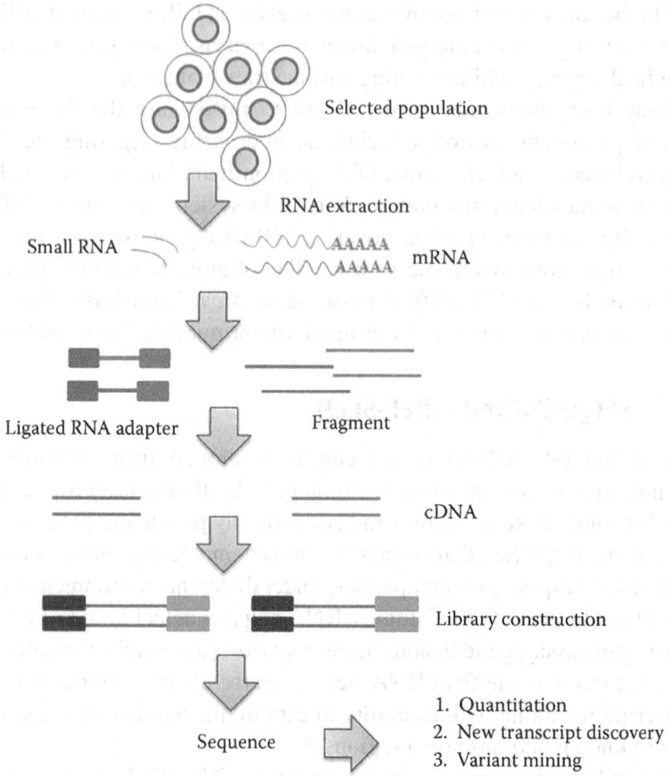

FIGURE 5.3 Schematic of RNA-seq experimental steps.

sequences. Read quality decreases towards 3′ end of the reads and as it becomes too low, the bases should be removed to improve the quality and mappability.

5.4.1.2 Read Mapping

After preprocessing of the sequence, the following step is to map the short reads to the reference genome or assemble the reads onto the contigs and align them to the reference genome. If the reads accumulate at the 3′ end of the transcripts in Poly (A) selected samples, it indicates that the RNA quality is low. Many bioinformatics program such as ELAND, SOAP, Bowtie, MAQ, BWA, ZOOM, STAR, and others are suitable to map reads that are not located at the poly (A) tails and exon-intron splicing junctions. Tools such as BLAT, TopHat, MapSplice, and GEM align the reads located in the exon-intron boundaries to help to determine the alternative splicing patterns. The alignment of sequence reads to multiple locations of the genome may occur due to polymorphisms.

5.4.1.3 Read Quantification and Counting

Once the initial quality control steps are completed, the reads should be checked for GC content and gene length bias. With the availability of well-annotated reference transcriptome, we can analyze the genetic constitution of the sample of our study.

Read counting is an essential step in RNA-Seq workflow. The number of RNA-Seq reads generated from a transcript is directly proportional to the transcript's relative abundance in the sample. In case of alternative splicing, the reads mapping to this region will map to constitutive or shared exons, complicating the process of read counting.

5.4.1.4 Transcript Identification and Alignment

When the reference genome is accessible, the RNA-Seq analysis maps the reads to the reference genome or transcriptome and determines which transcripts are expressed. The alignment of reads to the reference genome can be done either mapping to the genome or to the transcriptome. Regardless of the mapping reference, the reads may align to either one position or to multiple positions (multireads). It is difficult to align short reads since they span across several splice junctions. Varieties of tools such as GRIT can annotate the isoforms correctly.

5.4.1.5 Transcript Quantification

The most common application of RNA-Seq is the estimation of gene and transcript expression. It depends on the number of reads that map to each of the transcript. Numerous statistical methods are used to quantify transcript abundance based on the read coverage. The RPKM (reads per kilo base per million mapped reads) is a widely used method to account for expression and normalized read counts with respect to overall mapped read number and gene length. Other than read coverage, sequence depth, gene length, and isoform abundance also determine the transcript abundance. Many algorithms such as RSEM, eXpress, Sailfish, kallisto, and others are developed to estimate the transcript-level expression. Once the read counts are estimated, data normalization is a crucial step in the data processing since it is important to determine accurate results of the gene expression and further analysis. Multiple parameters such as transcript size, sequencing depth and error rate, GC-content, and many more should be considered while choosing normalization methods. For example quantile normalization can improve the quality of the mRNA-Seq data. EDASeq, an R package using within and between normalization methods can decrease the GC-content.

Most algorithms utilize mathematical functions such as probability distributions (e.g., Poisson distribution) followed by Fisher's exact test. The list of differentially expressed genes can be used to gain biological insights into the experiment. Various advanced analyses such as gene ontology, gene sets, network interference, and knowledge databases can be used to understand the biological context of the differentially expressed genes.

5.4.2 Quality Control

The quality of the RNA-Seq can be assessed using both low-level metrics (including read quality, duplication rate, GC content, nucleotide composition, and so on) and high-level metrics (including mapping statistics, coverage uniformity, saturation of the sequence depth, ribosomal RNA contamination, reproducibility of the replicates, read distribution, and strand specificity).

5.4.2.1 Low-Level Metrics

The raw sequence-based metrics checks the experiments at low-level as prior sequence alignments are not required. The raw sequence quality is assessed based on the Phred quality score (Q). Phred score measures the base-calling reliability from Sanger sequencing chromatograms. It is defined as $Q = -10 \times \log10(P)$, where P is the probability of erroneous base calling. GC content gives the percentage of either guanine or cytosine bases in a DNA sequence. It is a simple way to measure the nucleotide composition. Read duplication is often determined by read length, transcript abundance, PCR amplification, and sequence depth.

5.4.2.2 High-Level Metrics

Mapping statistics is an easy intuitive way to assess the success of the RNA sequencing. This includes the number of reads aligned to the reference genome, number of reads aligned to the unique location in the reference genome, number of reads mapped to mitochondria, and number of splice mapped reads. Assessing rRNA contamination in the sequencing is simple, and is accomplished by aligning the reads to the reference genome and counting how many reads are mapped to ribosome genes, or by aligning reads directly to the ribosomal RNA sequences. The saturation test is done to determine whether the current sequence depth is ideal to satisfy the purpose of the experiment. When the sequence is unsaturated, the RPKM will be unstable and the presence of low abundant isoforms will remain unknown.

5.4.3 Tools and Databases

The next generation sequence analysis largely depends on the bioinformatics tools to support the analysis process. Some of the tools and databases, which are commonly utilized for the processing and analysis are summarized in Table 5.6.

TABLE 5.6
List of Tools and Packages Used for Data Processing

Function	Tools
Preprocessing and quality assessment	FastQC, HTQC, Trimmomaatic, BBMap, FLASH, RSeQc
Mapping	ELAND, SOAP, SOAP2, MAQ, Bowtie, BWA, ZOOM, STAR, BLAT, HTSeq, Easy RNASeq, GenomicRanges, Feature-Counts
Expression quantification	Alexa-seq, Cufflinks, RSEM
Differential expression	Cuffdiff, DESeq, DESeq2, EdgeR, PoissonSeq, Limma-voom, MISO
Alternative splicing	TopHat, MapSplice, SpliceMap, SplitSeek, GEMmapper, SpliceR, Splicing-Compass, GliMMPS, MATS, rMATS
Variants detection	GATK, ANNOVAR, SNPiR, SNiPlay3
Pathway analysis	GSEA, GSVA, SeqGSEA, GAGE, SPIA, TAPPA, DEAP, GSAASeqSP
Coexpression network	GSCA, DICER, WGCNA

5.4.4 Applications in Health Care

RNA-Seq technology can be applied for various prognostic, diagnostic, and therapeutic needs implicated in various disease conditions. The multifaceted applications of RNA-Seq include identification of gene fusions, differential expression of the disease causing transcripts, and detection of diversity of RNA species.

RNA-Seq is currently being utilized to assess the expression profiles of genes involved in a number of complex disease such as metabolic disorders, cardiomyopathy, multiple sclerosis, and so on. Leti et al. analyzed the involvement of microRNAs in the regulation of metabolic disorder in patients suffering with nonalcoholic fatty liver disease (NAFLD). Around 30 and 45 miRNAs were found to be upregulated and downregulated, respectively. The genes were implicated in various significant pathways such as apoptosis signaling and growth factor beta signaling.[18]

Punt et al. identified biomarkers expressed in both cancer and immune cells involved in cervical cancer. The authors utilized samples from squamous cell cervical cancer. The RNA-Seq and differential gene expression analysis revealed T-cell leukemia/lymphoma 1A (*TCL1A*) as a novel immune cell expressed biomarker for predicting the survival in patients with cervical cancer.[19]

Ren et al. investigated the complex landscape of genomic alternations in prostate cancer by analyzing the transcriptomes of 14 pairs of prostate cancer and normal tissue. The RNA-Seq analysis revealed the involvement of various gene fusions, alternative splicing, expression of long noncoding RNAs and genes, and somatic mutations in the disease condition.[20]

Amyotrophic lateral sclerosis (ALS) is associated with degeneration of the upper and lower motor neurons in the brainstem and spinal cord. Brohwan et al. determined the major contributing factor by comparing the RNA-sequencing reads from the postmortem spinal section tissue samples from ALS and healthy individuals. Many differentially expressed genes were identified and the tumor necrosis factor (TNF) was identified as an upstream regulator involved in the disease and inflammatory processes, and the TNF signaling cascade. The investigators identified TNFAIP2 as a hub gene, which encodes intracellular proteins of the tumor necrosis factor family.[21]

Li et al. utilized transcriptome analysis to gain insights about the psoriasis disease mechanism. The investigators used 92 psoriatic and 82 normal punch biopsy samples to sequence polyadenylated RNA-derived complementary DNAs. The results identified differentially expressed genes were mostly enriched in immune mediated cascades. The weighted gene coexpression network analysis (WGCNA) revealed multiple modules of coordinately expressed epidermal differentiation genes, overlapping significantly with genes regulated by the long noncoding RNA. They also noted that the genes expressed in the dermal regions were significantly downregulated in psoriatic biopsies.[22]

5.5 FUTURE PROSPECTS

The high-throughput technologies provide a valuable pathway to study the transcripts of organisms. The technologies provide more insight about the genes expressed in different disease condition, disease progression, RNA editing, fusion transcripts, alternative splicing, and allele-specific expression. Transcriptomics research is moving forward towards single-cell transcriptomics studies, since the expression of the genes vary from one cell to other. Mining data from public repositories such as ENCODE, TCGA, GEO, and the Geuvadis project may provide new insight about gene regulation. The high volume of data requires high-end computational resources for transcriptome assembly and other downstream analysis. Hence, highly parallel data processing steps are required for better analysis and faster processing of the results.

REFERENCES

1. R.M. Fryer, J. Randall, T. Yoshida et al., Global analysis of gene expression: Methods, interpretation, and pitfalls, *Exp. Nephrol.* 10, 2002, pp. 64–74.
2. G. Barker and D. Stekel, Microarray bioinformatics, *Ann. Bot.* 93, 2004, pp. 615–616.
3. L.D. Miller, P.M. Long, L. Wong, et al., Optimal gene expression analysis by microarrays, *Cancer Cell* 2, 2002, pp. 353–361.
4. N.E. Olson, The microarray data analysis process: From raw data to biological significance, *NeuroRx J. Am. Soc. Exp. Neurother.* 3, 2006, pp. 373–383.
5. L. Kaufman and P.J. Rousseeuw, *Finding Groups in Data: An Introduction to Cluster Analysis*, Wiley series in probability and mathematical statistics, Wiley, Hoboken, N.J, 2005.
6. Sandrine, D. and F. Jane, Classification in microarray experiments. In *Statistical Analysis of Gene Expression Microarray Data.* CRC press. 2003; 93–159.
7. V. Rus, S.P. Atamas, V. Shustova et al., Expression of cytokine- and chemokine-related genes in peripheral blood mononuclear cells from Lupus Patients by cDNA Array, *Clin. Immunol.* 102, 2002, pp. 283–290.
8. T.R. Adib, S. Henderson, C. Perrett et al., Predicting biomarkers for ovarian cancer using gene-expression microarrays, *Br. J. Cancer* 90, 2004, pp. 686–692.
9. S. Sundarrajan, S. Lulu, and M. Arumugam, Insights into protein interaction networks reveal non-receptor kinases as significant druggable targets for psoriasis, *Gene* 566, 2015, pp. 138–147.
10. S. Sundarrajan and M. Arumugam, Weighted gene co-expression based biomarker discovery for psoriasis detection, *Gene* 593, 2016, pp. 225–234.
11. T.R. Golub, D.K. Slonim, P. Tamayo et al., Molecular classification of cancer: Class discovery and class prediction by gene expression monitoring, *Science* 286, 1999, pp. 531–537.
12. M.M. Kittleson, K.M. Minhas, R.A. Irizarry et al., Gene expression analysis of ischemic and nonischemic cardiomyopathy: Shared and distinct genes in the development of heart failure, *Physiol. Genomics* 21, 2005, pp. 299–307.
13. F.-L. Tan, C.S. Moravec, J. Li et al., The gene expression fingerprint of human heart failure, *Proc. Natl. Acad. Sci. USA.* 99, 2002, pp. 11387–11392.
14. M. Steenman, G. Lamirault, N. Le Meur et al., Distinct molecular portraits of human failing hearts identified by dedicated cDNA microarrays, *Eur. J. Heart Fail.* 7, 2005, pp. 157–165.
15. M.B. Miller and Y.-W. Tang, Basic concepts of microarrays and potential applications in clinical microbiology, *Clin. Microbiol. Rev.* 22, 2009, pp. 611–633.
16. U. Nagalakshmi, Z. Wang, K. Waern et al., The transcriptional landscape of the yeast genome defined by RNA sequencing, *Science* 320, 2008, pp. 1344–1349.
17. Z. Wang, M. Gerstein, and M. Snyder, RNA-Seq: A revolutionary tool for transcriptomics, *Nat. Rev. Genet.* 10, 2009, pp. 57–63.

18. F. Leti, I. Malenica, M. Doshi et al., High-throughput sequencing reveals altered expression of hepatic miRNAs in non-alcoholic fatty liver disease-related fibrosis, *Transl. Res. J. Lab. Clin. Med.* 166, 2015, pp. 304–314.
19. S. Punt, W.E. Corver, S.A.J. Van der Zeeuw et al., Whole-transcriptome analysis of flow-sorted cervical cancer samples reveals that B cell expressed TCL1A is correlated with improved survival, *Oncotarget* 6, 2015, pp. 38681–38694.
20. S. Ren, Z. Peng, J.-H. Mao et al., RNA-seq analysis of prostate cancer in the Chinese population identifies recurrent gene fusions, cancer-associated long noncoding RNAs and aberrant alternative splicings, *Cell Res.* 22, 2012, pp. 806–821.
21. D.G. Brohawn, L.C. O'Brien, and J.P. Bennett, RNAseq analyses identify tumor necrosis factor-mediated inflammation as a major abnormality in ALS spinal cord, *PloS One* 11, 2016, pp. e0160520.
22. B. Li, L.C. Tsoi, W.R. Swindell et al., Transcriptome analysis of psoriasis in a large case-control sample: RNA-seq provides insights into disease mechanisms, *J. Invest. Dermatol.* 134, 2014, pp. 1828–1838.

6 Comparative Study of Predictive Models in Microbial-Induced Corrosion

Nitu Joseph and Debayan Mandal

CONTENTS

6.1 INTRODUCTION

Iron is such a metal, whose origin itself is shrouded in mysteries. While some studies suggest that the earliest iron artifacts have their origin from meteoric iron deposits, others emphasize the existence of iron smelting processes as early as 5000 BCE. The oldest artifacts have been discovered from the cradles of ancient civilizations, like Mesopotamia and Egypt. But the iron wares were not as widely used as their silver and gold counterparts, owing to the single major problem of "corrosion." With the onset of Iron Age around 1200 BCE–1300 BCE, as the metal became more freely available and abundant, man began to venture into the inner workings of this troublesome menace called corrosion, along with ways to try and prevent it.

The phenomenon of anaerobic iron corrosion has mind-boggled scientists for centuries. The popular notion of metal corrosion occurring due to the interaction of the metal with air and moisture could not explain the mechanisms driving the corrosion of buried pipelines, pasteurization tanks, along with a variety of structures of mechanical, infrastructural, as well as cultural significance. It was not until the early 1900s scientists started looking at the gravity of the corrosion problem (Olmstead and Hamlin 1900). Scientists had always viewed the phenomenon with skepticism, as in most cases if an abiotic mechanism could be related to a particular case of corrosion, it was explained to have been caused by microbes. It was not until the latter half of the twentieth century that the frame of focus was shifted on to the persistent presence of microbial colonies, which were causing severe monetary losses.

6.2 BACKGROUND

The progressive deterioration of a substrate after being subjected to redox reaction, facilitated by biofilms formed by the action of colonizing bacteria, is known as MIC, alternatively known as biocorrosion (Xu et al., 2016). The conventional electrochemical corrosion involves the release of metal cations as a resultant of electron transfer from a zero-valent metal to an external electron acceptor. However, biocorrosion is synergistic in nature, involving the interactions between the metal substrate, abiotic corrosion products, microbial cells, and the EPS secreted by them.

With present-day machineries and infrastructure being heavily dependent on metals and related products, MIC poses an important challenge. Across all industrial sectors, 20%–30% of corrosion in metal structures could be attributed to MIC (Koch et al., 2010), amounting to 2%–3% of the GDP of developing nations (Koch et al., 2001). Its effect is more prominent in the energy industry, as the corrosion in the oil and gas pipelines cause a loss in the range of hundreds of billion dollars (Zhiyuan and Zongdi 2017).

This chapter attempts to analyze the complex links between bacteria and the corrosion process they induce. Various mathematical modeling methods used for studying MIC have been analyzed, and a set of feasible combinations are presented for better understanding.

6.3 BROADLY CLASSIFIED PHASES

6.3.1 Substrate Organism Interaction

From an electrochemical perspective, corrosion can be viewed as an interfacial process. The multiple parameters of the physicochemical environment are crucial in determining reaction kinetics. Parameters like pH, salt content, redox potential, etc. can be altered by the presence of a microbial colony at the interface (Wells and Melchers 2011).

This alters the otherwise purely abiotic nature of the redox reaction involving iron and suitable oxidants. In environments with little or no oxygen and a pH > 6, the normal chances of abiotic corrosion are abysmally low (Wells and Melchers 2011). But it was observed these areas exhibit a rather high corrosion rate (Gu 2014).

The basic requisite for bacterial proliferation is water. The presence of water along with an oxidized electron donor, a reduced electron acceptor, an energy, and a carbon source facilitates the accelerated growth of microbial consortia. A comprehensive broad-spectrum analysis on the genomics of bacterial colonies has revealed the existence of a multitude of anaerobic bacterial species instrumental in the corrosion process (Marangoni et al., 2013). Obligate anaerobic bacteria involved in sulfate respiration are one of the primary organisms responsible for the MIC. They are popularly known as sulfate reducing bacteria (SRB), metal reducing bacteria (MRB), metal depositing bacteria (MDB), and acid producing bacteria (APB) (Beech and Gaylarde 1999).

6.3.1.1 Sulfate Reducing Bacteria (SRB)

The SRB family includes bacteria like *Desulfovibrio desulficans, Desulfomaculum ruminis, Desulfonema limicola,* and facultative anaerobic bacteria like *Desulfuromonas acetoxidans* (Little et al., 2000). Biogenic sulfur compounds thus formed also pose serious health risks. From the early 1930s, investigations had begun on the pitting corrosion caused in cast iron, as SRB consortia were traced as the causative reason (Beech et al., 2000).

6.3.1.2 Metal Reducing Bacteria (MRB)

Naturally, iron objects form a protective passive layer of oxide on their surfaces. The bacterial consortium involving *Alteromonas putrefaciens, Shewanella putrefaciens,* and *Pseudomonas* genera are key causative species (Maluckov 2012). Organisms belonging to this category are predominantly involved in metabolic activities resulting in the reduction of iron and manganese metals (Beech et al., 2000).

6.3.1.3 Metal Depositing Bacteria (MDB)

These bacterial species are involved in the process of biologically metabolizing various metal oxides, commonly iron and manganese. The divalent ferrous ions (Fe^{2+}) present in the bulk-dissolved state are converted into a metabolized precipitate chiefly composed of trivalent ferric ions (Fe^{3+}). Bacterial species like *Gallionella, Leptothrix,* and *Geobacter sp* are involved in this process (Beech et al., 2000).

6.3.1.4 Acid Producing Bacteria (APB)

Inorganic acids like sulfuric acid, sulfurous acid, nitric acid, and carbonic acids are produced as metabolic wastes by some bacteria (Beech et al., 2000). Bacterial species like *Thiobacillus sp*, *Thiothrix,* and *Beggiatoa spp* are involved in the production of sulfurous acid while those like nitrogen oxidizing bacteria and ammonia oxidizing bacteria are involved in the production of nitric acid resulting corrosion of concrete and natural stone. Consequently, with the fluctuation in the alkalinity, there arises an increase in lead contamination and copper solubility.

6.3.1.5 Mycetes (Fungi)

The most widely occurring form of corrosion induced by the fungi is that of aluminum. There have been extensive studies on the corrosion of aircraft fuel tanks due to jet fuel contamination. The biocontaminant was identified as a consortium of *Hormoconis resinae, Aspergillus spp., Fusaruium spp.,* and *Penicillium spp.* The *Hormoconis resinae* biometabolized the hydrocarbons into organic acids. These metabolites form chelation compounds with copper, iron, aluminium, and zinc, thus forming weak spots in the fabric of the aircraft metal lining (Beech et al., 2000).

6.3.2 Biofilm Formation

Matrix layers are formed when bacterial species colonize a particular substrate and start the secretion of Extra cellular Polymeric Substances (EPS) (Beech and Sunner 2004), leading to the onset of corrosion (Alabbas et al. 2013). The EPS, along with the microbial colonies, constitute the biofilm commonly called "slime layer." The EPS layer are commonly composed of minor concentrations of lipids and nucleic acids engulfed in ploysaccharides and protiens. Formed as a monolayer in the initial stages, these have a potential of increasing their thickness up to several centimeters and tend to be highly hydrated ($\sim$97% water). One of the primary characteristics of biofilm is the exponential rate of bacterial cell growth exhibited; with the cell count in the colonies reaching as high as Log_{10} (6.0) cells/cm^2 (Bouman et al., 1982) thereby establishing synergistic communities commonly known as "Microconsortia." Another prominent characteristic is the free motility of the bacteria within the EPS layer, resulting in pronounced pH gradient across the matrix (Figure 6.1) (Gu 2012).

6.3.3 Biofilm Induced Corrosion

The primary mechanism of MIC involves the release of the dehydrogenase enzyme, which results in the formation of cathodic hydrogen as a byproduct at the metal surface. This further aids the growth of the SRB consortia. The bacterial reduction of sulfate is carried out by the hydrogenase positive (H+) SRB. This results in a cathodic depolarization process, as the hydrogen are scavenged away by the bacteria, thus leading to anodic corrosion (Figure 6.2).

This process culminates in the bacterial release of hydrogen sulfide, which in turn forms metallic sulfide (ferrous sulfide) as an end product (Videla 2010). Subsequently, a galvanic cell is formed between the metallic substrate and its sulfide promoting further corrosion. Here the metal sulfide acts as a cathode. Therefore, this implies

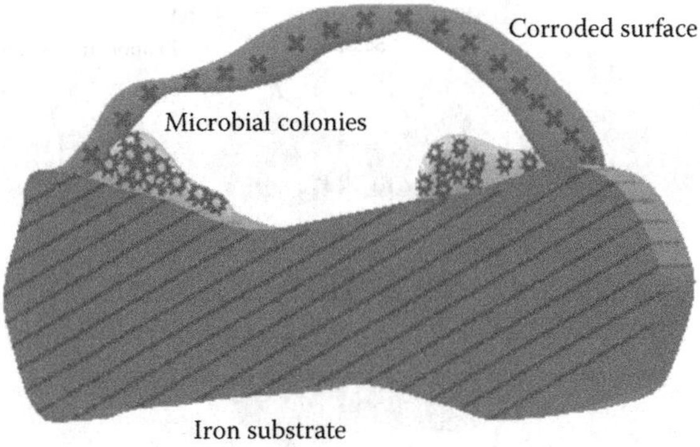

FIGURE 6.1 MIC in progress under the biofilm.

that the rate of corrosion is directly proportional to the concentration of metal sulfide present in the vicinity. The metal sulfide and hydroxide are products of the corrosion process. The rate of anodic reaction is accelerated by the adsorbed hydrogen-driven cathodic reaction (Loosdrecht et al., 2002).

The precipitated metal sulfide has a depolarizing effect, thus resulting in production of hydrogen. The thiosulfate and polythionate metabolites also aggravate the corrosion problem (Figure 6.3).

The process of biofilm formation involves the following steps:

a. Planktonic colonization of the vicinity of an exposed weak point.
b. Development of exopolymer chains anchoring the planktonic SRB to the reaction site (Sessile state) (Beech et al., 1994).
c. Formation of extensive network of exopolymers and consequent thickening of the biofilm matrix.
d. Initiation of the corrosion process under anoxic conditions of the biofilm.

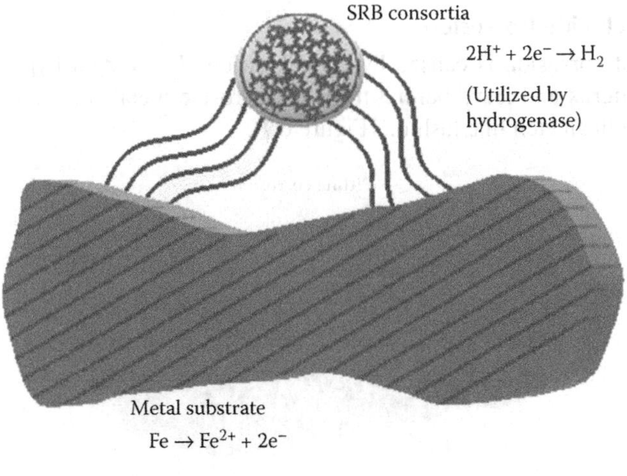

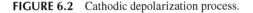

FIGURE 6.2 Cathodic depolarization process.

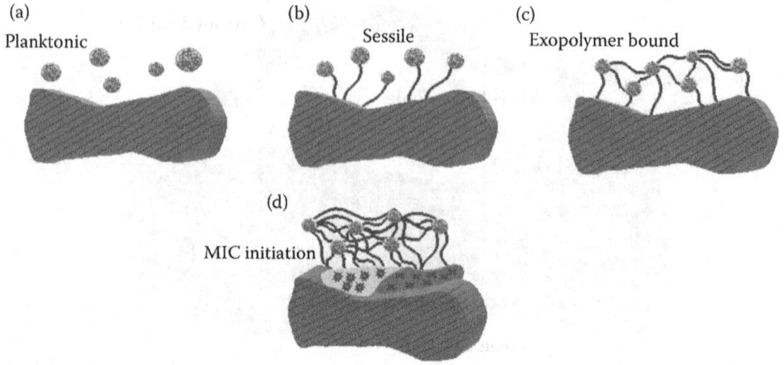

FIGURE 6.3 Stages of biofilm formation.

6.3.4 Nature of Corrosion

6.3.4.1 Pitting Corrosion

This form of corrosion is the resultant of passive breakdown of a metal caused by the metabolites of SRB forming multiple pits of varying sizes on the metal surface, as shown in the Figure 6.4 (Amadi et al. 2007). For example, stainless steel pipeline corrosion (Liang et al., 2016).

6.3.4.2 Flask Shaped Cavitation

This form of corrosion occurs with the SRB entering the hairline cracks formed as a result of the stress rupture of the substrate (Figure 6.5).

6.3.4.3 Uniform Layer Corrosion

This form of corrosion is caused by acid-producing bacteria. The metabolites of the biofilm cause the total erosion of the top layer of the substrate (Figure 6.6) (Wells et al., 2012).

6.3.4.4 Etch Line Corrosion

This form of corrosion is caused by fungal action. The fungal hyphae spread in irregular patterns and leave behind trails of corrosive metabolites corroding away the substrate in an etch line fashion (Figure 6.7).

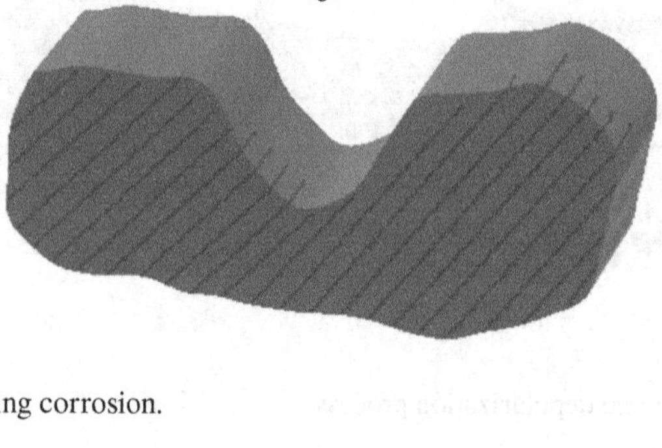

FIGURE 6.4 Pitting corrosion.

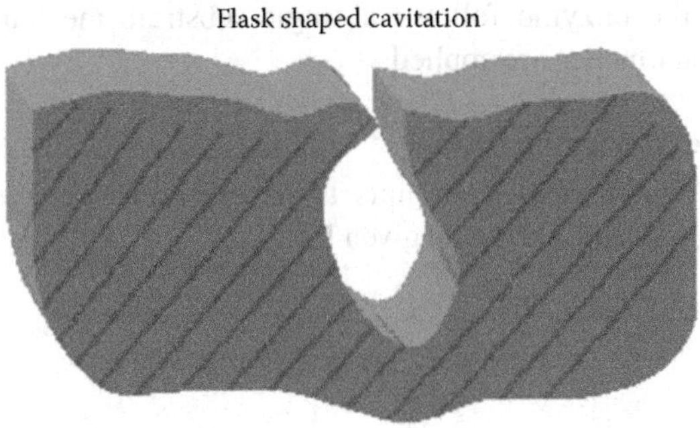

FIGURE 6.5 Flask-shaped cavitation.

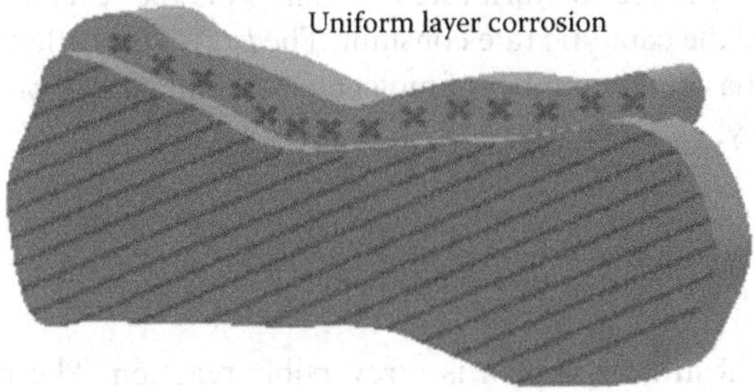

FIGURE 6.6 Uniform layer corrosion.

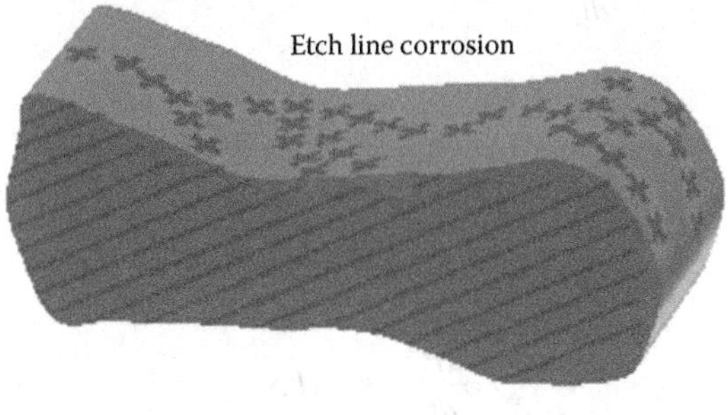

FIGURE 6.7 Etch line corrosion.

6.3.5 Kinetics Prediction Models

As biofilm formation in case of MIC phenomenon is essentially kicked off by the release of enzymes, as explained earlier, enzyme kinetics models are used. Enzyme kinetics encompasses the understanding on how the enzyme functions, formation of the enzyme substrate complex, and the factors effecting enzyme substrate complex. One can classify these mechanisms into:

a. Single-substrate mechanisms.
b. Multiple-substrate mechanisms.

In this case, the enzyme follows a single-substrate mechanism. Therefore, Michaelis–Menten kinetics are applied.

6.3.5.1 Michaelis–Menten Model

The Michaelis–Menten model attempts to describe ideal enzyme kinetics. An enzymatic reaction in equilibrium is given by:

$$E+S \underset{k_r}{\overset{k_f}{\rightleftharpoons}} ES \xrightarrow{k_{cat}} E+P$$

where E is the Enzyme, S is the Substrate, ES is the enzyme-substrate complex, P is the final product, k_f is the forward rate constant, k_r is the reversed (backward) rate constant and k_{cat} is the catalytic rate constant. The k_{cat} is also called turnover rate and is defined as the maximum number of molecules an enzyme can produce per second. It is represented by:

$$k_{cat} = \frac{V_{max}}{E_T}$$

The enzyme-substrate complex is a reversible reaction. The rate laws of each reactant from this equation would be:

$$\frac{dE}{dt} = -k_f \cdot E \cdot S + k_r \cdot ES + k_{cat} \cdot ES$$

$$\frac{dS}{dt} = -k_f \cdot E \cdot S + k_r \cdot ES$$

$$\frac{dES}{dt} = k_f \cdot E.S - k_r \cdot ES - k_{cat} \cdot ES$$

$$\frac{dP}{dt} = k_{cat} \cdot ES$$

while solving this equation, the steady state assumption is relied upon, which in an equation form is:

$$E + ES = E_o$$

This assumption states the concentration of the free or bound enzymes is constant. After solving the rate of product formation (P), we obtain:

$$v = \frac{dP}{dt} = \frac{(V_{max}) \cdot S}{k_m + S}$$

This is the Michaelis–Menten kinetics equation. Here k_m represents the substrate concentration when the reaction rate is half of its maximum value. It is given by the following equation:

$$k_m = \frac{k_r + k_{cat}}{k_f}$$

In biofilm reactions, this same equation is used to derive the following formula:

$$D\frac{d^2C}{dz^2} = q\frac{C}{k+C}X_f$$

with boundary conditions of: At $z = 0$, $dC/dz = 0$, and at $z = 1$, $C = C_1$.

The annotations are explained in the Adhomian decomposition method, in the following section.

6.3.6 Formulating Biofilm Formation

Biofilm structure formation is governed by various physical and biological factors. Common factors, which come into interplay here, are growth yield, mass transfer, conversion rates (substrate), detachment, etc. These are further micromanaged by specific strain dependent factors. They come in two forms—porous, heterogenous structures, and compact ones.

Mass transfer is affected by pores only at a local level. So, at lower flow rates and detachment forces, the porous and heterogeneous structures are formed. While at higher flow rates, the sheer increase of shear stress lead to the formation of compact biofilms. Through all these factors, one seeks to assert substrate flux conditions by use of various analytical expressions.

6.3.6.1 Adhomian Decomposition

Muthukaruppan et al. (2013) has used this method for analyzing biofilm for a square law of microbial death rate. Taking substrate consumption, as defined by Michaelis Menten Kinetics, the nonlinear equations thus obtained with boundary values can be both solved by the Adomian decomposition method and other numerical methods, which are prevalently used in programming software. The main advantage of Adomian decomposition—and why it is preferred—is because it procures a fast converging series with repeated iterations. So much so that even the first iteration itself, as shown in various case studies, has a high degree of accuracy.

The nonlinear ODE obtained is as follows:

$$D\frac{d^2C_f}{dz^2} = \frac{r^2M}{d}\left(\frac{C_f}{k+C_f}\right)^2$$

where k is the Michaelis–Menten constant, C_f is the biofilm's substrate concentration, D is biofilm's diffusion coefficient, M being biomass yield per substrate unit, d is Microbial death constant, z is taken as a coordinate value.

Making it dimensionless by including the following parameters, the equation can be solved using Adomian decomposition:

$$C = \frac{C_f}{k};\ x = \frac{z}{l};\ t = \frac{Mr^2l^2}{dkD};\ C_l = \frac{C_1}{k}$$

Where l is biofilm thickness, t is the dimensionless version of biofilm thickness, C_1 is the substrate concentration outside the biofilm. The two equations we obtain on substrate concentration, C, and concentration flux, $\varnothing$, are:

$$\frac{d^2C}{dx^2} = t\left(\frac{C}{1+C}\right)^2$$

$$\varnothing(x) = \frac{1}{\sqrt{t}}\left|\frac{dC}{dx}\right|_{x=1}$$

Now, Adomian decomposition method essentially depends on the nonlinear differential equation, suppose;

$$F(x, y(x)) = 0$$

Breaking it into its linear (let L) and nonlinear (let N) parts, one gets:

$$L(y(x)) + N(y(x)) = 0$$

L is taken as an invertible operator and solving for $L(y)$ gives us:

$$L(y) = -N(y)$$

As L is invertible, taking inverse on both sides, we obtain:

$$y = -L(N(y)) + \varphi(x)$$

where $\varphi(x)$ is the constant of integration. This y can now be represented in the form of an infinite series:

$$y = \sum_{n=0}^{\infty}(y_n)$$

Following this, the nonlinear term (N) can be written in terms of Adomian polynomials (A_n):

$$N(y) = \sum_{n=0}^{\infty}(A_n)$$

where $(A_n(x)) = \left|\frac{1}{n!}\frac{d^n}{d\lambda^n}N\sum_{n=0}^{\infty}(\lambda^n y_n)\right|\lambda = 0.$

Substituting these values y can be solved from the previous equation after some truncations in the infinite series.

Coming back to the substrate concentration equation, when solved, the resulting equation is:

$$C(x)=C_l+\frac{t}{2}\left(\frac{C_l}{1+C_l}\right)^2(x^2-1)+\frac{t^2C_l^3}{(1+C_l)^5}\left(\frac{x^4}{12}-\frac{x^2}{2}+\frac{5}{12}\right)$$
$$+\frac{t^2C_l^2}{4(1+C_l)^6}\left(\frac{x^6}{30}-\frac{x^4}{6}+\frac{x^2}{2}\right)-\frac{t^2C_l^5}{(1+C_l)^7}\left(\frac{x^6}{30}-\frac{x^4}{6}+\frac{x^2}{2}\right)$$
$$+\frac{2t^2C_l^2}{(1+C_l)^7}\left(\frac{x^6}{360}-\frac{x^4}{24}+\frac{5x^2}{24}\right)+\frac{3t^2C_l^6}{4(1+C_l)^8}\left(\frac{x^6}{30}-\frac{x^4}{6}+\frac{x^2}{2}\right)$$
$$-\frac{2t^2C_l^5}{(1+C_l)^8}\left(\frac{x^6}{360}-\frac{x^4}{24}+\frac{5x^2}{24}\right)+\frac{t^2C_l^4}{360(1+C_l)^8}(-155+66C_l)$$

The concentration flux equation yields the following solution:

$$\varnothing=\frac{1}{\sqrt{t_l}}\left[t\left(\frac{C_l}{1+C_l}\right)^2-\frac{2}{3}\frac{t^2C_l^3}{(1+C_l)^5}+\frac{2}{15}\frac{t^2C_l^2}{(1+C_l)^6}-\frac{8}{15}\frac{t^2C_l^5}{(1+C_l)^7}\right.$$
$$\left.+\frac{8}{15}\frac{t^2C_l^2}{(1+C_l)^7}+\frac{2}{5}\frac{t^2C_l^6}{(1+C_l)^8}-\frac{8}{15}\frac{t^2C_l^5}{(1+C_l)^8}\right]$$

These equations give the needed value of concentration flux factor and thereby help in finding the property of biofilm formed through its microbial death rate.

6.3.7 Decomposition Rate Evaluation

The weight loss of a metal due to decomposition, or more accurately in this case—corrosion, is dependent on various factors pertaining to immediate or absolute environments. As the corrosion rate is determined the amount of loss can be quantified. This weight loss is seen to be increasing with increased periods of exposure; more in saltwater environments as opposed to their freshwater counterparts.

Corrosion of metals is heavily influenceby redox reactions, environmental physicochemical properties, oxygen diffusion rate, and microbial growth in addition to the previously mentioned factors. The analytical models for this are solely developed with the purpose of predicting the corrosion rate of a metal exposed to a certain environment.

6.3.7.1 Laplace Transformation of Plank–Nerst Equation

Ukpaka et al. (2011) has developed an analytical model for corrosion rate with respect to oxygen diffusion transport in a water environment. To achieve this, at first the model is taken to represent the steel's corrosion in the presence of water, shaped by oxygen diffusion, while neglecting the convective motion of oxygen. Therefore, the Plank–Nernst law is taken as:

$$\overrightarrow{\varnothing_k} = -D_r^o \vec{\nabla} C_k + \left(\frac{-z \cdot D_r^o \cdot F}{RT} C_k \right) \vec{\nabla} E$$

where, $\overrightarrow{\varnothing_k}$ is the mass flux, D_r^o is the oxygen diffusivity constant in the rust layer, C_k is the species concentration, F is Faraday constant, R is Ideal Gas constant, T is absolute temperature, and E is the electrolyte's electric potential.

Taking into consideration the migration and diffusion, the mass transport equation becomes

$$\frac{\partial C_k}{\partial t} + \vec{\nabla} \overrightarrow{\varnothing_k} = S_k$$

where, S_k is an oxygen production/consumption source term.

Assuming oxygen consumption occurs only at the metallic surface, due to biochemical reactions; substituting the second equation into the first we get the migration diffusion equation:

$$\frac{\partial C_k}{\partial t} + \frac{\partial}{\partial x} \left(\frac{-z \cdot D_r^o \cdot F}{RT} \frac{\partial E}{\partial x} C - D_r^o \frac{\partial C}{\partial x} \right) = 0$$

Taking only the diffusion equation into account, neglecting the migration part;

$$E = \frac{\partial C}{\partial t} - \frac{\partial}{\partial_X} [\partial C / \partial x] = 0$$

Assuming D_r^o to be dimensionless, and $D_r^o = W(x) \cdot D_*^o$; where D_*^o is independent of position.

So, the previous equation can be written as

$$\frac{\partial C}{\partial t} - D_*^o \frac{\partial}{\partial_X} \left[W(x) \frac{\partial_C}{\partial_X} \right] = 0$$

$$\underset{\text{or}}{\Rightarrow} \frac{\partial_C}{\partial t} - D_*^o \cdot W(x) \frac{\partial^2 C}{\partial_{X^2}} = 0$$

Thus, a second order differential is obtained. So, Laplace transformation and separation of variables can be applied to solve it

$$\frac{\partial c}{\partial t} = D_*^o \cdot W(x) \frac{\partial^2 C}{\partial_{X^2}}$$

So, $C = TX$; where C is the oxygen concentration, T is the time coefficient, and X is the position coordinate. Differentiating it twice with respect to t and x, two *equations can be obtained*:

$$\frac{\partial c}{\partial t} = T^1 X$$

$$\frac{\partial^2 C}{\partial_x^2} = TX^{II}$$

Substituting them in the primary equation that was obtained:

$$T^1 X = D_*^o \cdot W(x) TX^{II}$$

$$\text{or} \quad \frac{T^1 X}{TX} = D_*^o \cdot W(x) \frac{TX^{II}}{TX} = \lambda^2$$

$$\text{or} \quad \frac{T^1}{T} = \lambda^2 \quad \text{and} \quad D_*^o \cdot W(x) \frac{X^{11}}{X} = \lambda^2$$

The first equation becomes a first order differential equation if we resolve further;

$$T^1 - \lambda^2 T = 0$$

Solving it using Laplace, the resulting equation is:

$$T_{(t)} = t_o e^{\lambda^{2} f}$$

The second equation obtained can be written as;

$$D_*^o \cdot W(x) X^{11} = \lambda^2 x$$

Solving through Laplace transform and application of partial law fraction, gives:

$$X_{(t)} = \frac{X_o}{\lambda}(e^{\lambda} - e^{-\lambda})$$

Among the two equations at hand, the second one is used to obtain the biocorrosion reaction's functional parameters. As for the first equation, it can be used to modify the general formula of corrosion rate, which is;

$$c_R = \frac{534W}{DAt}$$

to:

$$C_R = \frac{534W}{DAt_o e^{\lambda^2 f}}$$

This corrosion rate is now tailored to the factor of oxygen diffusion transport in water, making it easier to apply.

Previously, $C = TX$ was taken. Substituting the resolved equations now

$$C_{(t)} = t_o e^{\lambda^2 f}\left[\frac{X_o}{\lambda}(e^{\lambda} - ^{-\lambda})\right]$$

Putting the constant values yield

$$C_{(t)} = 0.679\frac{X_o}{\lambda}t_o e^{\lambda^2 f}$$

This equation gives us the oxygen diffusion transport rate at any given time and space.

6.3.7.2 Butler–Volmer Equation Modification

The corrosion current has a lukewarm relationship with the potential and pH. Usual iron dissolution comprises active, transition, passive, and pitting stages. The biofilm spread is heavily influenced by the amount of dissolved oxygen through differential aeration. Localized corrosion shifts to a broader area just by an increase in the amount of dissolved oxygen. In the presence of saline water, however, if chloride ions are present, a biofilm would slow down the corrosion rate.

Picioreanu and van Loosdrecht (2002) have formulated a mathematical model based on the factor of differential aeration and saline presence; while conjuring out a relation between corrosion rate and corrosion potential. A major positive aspect of this model is that the cathodic and anodic regions need not be defined. However, as there are a lot of passive undercurrents affecting the metal pertaining to the surroundings, sometimes due to these the potential cannot be directly correlated with rate; therefore, the model should be employed with caution.

As mentioned before, the anodic current is affected by its four phases.

$$j_1 = \left[(1-\sigma)j_{1,a} + \sigma j_{1,p}\right](1-\sigma_p) + \left[(1-\sigma_{pit})j_{1,apit} + \sigma_{pit}j_{1,p}\right]$$
$$\times \sigma_p \exp\left[\frac{-\alpha_1 F}{RT}(\Phi_M - \Phi - E_{pit})\right]$$

where j_1 is the anodic current density, σ is the shift of regions, Φ is the solution potential, Φ_M is the metal potential, α_1 is the charge transfer coefficient ranging from 0 to 1, F is Faraday's constant, R is ideal gas constant, and T is the absolute temperature.

The Butler–Volmer equation is used in depicting the current density of the active region:

$$j_{1,a} = j_{1,0}\left\{\exp\left[\frac{\alpha_1 F}{RT}(\Phi_M - \Phi - E_{e,1})\right] - \exp\left[\frac{\alpha_1 F}{RT}(\Phi_M - \Phi - E_{e,1})\right]\right\}$$

where $j_{1,0}$ is current exchange density, and the given fraction can be calculated from the Tafel slope. Some of the values can be calculated by the following formulas:

- At local conditions, the equilibrium potential is: $E_{e,1} = E^0_{e,1} + \frac{0.059}{2}\log(C_2)$
- $\sigma = \exp\frac{\left[2\frac{\alpha_1 F}{RT}(\Phi_M - \Phi - E_p)\right]}{1+\exp\left[-2\frac{\alpha_1 F}{RT}(\Phi_M - \Phi - E_p)\right]}$; as proposed by Hines.
- Flade potential is given by: $E_p = E^0_p - 0.059\,pH$
- $\sigma_p = \exp\frac{\left[4\frac{\alpha_1 F}{RT}(\Phi_M - \Phi - E_{pit})\right]}{1+\exp\left|-4\frac{\alpha_1 F}{RT}(\Phi_M - \Phi - E_{pit})\right|}$.
- $\sigma\, pit = \exp\frac{\left|2\frac{\alpha_1 F}{RT}(E_{pit} - E_p)\right|}{1+\exp\left|-2\frac{\alpha_1 F}{RT}(E_{pit} - E_p)\right|}$

According to these equations current density above pitting potential experiences a sudden surge:

$$j_{1,apit} = j_{1,0}\left\{\exp\left|\frac{\alpha_1 F}{RT}(E_{pit} - E_{e,1})\right| - \exp\left[\frac{-\alpha_1 F}{RT}(E_{pit} - E_{e,1})\right]\right\}$$

However, the pitting potential linearly decreases

$$E_{pit} = -0.29 + 0.125p$$

Apart, from the anodic reactions, the cathodic reactions too can be given by Butler–Volmer Equations:

$$j_2 = j_{2,0}\left\{\exp\left|\frac{\alpha_2 F}{RT}(\Phi_M - \Phi - E_{e,2})\right| - \exp\left[\frac{-\alpha_2 F}{RT}(\Phi_M - \Phi - E_{e,2})\right]\right\}$$

The $j_{2,0}$ usually depends on molar concentrations and is calculated as

$$i_{2,0}(C_1, C_4) = i_{2,0}^{\infty}(C_1)^{\gamma_1}(C_4)^{\gamma_4}\,[1S]$$

In this case, the equilibrium potential (at 298 K) is given by

$$E_{e,2} = E_{e,2}^0 + \frac{0.059}{4}\log\left(\frac{C_1}{C_1^*}\right) - 0.059\log(C_4)$$

The Tafel slope is similarly used for the fraction part. Apart from these, oxygen intake can be calculated by Monod reactions (Meena et al., 2014). The ferrous ion oxidations are calculated depending on ferrous ion concentrations, pH value and dissolved oxygen. The ferric ion precipitation and ionic product of water, being homogenous equations, which can be taken at equilibrium, can be represented by respective equilibrium constants.

While solving this model, the assumptions taken are:

a. System is taken at steady state. This is pertaining to the fact that chemical reactions occur at a relatively faster rate than biofilm phase changes.
b. Transport terms for chemicals are defined by molecular diffusion and migration in the mass flux equations.
c. Nernst–Einstein equations are used to have a relation between ionic mobility and diffusion coefficient, assuming the solution to be infinitely diluted.
d. System was taken to be isothermal in nature wherein the physical parameters are constant.

So the volume element can be now given by

$$D_i\nabla^2 C_i + D_i\frac{z_i F}{RT}\nabla(C_i\nabla\Phi) + \sum_{j=1}^{1} s_{ij}r_j = 0$$

where D_i is the coefficient of diffusion, S_{ij} gives the coefficient of stoichiometry.

While solving this some boundary conditions are set: electroneutrality condition is maintained, potential value on the outside boundary is taken as 0, boundary conditions of each reacting chemicals must have their rates conform to the flux of intake, and the cathodic current must counterbalance the anodic one.

After this stage, simple arithmetic methods to calculate boundary value problems are incorporated. At a 3D spatial distribution, the final relation between concentrations, total ionic currents (j), and potentials of solution, as well as metal can be obtained as

$$j = -\frac{F^2}{RT}\nabla\Phi\sum_{i=2}^{7} z_i^2 D_i C_i - F\sum_{i=2}^{7} z_i D_i \nabla C_i$$

where i represents the number of chemical species involved, which in this case is 7. Therefore, this equation obtained can be used in several case studies to formulate a corrosion rate if the electrode and solution potential is obtained.

6.3.8 Relevant Data Handling Techniques

Data obtained during experimental procedures may be broadly classified into the following types on the basis of their source:

a. *Structured data*: These are the data sets which comply with the predefined type, format, and structural criteria. For example, the analytical and experimental data obtained under controlled conditions.
b. *Semistructured data*: These are the data sets that are self-describing in nature and possess minimal structural features. For example, sensor generated data on biofilm behavior.
c. *Unstructured data*: These are the data sets lacking an inherent structure. For example, qualitative parametric outputs, random variables, etc.

To ensure a hassle-free handling of the obtained data it is quintessential to divide the data into three groups. The divisions are as mentioned hereafter:

a. *Training data*: A model is built using individual tests or model selection techniques. Optimal parameters are identified using a trained data-analyzing model.
b. *Test data*: The model built with the training data is evaluated followed by testing using hypothesis data.
c. *Confirmation data*: The model built with the training data is evaluated followed by confirmation of finding by test data.

The prevalent data handling techniques used in analysis of laboratory data are as follows:

6.3.8.1 Classification and Regression Trees (CART)

It is a model which predicts the target output value using multiple input values. The algorithmic structure involves a series of conditional and absolute questions and answers, which will link the input data points to the next level of question. The resultant of this algorithm will be a tree structure terminating in points called nodes. The principal components of this algorithm include data splitting rules, stopping rules, and prediction criteria at the end of each terminal node. This method renders an overall accuracy of 97.35% (Nisbet et al., 2009).

Characteristic features of this method are

i. *Nonparametric nature*: The algorithm does not heavily rely on data belonging to certain distribution types.

ii. The impact of outliers (any sampling unit that does not belong to the population of interest) in the input variable is minimum.
iii. Incorporates both testing with the data sets as well as cross checking to determine the goodness of fit.
iv. Use of same variable is allowed in different parts of the tree.
v. This method can be used in conjunction with other methods.

6.3.8.2 K-Nearest Neighbor Method (KNN)

This algorithm classifies a new data point on the basis of attributes and training samples. For a given query point *K*, a number of closest neighborhood points are identified and are assigned to the same class. In other words, it uses a majority vote in the classification of *K* objects. This method is solely based on memory of the algorithm and does not involve any model to fit. Hence, any ties can be broken at random. The average accuracy of this method is 71.28% (Danades et al., 2016).

Characteristic features of this method are

i. All the data points feature in the n dimensional Euclidean space.
ii. Classification is withheld till a new instance arrives.
iii. Comparison of feature vectors of each point forms the basis of classification.
iv. Real or discrete value can be assumed by the target function.

6.3.8.3 Clustering

The method of identifying similar groups of data in a data set is called clustering. This method attempts to fit a data set with a particular model assuming the data are the resultant of a mixture of underlying probability distributions. The method is broadly classified into hard clustering where each data point belongs completely to a cluster or not, and soft clustering where each data point is assigned with a probability to be grouped into a cluster. Two major techniques involved under this approach are:

i. *Expectation maximization*: In this approach a parametric probability distribution represents each particular cluster. This is followed by parameter estimation by using an iterative refinement algorithm. This algorithm comprises of an expectation step and a maximization step. The former step involves probability assignment to each data point while the latter step involves reestimation of model parameters.
ii. *Conceptual clustering*: In this approach a classification scheme is produced for a set of unlabeled data points. A popular method of COBWEB is used for the purpose of incremental conceptual learning. It renders a hierarchic clustering as a classification tree. Each node represents a cluster and its corresponding probabilistic description.

Characteristic features of this method are:

i. Most methods biased towards finding clusters with features such as size, shape, or dispersion.
ii. Methods with least square, as criteria tend to find equal number of data points in each cluster.
iii. Methods based on nonparametric density estimation tend to be least biased.

The efficiency of this method is highly variable on the basis of the basic criteria used for clustering.

6.3.8.4 Artificial Neural Networks

The abstract mathematical method built on the foundations of the modern neuroscience employed for solving complex mathematical data problems is known as neural networks. The most widely used form of this method is that of BPANN. It is a multilayer feed forward method with error back propagation-trained algorithm. The average efficiency of this method is 84.46% (Safi and White 2017).

Characteristic features of this method are

i. Massively parallel distributive structure
ii. Nonlinearity
iii. Input/output mapping
iv. High degree of adaptability and fault tolerance

6.3.8.5 Support Vector Machines (SVN)

In this method, each data item is plotted as a point in an n-dimensional space. Value of each of the n features is assigned as the coordinate value. The classification solution is arrived at by identifying a hyper plane clearly differentiating the two classes. The average efficiency of this method is found to be around 70% (Danades et al., 2016).

Characteristic features of this method are:

i. Effective in high dimensional spaces.
ii. Cases with number of dimensions exceeding the number of samples can be solved well.
iii. Performance decreases with the increase in the size of the data sets.

6.3.9 Optimal Combination

As this area of research has been taken into the limelight recently, it is at the nascent stage where not all of its processes can be analytically determined. Pre-existing biological models have been predominantly used and then modified based on initial, intermediary, and final conditions as boundary values and tailored equations are obtained. One menacing factor about this whole process is most of these analytical equations are based on one or two factors; but in reality, this type of corrosion is a very complex process encompassing many undercurrents of physical, biological, and environmental process. However, since the laboratory experimental data for many a case study have shown that these equations give reasonable accuracy in the resulting values, they can be used.

An optimal combination would be a universal equation encompassing all the possible factors. However, as we increase the number of factors, human-made errors during the data intake also increase leading to an unfavorable result. So, it is advisable to choose equations that correlate the obtained data point to the desired factor. As far as kinetics, due to this being a single substrate mechanism, the Michaelis–Menten model is preferred. In the phase of biofilm formation, Min'kov et al. (2006) have modified and presented an analytical relation between substrate flux concentration and volumetric flow rate. In the same phase, Muthukaruppan et al. (2013) has presented a relationship of substrate flux concentration to biofilm thickness, substrate concentration on the outside of the biofilm. As the second issue has high accuracy due to the application of the Adhomian method, it might be preferred but it will still depend on available data sets. For the corrosion rate, however, the Plank–Nernst equation solved using the Laplace transform seems more reliable, as the other available option of Butler–Volmer does not always hold true. However, if the specific combination of metal is known to have followed a correlation of potential to rate, the Butler–Volmer equation would yield better results than the generalized version. So, choosing of equation would be mostly circumstantial.

6.4 CONCLUSION

As nature would have it, microbes always find an exception to every generalized theory known to man. So is the case with theorizing the process of MIC. An all-inclusive approach perfectly dealing with all the variables has yet to be found. However, in this chapter an attempt has been made to put together simple techniques of biological experimentation models coupled with data handling methods to formulate optimal combinations. The Kinetic modeling of the biocorrosion process can be effectively accomplished by the Michaelis–Menten model, while that of corrosion rate by Plank–Nernst equation solved using Laplace transform. Usually, MATLAB is used for solving them, as it has many in-built functions for boundary value problems. For example, in the Adhomian Decomposition method in Matlab, for two-point boundary value problems consisting of ordinary differential equations (ODEs)—a built-in function called bvp4c is popularly used. Similarly, the symbolic package of Matlab is used to calculate the analytical derivatives of the Butler–Volmer model. As the Newton–Rhapson system was followed, the linear system in it had been subjected to the BiCG (biconjugated gradient method) with Jacobi preconditioning. This method is frequently used for large systems of linear equations. Any large system of data points are smoothly iterated through predefined programs—significantly reducing calculation times. Artificial neural network was found to be the optimally suited handling method for data inputs of the experimentation processes and subsequent modelings, owing to its high adaptability to variations and superior efficiency. Moreover, if the species is of a rare kind and the studies need to be focused on it's definitive characteristics, an individual based model, like BacSim (Kreft et al., 2001) is more preferable than analyzing them in bricks. A combination of the aforementioned methods is found to be capable of providing a qualitative and quantitative explanation to the process of corrosion induced by microbial infestation.

REFERENCES

Alabbas, F. M., Williamson, C., Bhola, S. M., Spear, J. R., Olson, D. L., Mishra, B., & Kakpovbia, A. E. 2013. Microbial corrosion in linepipe steel under the influence of a sulfate-reducing consortium isolated from an oil field. *Journal of Materials Engineering and Performance*, 22(11), 3517–3529. https://doi.org/10.1007/s11665-013-0627-7

Amadi, S. A., Ukpaka, C. P., & Neeka, J. B. 2007. Mechanisms of the microbial corrosion of aluminum alloys. *Journal of Industrial Pollution Control*, 23(2), 197–208. https://doi.org/10.5006/1.3580810

Beech, I. B., & Gaylarde, C. C. 1999. Recent advances in the study of biocorrosion: An overview. *Revista de Microbiologia*, 30(3), 177–190. https://doi.org/10.1590/S0001-37141999000300001

Beech, I. B., & Sunner, J. 2004. Biocorrosion: Towards understanding interactions between biofilms and metals. *Current Opinion in Biotechnology*, 15(3), 181–186. https://doi.org/10.1016/j.copbio.2004.05.001

Beech, I., Bergel, A., Mollica, A., Flemming, H., Scotto, V., & Sand, W. 2000. Simple methods for the investigation of the role of biofilms in corrosion. *Brite-Euram III Themat.* Netw. N° ERB BRRT-CT98-5084; 5084.

Beech I. B., Cheung C. W. S., Chan C. S. P., Hill M. A., Franco R., & Lino A. R. 1994. Study of parameters implicated in the biodeterioration of mild steel in the presence of different species of sulfate-reducing bacteria. *Int. Biodeterior. Biodegradation*, 34:289–303. 10.1016/0964-8305(94)90089-2

Bouman, S., Lund, D., Driessen, F. M., & Schmidt, D. G. 1982. Growth of thermoresistant streptococci and deposition of milk constituents on plates of heat exchangers during long operating times. *J. Food Prot*, 45, 806–12, 815.

Danades, A., Pratama, D., Anggraini, D., & Anggriani, D. (2016). Comparison of accuracy level K-Nearest Neighbor algorithm and support vector machine algorithm in classification water quality status. 137–141. 10.1109/FIT.2016.7857553.

Gu, T. 2012. New understandings of biocorrosion mechanisms and their classifications. *Journal of Microbial & Biochemical Technology*, 4(4). https://doi.org/10.4172/1948-5948.1000e107

Gu, T. 2014. Theoretical modeling of the possibility of acid producing bacteria causing fast pitting Biocorrosion. *Journal of Microbial & Biochemical Technology*, 6(2), 68–74. https://doi.org/10.4172/1948-5948.1000124

Koch G. H., Brongers M. P. H., Thompson N. G., Virmani Y. P., & Payer J. H. 2001 *Corrosion Cost and Preventive Strategies in the United States*. FHWA-RD-01-156.CC Technologies Laboratories, NACE International, Dublin, OH.

Koch, G. H., Thompson, N. G., Brongers, M. P. H., Virmani, P. Y., & Payer, J. H. 2010. Cost of corrosion, study unveiled. *Nace International*, (NACE), 65(1), 10.

Kreft, J. U., Picioreanu, C., Wimpenny, J. W. T., & Van Loosdrecht, M. C. M. 2001. Individual-based modelling of biofilms. *Microbiology*, 147(11), 2897–2912. https://doi.org/10.1099/00221287-147-11-2897

Liang, R., Aktas, D. F., Aydin, E., Bonifay, V., Sunner, J., & Suflita, J. M. 2016. Anaerobic biodegradation of alternative fuels and associated biocorrosion of carbon steel in marine environments. *Environmental Science and Technology*, 50(9), 4844–4853. https://doi.org/10.1021/acs.est.5b06388

Little, B. J., Richard I, R., & Pope, R. K. (2000). The relationship between corrosion and the bilogical sulfur cycle. *Corrosion NACE International*, 56(394), 194–196.

Maluckov, B. S. (2012). Corrosion of Steels Induced by Microorganism. *Association of Metallurgical Engineers of Serbia (AMES)*, 223–231.

Marangoni, P. R. D., Robl, D., Berton, M. A. C., Garcia, C. M., Bozza, A., Porsani, M. V., … Pimentel, I. C. (2013). Occurrence of Sulfate Reducing Bacteria (SRB) associated with biocorrosion on metallic surfaces in a hydroelectric power station in ibirama (SC)-Brazil. *Brazilian Archives of Biology and Technology*, 56(5), 801–809. https://doi.org/10.1590/S1516-89132013000500011

Meena, V., Indira, K., Kumar, S., & Rajendran, L. 2014. A new mathematical model for effectiveness factors in biofilm under toxic conditions. *Alexandria Engineering Journal*, 53(4), 917–928. https://doi.org/10.1016/j.aej.2014.09.003

Min'kov, L. L., Pyl'nik, S. V., & Dueck, L. H. 2006. Steady-state problem of substrate consumption in a biofilm for a square law of microbial death rate. *Theoretical Foundations of Chemical Engineering*, 40(5), 496.

Muthukaruppan, S., Eswari, A., & Rajendran, L. (2013). Mathematical modelling of a biofilm: The adomian decomposition method. *Biomedical & Life Sciences*, 5(4), 456–462. https://doi.org/10.4236/ns.2013.54059

Nisbet, R., Elder, J. F., & Miner, G. 2009. *Handbook of Statistical Analysis and Data Mining Applications*. Amsterdam: Elsevier.

Olmstead, W. M., & Hamlin, H. 1900 Converting portions of the Los Angeles outfall sewer into a septic tank. *Engineering News* 44, 317–318.

Picioreanu, C., & van Loosdrecht, M. C. M. 2002. A mathematical model for initiation of microbiologically influenced corrosion by differential aeration. *Journal of The Electrochemical Society*, 149(6), B211. https://doi.org/10.1149/1.1470657

Safi, S., & White, A. 2017 Short and long-term forecasting using artificial neural networks for stock prices in Palestine: A comparative study. *Electronic Journal of Applied Statistical Analysis*, North America.

Ukpaka, C. P., Amadi, S. A., Ahuchogu, I. G., & Odharo, J. 2011. Modeling the rate of biocorrosion and the effects of redox-reactions of metals in water environment. *Journal of Engineering and Technology Research*, 3(13), 371–380. https://doi.org/10.5897/JETR11.058

van Loosdrecht, M. C. M., Heijnen, J. J., Eberl, H., Kreft, J., & Picioreanu, C. 2002. Mathematical modelling of biofilm structures. *Antonie van Leeuwenhoek*, 81, 245–256.

Videla, H. A. 2010. Biocorrosion and biofouling of metals and alloys of industrial usage. Present state of the art at the beginning of the new millennium. *Revista de Metalurgia de Madrid*, 39(Extra), 256–264. https://doi.org/10.3989/revmetalm.2003.v39.iExtra.1128

Wells, P. A. (Tony), & Melchers, R. E. 2011. Microbial corrosion of sewer pipe in Australia - Initial field results. *18th International Corrosion Congress 2011*, 1–12.

Wells, T., Melchers, R., Joseph, A., Bond, P., Vitanage, D., Bustamante, H., … Evans, T. 2012. A collaborative investigation of the microbial corrosion of concrete sewer pipe in Australia. *OzWater-12 Australia's National Water Conference and Exhibition*, May, 8–10.

Xu, D., Li, Y., & Gu, T. 2016. Mechanistic modeling of biocorrosion caused by biofilms of sulfate reducing bacteria and acid producing bacteria. *Bioelectrochemistry*, 110, 52–58. https://doi.org/10.1016/j.bioelechem.2016.03.003

Zhiyuan, L., & Zongdi, S. 2017. The carbon trading price and trading volume forecast in Shanghai City by BP neural network, *International Journal of Economics and Management Engineering*, 11(3), 623–629.

7 Application of Data Mining Techniques in Autoimmune Diseases Research and Treatment

Sweta Bhattacharya and Sombuddha Sengupta

CONTENTS

7.1 INTRODUCTION

The human body is indeed an interesting yet complex unit, formed through million years of evolution, transcending the unicellular to arrive at the complex multicellular. For proper functioning, the body comprises several inherent systems needing to work in harmony to facilitate homeostasis and normal behavior. If a system, or several systems, go awry from their normal function it can lead to severe detrimental or life threatening complications. One of the key systems is the immune system. To explore the immune system or its nuances is beyond the scope of this chapter. Hence, we shall be concentrating on a portion of this vast topic.

Autoimmunity is a situation in which the immune system goes off-course, resulting in an inappropriate response to self elements or self antigens.[1] Earlier, it was proposed that the self reactive lymphocytes were screened and eliminated from the body, preventing an autoimmune response. However, later studies have shown that not all T-cells and B-cells that have the potential to target self antigens are successfully removed. This leads to a situation where the "army is fighting against the king." The damage induced to cells and organs may be due to the antibodies or the T cells of the body. Some common autoimmune diseases have been listed in Table 7.1.

Data mining simply means the process of extracting data, or a mining process to squeeze out valuable information from a collection of existing data. Other terms commonly used in place of data mining, although having a slight different meaning,

TABLE 7.1
Some Common Autoimmune Diseases in Humans

Disease	Self Antigen	Immune Response	Reference
	Organ Specific Diseases		
Hashimoto's Thyroiditis	Thyroid cells	T_{DTH} cells and auto- antibodies	Caturegli et al. 2014[3]
Pernicious Anemia	Gastric parietal cells, intrinsic factors	Auto-antibodies	Bizzaro et al. 2014[4]
Myocardial Infarction	Heart	Auto-antibodies	Li et al. 2013[5]
Myasthenia Gravis	Acetylcholine receptors	Auto-antibodies	Lindstrom et al. 1998[6]
Grave's disease	Thyroid stimulating hormone receptor	Auto-antibody	Weetman et al. 2000[7]
Goodpasture's syndrome	Lung and renal basement membrane	Auto-antibodies	Greco et al. 2015[8]
	Systemic Autoimmune Disease		
Systemic lupus erythematosus (SLE)	DNA, nuclear protein, RBC	Auto-antibodies, complexes	Tan et al. 1982[9]
Multiple sclerosis	Brain	T_{H1} and T_C cells, auto-antibodies	Lublin et al. 1996[10]
Rheumatoid arthritis	Connective tissues	Immune complex, auto-antibody	McInnes et al. 2017[11]
Ankylosing sponkylitis	Vertebrae	Immune complex	Chen et al. 2016[12]

are: data dredging, data archaeology, data analysis, data extraction, etc. Nowadays, people synonymously use this term with another popular term called knowledge discovery from data (KDD).[2] This process of knowledge extraction of useful data from the bulk data has several steps, as listed below:

- *Data cleaning*: To remove noise or inconsistency or misleads in the data collected or stored.
- *Data integration*: A step where multiple sources of data are combined.
- *Data selection*: A step where which analysis is to be done is determined, and the data pertaining to that cause is screened.
- *Data transformation*: Data is consolidated to make it suitable for mining purposes. This is done by summary or aggregation operations.
- *Data mining*: Intelligent methods are used to extract data patterns.
- *Pattern evaluation*: Done to identify truly fascinating patterns.
- *Knowledge presentation*: Various visualization procedures are employed to display the mined knowledge to the interface user.

The steps of data cleaning, data integration, data selection, and data transformation come under the category of preprocessing of the data collected.[13] Data mining is a single and important step in the entire process, as it helps in the retrieval of important patterns to make the vast data collection potentially useful to the interface user. A flowchart of the process of knowledge retrieval has been shown in Figure 7.1.

The incidence of autoimmune diseases is great and vast in number. Through the years, the data pertaining to these diseases is increasing at an exponential rate. Data mining offers a suitable and easy method of dealing with this avalanche of data, and hence, pattern recognition and deciphering becomes easier.[2] The later sections of this chapter explain in great detail the crux of the concept of data mining, along with its application, in order to diagnose or treat some of the commonly occurring autoimmune diseases in the world.

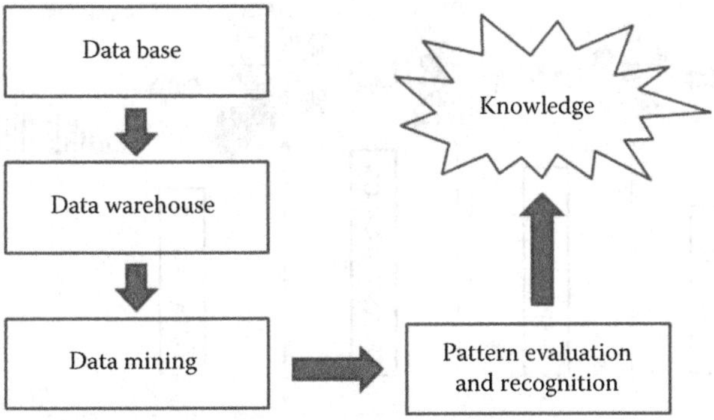

FIGURE 7.1 Knowledge retrieval process.

7.2 MACHINE LEARNING AND KNOWLEDGE DISCOVERY TASKS

Machine learning is an application of artificial intelligence, which enables systems to learn information automatically and improve based on training experience, without being programmed explicitly. The basic objective is to develop computer programs able to access data and learn by themselves, and help in predictions based on examples provided during training. The machine learning technique helps to create systems that can adapt and learn on their own—based on the experience of analyzing data patterns.[14]

Knowledge discovery, on the other hand, is a concept including methods and techniques to extract useful information and understanding from data. The knowledge discovery process comprises multiple stages to be followed sequentially. The steps of knowledge discovery are selection, preprocess, transformation, data mining, interpretation, and evaluation, as shown in Figure 7.2. The crucial part of the knowledge discovery process is the application of machine learning algorithms when analyzing data for mining important information to understand correlations and dependencies among variables.[15]

Machine learning is broadly categorized in to three types, which are supervised learning, unsupervised learning, and reinforced learning.

7.2.1 Supervised Learning

In supervised learning, the system learns a function called "target function" from a labeled training data set. The training data set is analyzed and an inferred function is produced, which is used to predict the value of output variables or dependent variables. The input values of the function consist of a set of attributes and features. The subset of the input variables, for which the output is already known, is considered as training data used to make predictions for the values of unknown output variables.

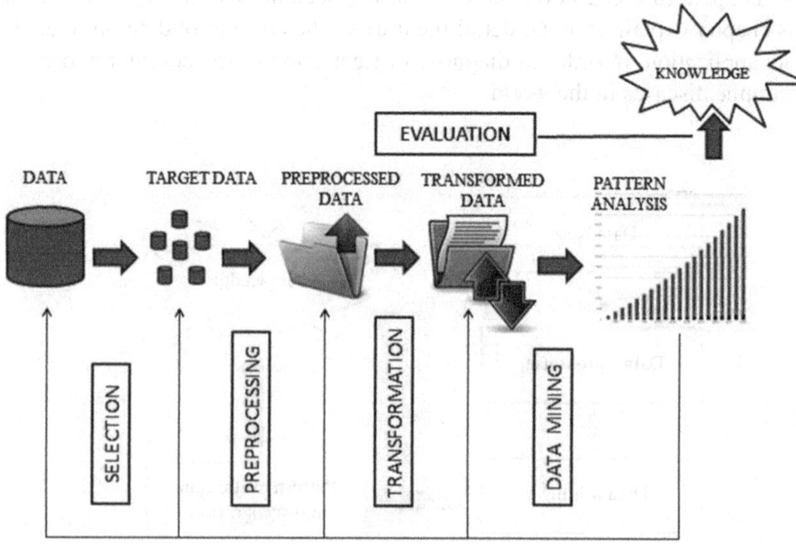

FIGURE 7.2 Steps involved in knowledge discovery.

The supervised learning method is also categorized into two types—classification and regression, on the basis of learning techniques. The popular techniques used in supervised learning are decision trees, instance based learning, k-nearest neighbors (k-NN), genetic algorithm, artificial intelligence, and support vector machines. The details of these techniques are briefly discussed in the following section.[16]

- *Decision Tree*: A decision tree is basically a tree structure having root, branches, and leaves. The tests conducted on an attribute is denoted by the internal nodes, and the branches represent the outcome of the tests. The leaf nodes represent the classes, and the highest node is considered as the root note with the utmost significance.[17]
- *Instance Based Learning* (IBL): It is a learning algorithm where new problem instances are compared with instances found in training data sets to adapt the model to formerly unseen data.[18]
- *k—Nearest Neighbors* (k-NN): It is a type of IBL used for classification of new objects based on attributes and training sample data. The *k* number of objects closest to a given query points are identified and classified in this technique.[13]
- *Genetic Algorithm*: Is a search-based algorithm founded on the theory of natural selection and genetics. In genetic algorithm there are multiple options or choices of solutions for a particular problem. These solutions are recombined and mutated to get the optimal solution, which would be extremely time-consuming using traditional approaches.[19]
- *Artificial Intelligence (AI)*: As said by the father of AI—John McCarthy, it is "The science and engineering of making intelligent machines, especially intelligent computer programs." It is a technique to program a computer or a computational robot to think logically and intelligently, similar to the working of a human brain. As human beings act based on experience gained, AI also uses knowledge gained (through training of data sets) to solve problems, perform predictions, and generate outputs from respective inputs.[20]

7.2.2 Classification

The concept of classification is similar to categorization, wherein a machine-learning technique is used on known data to categorize new data into distinct classes. As an example of class, we can consider blood groups or classification of mails as spam.[13]

7.2.3 Regression

Here the objective is to predict continuous or discrete numerical values and identify the linear relationships existing among attributes in the dataset. The popularly used regression algorithm types are "basic," "linear," and "polynomial."[13]

7.2.4 Unsupervised Learning

In unsupervised learning, the system tries to discover information from unlabeled data sets and find associations between variables. The training data consists of only instances without any data labels. "Association rule mining" and "clustering" are forms of unsupervised learning algorithms.

7.2.5 Association Rule Mining

This technique aims to discover interesting relationship of correlations ,which exist among variables in data sets. The most popular association rule mining algorithm "apriori," is an exploratory data analysis technique used in a wide range of research in biological science and bioinformatics.[21]

7.2.6 Clustering

Clustering is a technique to form groups, or clusters, of data based on common attributes or characteristics. The method analyses the input data set and—based on the nature of the data—forms clusters or groups depending on informative patterns existing in the data set.

7.2.7 Reinforced Learning

It is a family of machine learning techniques wherein the system learns through direct interaction with the environment in order to maximize the notion of cumulative award. The system is completely unaware of the behavior of the environment, and hence, applies trial and error methods to find plausible solutions to a problem.[22]

7.3 AUTOIMMUNE DISEASES: AN OVERVIEW

The immune system has developed over the course of evolution to protect the body from invading pathogens. However, the immune system can lead to pathogenesis in two cases: in the case of immunodeficiency diseases and autoimmune diseases.[23] Immunodeficiency diseases results from the failure of one part, or more than one part, of the immune system to protect the body. An autoimmune disease, on the other hand, is a case where the body's immune system fails to distinguish between itself and foreign antigens. Autoimmune diseases are rare and have been seen to affect 3%–5% of the world population, with thyroid diseases and diabetes among the most common of these conditions.[23] Statistical records have shown there are more than 100 different types of autoimmune diseases, some being organ-specific while others are systemic in nature.

Immune tolerance has been seen to be an innate characteristic, which was reported by Macfarlane Burnet in 1978.[24] Tolerance denotes the state where the immune system is accustomed to the innate antigens present in the body, and hence does not attack the cells or organs from the same body. The concept of autoimmunity was still not fully accepted, even though the concept was reported as early as the twentieth century by Paul Ehrlich who coined the term "horror autotoxicus."[25] Autoimmune diseases can be either organ-specific or systemic in nature. In organ-specific disease, the immune response is induced in response to a single antigen on a particular organ.[1] The cells of that organ are damaged and overproduction or overstimulation of the antibodies may even block further the functioning of the organ. In case of systemic autoimmune diseases, the immune response is towards a wide range of antigens. A large number and types of tissue can be damaged, and the disease is often marked by the hyper stimulation of the T and B cells.[23]

Central tolerance in the thymus and the bone marrow help shape the immune system of the body. This way, the self reacting antibodies are removed in the thymic medulla so that the immune system does not target the self antigens. Even in the presence of the strict surveillance of the central and peripheral tolerance, some of the self reactive antibodies escape leading to the manifestation of autoimmune disease.[26]

The occurrence of autoimmune disease has been seen to be greater in females than males; however, the geoepidemiology becomes even more complex with the consideration of factors like age, ethnicity, gender, etc. The exact reason for the occurrence of autoimmune disease is still under much debate, however, a few main points need to be mentioned at this point (Wang 2015).[23]

Genetics can be a plausible cause for the onset of autoimmune diseases. Diseases like Addison's disease or autoimmune polyendocrinopathy syndrome type I, etc. are example of this. Mutation in the autoimmune regulator gene has been seen, which affects the negative selection in the thymus and bone marrow. Mutation in the tumor necrosis factor (TNF) receptor superfamily member 6 (TNFRSF6), TNFRSF6 bound membrane ligand, and caspase 10 can also lead to apoptosis of lymphocytes, which can lead to the accumulation of a large number of mononuclear cells in the lymphoid tissue, resulting in failure to delete the autoreactive cells.[26] Linkage variants of human leukocyte antigens (HLA), which is the gene product of the major histocompatibility complex (MHC), have also been held responsible for the onset of autoimmune disorders. Genome wide association studies (GWAS) have documented several of these variants.[27] Lastly, epigenetic and stochastic factors have also been known for causing autoimmune disorders. Environmental factors, like infectious agents and molecular mimicry of the pathogens, also hold important roles in these diseases.

The list for the causes of autoimmunity is exhaustive. Substances like glucose and cellular apoptosis have also been designated as causal agents in recent studies. For reference, the main causes have been given in Figure 7.3. Early diagnosis is possible nowadays with the advent of new technology and better diagnosis procedures. One such tool, as mentioned before is data mining. The following sections now elaborate

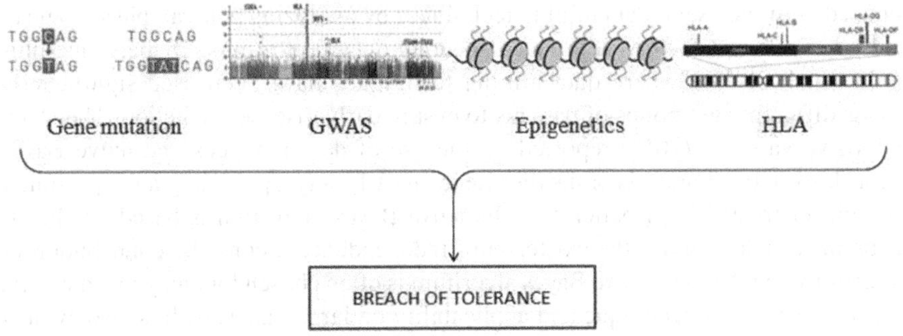

FIGURE 7.3 Causes of autoimmunity.

on the role of data mining in autoimmune disease research, and further show how biology and information technology may complement each other.

7.4 ROLE OF DATA MINING IN AUTOIMMUNE DISEASES

7.4.1 Diabetes Mellitus

The increase in the sedentary lifestyle due the advancement of technology has given diabetes the name, "disease of modern society." Lack of exercise, obesity, and consumption of foods rich in trans fat and saturated fats are some of the key reasons why an individual might be affected with diabetes. According to the data published by the CDC in 2017, diabetes affects 30.3 million individuals of which 7.2 million remain undiagnosed.[28] Biologically, diabetes is a disease marked by a spike in the blood glucose level of an organism.[29] This has been attributed to two reasons: (1) the body cells become unresponsive to insulin, and (2) the body does not produce insulin. Based on these characterizations we can classify them as Type I and Type II.[30] Overtime, an increase in blood sugar level can have serious complications such as imparting damage to eyes, kidney, and nerves, provoking heart attacks, strokes, etc. There are many tests to diagnose diabetes and some are listed below.[28,31]

- *Fasting plasma test*: The blood glucose level at a point in time is checked after the individual has fasted for more than 8 hours.
- *A1C test*: This test gives the average amount of blood glucose over a period of 3 months. Other names of this test include, HbA1C, glycated hemoglobin, and glycosylated hemoglobin test.
- *Random plasma glucose test*: Done for the diagnosis of the disease when the individual has not fasted. Blood for this analysis can be drawn at any time.

7.4.1.1 Application of Data Mining in Diabetes Mellitus

The basic objective behind using data mining is to extract valuable information from data sets using classification, association, and clustering techniques in order to aid in the prediction of diseases. Generally, multiple diagnostic tests are conducted on patients, which are often time consuming and painful. The number of tests could be reduced with the use of data mining techniques by analyzing general, physiological, and other relevant factors to produce accurate detection results. Diabetes mellitus is one such disease where data mining techniques have been used significantly across different age groups of patients to ensure early disease prediction. The study by Aiswarya et al. (2015) reported on the use of decision trees and naïve Bayes algorithms in the diagnosis of the disease at an early stage, providing an opportunity for quicker treatment of patients.[32] The naïve Bayes algorithm is based on Bayes' theorem, which assumes the existence of independence among the contribution or predictor variables. The naïve Bayes algorithm is often chosen for analyzing data sets due to its simplistic technique and applicability on large data sets. It is also evident that the detection of diabetes mellitus is often based on physical tests, which are prone to errors leading to falsified results. The indecisions related to these tests could be rectified with the use of data mining algorithms. The k-NN and ANFIS algorithms have proved to be more accurate in classification of the disease in comparison to other existing approaches.[33] In such cases, the k-NN algorithm helps in classifying the objects and produces predictions based on training data sets in the attribute space. The k means an algorithm uses a partitioning method depending on the input parameters for clustering of the dataset. The ANFIS algorithm uses a combination of the adaptive neural network and fuzzy inference system for prediction of the disease from a classified dataset.

Diabetes mellitus is often detected among patients with irregular lifestyles. The study by Pardha Repalli (2011) highlighted the likelihood of people being affected by the disease based on their lifestyle-related activities. The various factors contributing towards the disease have also been analyzed using decision tree algorithms. The results revealed that age is a significant factor in patients aged >65 years, who are more prone towards the disease in comparison to people aged <20 years.[34] Breault (2001) used naïve Bayes data mining algorithms in predicting the diabetic status among patients, and results generated had a 65% to 85.5% accuracy in the prediction of the disease considering eight variables.[35] Data mining techniques have also been used to predict the chances of heart disease among diabetic patients. In his study, Parthiban (2011) used a naïve Bayes data mining classification algorithm on a data set of patients of Indian origin, while also using the WEKA tool for implementation purposes. The study helped to generate an optimized prediction model using a minimal training data set while considering four major attributes namely age, gender, blood pressure, and blood sugar level of the patients. The results of the study helped to classify the age groups, ranges of body weight, and blood pressure and sugar levels of patients suffering from diabetes.[36]

In cases where medical diagnosis is not evident to detect diabetes mellitus among patients, clustering and classification techniques have proven to provide more accurate results. The use of machine learning techniques have helped in this regard using various tools such as WEKA, Tangra, and MATLAB. Also, these tools help to use various classification techniques on the same data set in minutes, yielding more accurate predictions than the conventional time-consuming methods used in the health care sector. The study by Rakesh (2013) used hybrid data mining techniques combining ANFIS and PCA algorithm, neural network and PCA Algorithm on the data set. The results provide an interesting insight in analyzing the contributions of the various factors towards diabetes mellitus disease, and the predictions have shown more accuracy than the traditional algorithmic approaches.[37]

Diabetes mellitus has become one of the major health issues in today's world, and the major challenge is to prevent the disease rather than cure it. Also, since the disease already affects large numbers of patients there exists an enormous data set in the form of medical and health records. Computer aided diagnostic (CAD) systems could be used successfully for knowledge discovery on these patient records, followed by application of machine learning and data mining algorithms for the detection and prediction of the disease.[38]

Diabetes is an everlasting disease, which has affected numerous lives making it a societal and health care challenge among the present generation. The confusion in detection of this disease is often guided by a lack of awareness and poor detection mechanisms. Data mining approaches help in this regard by investigating large data sets and discovering knowledge with the aim of disease prediction. The patterns found during this process also help in the detection of various other diseases related to diabetes mellitus, thus providing the ability for better treatment planning and risk management.

7.4.2 Lupus Disease

Lupus is a debilitating disease, where the body targets a wide variety of its own cells and organs. It can attack the joints, skin, kidney, heart, brain, lung, etc. Lupus is more common in females than males and statistics have shown that African American women are more prone to it than other ethnicities such as Caucasian or Asian. Hispanic women have been documented to present a more severe form of this disease.[39] Several kinds of lupus exist and their categorizations are given as follows:[39]

- *Systemic lupus erythematosus (SLE)*: The most common form of lupus; it can affect many parts of the body. It can manifest in a severe or a mild form.
- *Discoid lupus*: These give rise to red rashes that do not fade or go away.
- *Subacute cutaneous lupus*: Results in sores when the body has been exposed to sunlight.
- *Drug induced lupus*: Certain medicines can cause an immune response inside the body. These symptoms usually go away after medication intake stops.
- *Neonatal lupus*: It is an extremely rare case of lupus and affects newborns due to antibodies present in the mother.

The common symptoms include pain or swelling in joints, fevers with no known causes, mouth ulcers, swelling in the face and around the eyes, red rashes (butterfly rashes), hair loss, etc. Also, these symptoms overlap with other diseases making detection rather difficult. The symptoms have also been seen to "come and go." These advents and departures are said to be "flares," and common medical practice aims to make these flares less painful.[40] Till now there is no cure for lupus, however, changes in lifestyle and medicines can help to control this disease. The commonly available diagnostic test is the antinuclear antibody testing (ANA).

7.4.2.1 Role of Data Mining in Lupus Disease

Lupus disease has been commonly detected among patients in American, African, Caucasian, and Hispanic countries. The patients suffering from the disease are also prone to cardiovascular and osteoporosis diseases. The disease gets mostly detected at a later stage, due to lack of awareness, leading to high risk of infection and depression. Hence, data mining techniques could be used in the detection of this disease at an early stage to provide the chance of treatment among patients. The study by Gomathi (2015) uses the decision tree approach for the prediction of lupus among patients based on eleven criteria pertinent to the disease, as derived by the American College of Rheumatology. The criteria considered in the study are malar rash, photosensitivity, discoid rash, nonerosive arthritis, oral ulcers, pericarditis, neurological disorders, hematologic disorders, positive antinuclear antibody, and immunologic disorder. Being a technique without any requirement of parameter setting, the decision tree has helped to achieve results which are easier to read and comprehend. The Rapid Minor tool was used in the study where 15 patient data sets were used for creating the decision tree model. The data sets consisted of the clinical profiles of the patients and the generated decision tree revealed age as the most significant attribute, followed by gender. Caucasians were more affected by the disease than Hispanics, Americans, and other groups.[41]

A similar study was conducted by John (2013) highlighting the pattern of Lupus disease among patients based on age, gender, ethnicity and social class factor.[42]

Classification and regression tree (CART) algorithms have also been used in the analysis and prediction of lupus disease. The CART algorithm uses a sequence of questions whose answers build subsequent questions, thus forming a tree structure. The algorithm computation terminates when all questions are answered and no questions are left. The computation of specificity, sensitivity, and accuracy in the study of lupus disease has justified the use of CART algorithms to generate more accurate results in the prediction of the disease.[41] Another study on lupus disease used a data set consisting of seven factors pulled from the case sheets of patients suffering from the disease. The factors considered in the study were age, gender, sample type of serum and plasma, ACR criteria, disease activity (mild, moderate, and severe), and tests conducted. The performance of the algorithm used was validated against the results generated using decision tree, support vector machine, and ID3 algorithm. ID3 is a decision-making algorithm invented by Ross Quinlan, which helps to create a decision tree from an input data set.

7.4.3 Myasthenia Gravis

This is a chronic autoimmune disease demarcated by extreme weakness and the degeneration of skeletal muscles.[43] The name is derived from the Latin "grave muscle weaknesses." This disease has no known cure; however, treatment can lead to control of symptoms, thus improving the quality of life for the person afflicted. Nerve cells run through the length of the body and the transmission of impulses through them makes the day-to-day function of our body relatively simpler. In Myasthenia gravis, there is an error in nerve signal transduction via the neurons due to the generation of antibodies against the acetylcholine receptors.[44] The generated antibodies bind to the receptor ligand interaction sites and prevent the interaction of the receptors with acetylcholine. This hinders the transmission of the impulse and the polarization cannot cross the synapse, resulting in an interruption in the transmission network. Reports have shown that people with myasthenia gravis can develop thymomas, which may result in cancer.[43]

The common symptoms seen are drooping shoulders, difficulties in swallowing, weaknesses in arms and legs, and shortness of breath and blurred vision due to weaknesses in muscles controlling eye movements.[44] Years of dealing with the disease has brought out a number of diagnostic methods such as:

- *Edrophonium test*: This test involves injection of edrophonium chloride, which relieves muscle pain by blocking the breakdown of acetyl choline and increasing the neurotransmitters concentration at the neuromuscular junction.

- *Blood test*: Done to find out the presence of antibodies against the acetylcholine receptors. Anti MuSK is another antibody checked for in suspected cases of myasthenia gravis.
- *Diagnostic imaging*: Done using CT-Scans or magnetic resonance imaging to look for thymomas.
- *Electrodiagnostics*: This involves repeated stimulation of nerves with small pulses to tire the muscles. If the person suffers from myasthenia gravis, he or she will not respond to this treatment as the nerve transmission is hindered.

7.4.3.1 Application of Data Mining in Myasthenia Gravis

Myasthenia gravis affects nearly thirty million patients a year. Computational modeling techniques using 3D print technology have helped immensely in accurate diagnosis of the disease. There are many diseases with similar symptoms. As an example, tumors, strokes, nerve and muscle diseases have common symptoms of distorted speech and difficulty in swallowing similar to myasthenia gravis. Hence, often diagnostic tests fail to detect the disease early enough often because of physicians' erroneous judgments. Tumors and strokes are detected through MRI scans of the brain, muscular problems are detected through blood tests, and myasthenia gravis is detected through EMGs or nerve conduction tests. In order to eradicate confusions related to such expensive tests, studies suggest the use of machine learning, neural networks, and data mining algorithms for better predictive accuracy of the disease.

The graph theoretical method is an approach to help for data visualization and solution planning of the disease. Medical practitioners have used such an approach to map different symptoms with the hypothesis pertaining to the disease and perform multiple levels of investigations in parallel. The Bayesian network algorithm is one such graph theoretical algorithm where the probabilities of occurrence of the disease are mapped with the symptoms to provide more accurate and early detection of the disease.[45]

7.4.4 Grave's Disease

This particular autoimmune disease affects the thyroid leading to a symptom called hyperthyroidism. In this case, there is over stimulation of the thyroid gland leading to complications via the overproduction of thyroxine. People who are afflicted with complications like pernicious anemia, Addison's disease, celiac disease, and vitiligo are prone to develop Grave's disease. Here, the body produces antibodies against the hormone-producing cells of the body.[46] These antibodies mimic the thyroid-stimulating hormone, which is produced by the pituitary gland in the brain. Greater amounts of thyroxine are secreted due to overactivation of the gland cells in the thyroid, which leads to hyperthyroidism.

The common symptoms for hyperthyroidism are, shaking hands, a fast and irregular heartbeat, goiters, insomnias, muscle weaknesses, weight loss, and even sterility or erectile dysfunction. Females have been observed to get the disease more frequently than men, and even individuals aged below 40 years tend to get the disease.[47,48] It can lead to severe complications during pregnancy leading to preterm births, fetal thyroid dysfunctions, or even maternal heart failures and preeclampsias. It tends to be even more fatal when it deregulates the heartbeat rhythm, causing congestive heart failure.[46]

There are diagnostic tests able to determine whether a person is inflicted with Grave's disease. Some of these are listed below:

- *Thyroid functioning test*: A sample of the blood is drawn from the patient and the levels of thyroxine or TSH are checked. A high level of thyroid hormone plus a low level of TSH signifies the presence of Grave's disease.
- *Radioactive Iodine uptake (RAIU)*: This test tells us how much iodine the thyroid takes up. A high uptake of iodine suggests Grave's disease.
- *Antibody testing*: Blood samples drawn from the patient are checked for antibodies against specific receptors of cells producing thyroxine. If such antibodies are found they can suggest the presence of Grave's disease.

7.4.4.1 Application of Data Mining in Grave's Disease

Various studies have been conducted for the prediction of Graves' disease among patients. One of the methods used is the application of neural networks to predict the disease among patients treated with antithyroid medications. A study by Orunesu (2004) treated patients with methimazole for 18 months and found that in the majority of the patients there was a relapse after 2 years of therapy. The factors responsible for the relapse were difficult to predict. Artificial neural network approaches were used to predict the chances of relapse. Initially, 27 variables were considered as part of the therapy, which were then reduced to only 7 variables, and their combinations were finally considered to be more effective in predicting the factors responsible for the relapse. The optimized seven variables contributing significantly were heart rate, presence of thyroid bruits, psychological aspects, serum TGAb and fT4 levels at presentation, thyroid USG results, and cigarette smoking.[49,50]

Multilayer neural networks and learning vector quantization techniques have also been used in thyroid disease detection by classifying thyroid data. The interpretations of thyroid data using algorithms such as naïve Bayes, decision trees, back propagation neural networks (BPNNs), and support vector machines along with medical examinations have enabled a more accurate diagnosis and detection of the disease in comparison to traditional test-based approaches.[49–51]

7.4.5 Inflammatory Bowel Syndrome

In this particular autoimmune disease, the intestine gets inflated due to the action of the immune system on the intrinsic antigens associated with this region.[52] This condition is surrounded by a lot of controversy as to whether it really is an automimmune condition, or simply the action of the human immune system on pathogens, which might induce an inflammation as a side effect. However, both autoimmune and immune-mediated phenomena have been linked to this condition.[53] It has two main forms of manifestations: Crohn's disease and ulcerative colitis.[54] If we consider it as an autoimmune condition, then the body produces antibodies against the intestinal epithelial cells in both cases; however, antibodies against human tropomyosin fraction five has been seen in the case of ulcerative colitis. Furthermore, in the case of Crohn's disease, antibodies against *Saccharomyces cerevisiae* have been found and against perinuclear antineutrophil cytoplasmic antibodies have been seen to be common in the case of ulcerative colitis.[52]

Symptoms of this disease include cramps, diarrhea, weight loss, iron deficiency anemia, severe urgency to have bowel movements, etc. Other complications associated with this condition include a perforated or ruptured bowel, fistulae and perianal disease, toxic megacolon, and narrowing of the intestines.[54] Common forms of examinations for diagnosis include:

- *Complete blood count*: An increase in the number of white blood cells indicate inflammatory bowel syndrome.
- *Barium X-ray*: The patient is asked to intake the chalky white solution so that the upper part of the intestines can be scanned using X-rays. This method, however, is seldom used due to potential health complications.
- *Sigmoidoscopy*: A flexible tube with a camera attached is used to scan the last one-third region of the patient's large intestine. The anus and the intestinal walls are carefully scanned for ulcers, bleeding, or inflammations arising from the condition.
- *Capsule endoscopy*: The patient is made to swallow a small capsule with an inbuilt camera. The receiver system is worn as a belt by the patient. As the camera takes pictures of the gut scanning for symptoms of the disease, the pictures are sent to the receiver for subsequent download and analysis.

7.4.5.1 Application of Data Mining in Inflammatory Bowel Syndrome

Crohn's disease and ulcerative colitis (UC) are types of inflammatory bowel diseases, which are caused by abnormal responses of the body's immune system. In case of Crohn's disease, the gastrointestinal tract gets affected, which requires immediate medical attention. Studies have suggested that the diet plays a major role in this disease, and hence a study at Duke University created a web-based application to track food intake among patients. The participants would log their regular food intake and their fitness factors pertaining to the disease. The objective was to identify if any correlation existed among both of these factors. A correlation analysis was performed using advanced data mining approaches to understand the dependencies among the factors potentially contributing towards the disease. Due to the availability of EHRs and large data sets, the predictive analysis of factors impacting occurrence of such diseases have become extremely important to improve the quality of patients' lives.[55] A study by Eric Johnson (2012) highlighted the use of logistic regression using Bismarck on patients' data sets to analyze risk factors related to inflammatory bowel disease. The correlative factors were identified using a spatial data analysis technique and presented to researchers using data visualization tools for better insight. The Bismarck framework has proved to be successful in developing predictive models for the detection of the disease. Data mining approaches were successfully used to reduce irrelevant data in the huge and heterogeneous multidimensional database related to the disease, thus making predictions more accurate, specific, and relevant.[56,57]

7.4.6 Multiple Sclerosis

In this disease, the inherent immune system attacks the body's central nervous system (CNS), damaging the myelin sheath underlining nerve cells and oligodendrocytes. The main damage is done by the T cells, which enter the CNS through the blood vessels and produce inflammatory and degenerative changes.[58,59] The exact reason as to why these T cells become active against the CNS is still unknown and intense research is still in progress. This ailment comprises four main types: relapsing remitting multiple sclerosis (RRMS), secondary progressive multiple sclerosis (SPMS), primary progressive multiple sclerosis (PPMS), and progressing relapsing multiple sclerosis (PRMS).[60] This is indeed a complex disease as there seems to be no solid reason as to why this disease appears. Due to its mysterious origin, there are no specific diagnostic tests allocated for this disease. The main tests employed to detect this are MRI, lumber puncture, evoked potential testing, etc.[61] The common symptoms include paralysis, numbness, erectile dysfunction, pregnancy problems, slurred speech, muscle spasticity, etc.

The common treatment methods available include IV steroids, interferon injection, dimethyl fumarate, and copaxone.

7.4.6.1 Application of Data Mining in Multiple Sclerosis

Multiple sclerosis is an extremely unpredictable disease, which causes it to be very difficult to predict accurately. There are various clinical and biological markers related to neuroimaging and neurological aspects of the disease, nevertheless diagnostic accuracy in prediction remains a challenge. Various data mining approaches have proven to be extremely helpful in such complex predictions. Neural and Bayesian networks, decision trees, and linear regression models have been successfully used in such cases to compute accurate predictive outputs from complex data sets. Predictions are possible using both supervised and unsupervised learning methods. In supervised learning, predictions are made from known input/output samples, whereas in unsupervised learning the objective is to discover knowledge input data. Classification and regression are the two types of supervised learning methods. Clustering, summarization, and perception are types of unsupervised learning methods. Classification techniques have been used in the diagnosis of multiple sclerosis. The initial approaches considered in studies have eliminated irrelevant attributes from the data set with the help of feature selection such as the wrapper method. The input variables considered in the study were clinical, magnetic resonance imaging, and neurophysiological variables, which were further ranked using the wrapper algorithm. This led to the most significant input variables having an impact on the dependent variables of the disease using the Expanded Disability Status Scale. The accuracy of the prediction is calculated using a cross validation method, which revealed multilayer perceptron algorithm as the most accurate method in prediction of the disease. Neural networks have also been used to classify the neurophysiological pattern of the disease, namely the brainstem auditory evoked abilities, brain stem trigeminal evoked potential, visual evoked capabilities, and event-related capabilities. The classification of pathological gait patterns using machine-learning methods have also helped in the diagnosis of multiple sclerosis with the optical coherence tomography method. Advanced computational methods have been used on gene expression data related to multiple sclerosis. Feature set and design classifiers have been used to predict multiple sclerosis in the control set of patients participating in the diagnosis of the disease followed by linear classifiers. The application of classifiers has helped to highlight biomarker genes, dysregulated genes, and discriminatory genes pertinent to the disease. The accuracy of the predictions have been further validated using the leave-one-out cross validation method, which validates the applicability of data mining approaches in the prediction of multiple sclerosis.[62–64]

7.5 CONCLUSION

The various case studies thus reinforce the fact that applications of data mining approaches have become extremely essential and popular for the detection, prevention, and evaluation of treatment effectiveness in autoimmune diseases. Although clinical and physiological tests are available for diagnosis, often these become overwhelming for patients while incurring huge financial expenses. The use of machine-learning approaches in data mining has helped to analyze patterns related to the clinical and physiological symptoms of patients. It has also allowed computing an accurate diagnosis of the disease, thus minimizing the level of confusion and errors. Also, risk analysis and prediction of the disease, based on demographic and social backgrounds, has become possible through mining of information and knowledge discovery. There exists an enormous asset of medical records data, which needs to be utilized and explored more appropriately and optimally. The discovery of knowledge from these health records, using hybrid approaches, should be practiced more extensively for the prediction and analysis of diseases, therefore saving millions of human lives across the world.

REFERENCES

1. Kindt, T. J., R. A. Goldsby, B. A. Osborne, and J. Kuby. *Kuby immunology.* New York: W.H. Freeman, 2007.
2. Han, J., J. Pei, and M. Kamber. *Data mining: Concepts and Techniques.* Waltham, Massachusetts: Elsevier, 2011.
3. Caturegli, P., A. De Remigis, and N. R. Rose. "Hashimoto thyroiditis: Clinical and diagnostic criteria." *Autoimmunity reviews* 13, no. 4, 2014: 391–397.
4. Bizzaro, N., and A. Antico. "Diagnosis and classification of pernicious anemia." *Autoimmunity Reviews* 13, no. 4, 2014: 565–568.
5. Li, J., S. Göser, F. Leuschner, H. Christian Volz, S. Buss, M. Andrassy, R. Öttl, G. Pfitzer, H. A. Katus, and Ziya Kaya. "Mucosal tolerance induction in autoimmune myocarditis and myocardial infarction." *International Journal of Cardiology* 162, no. 3, 2013: 245–252.
6. Lindstrom, J. M., M. E. Seybold, V. A. Lennon, S. Whittingham, and D. D. Duane. "Antibody to acetylcholine receptor in myasthenia gravis Prevalence, clinical correlates, and diagnostic value." *Neurology* 51, no. 4, 1998: 933–933.
7. Weetman, A. P. "Graves' disease." *New England Journal of Medicine* 343, no. 17, 2000: 1236–1248.

8. Greco, A., M. I. Rizzo, A. D. Virgilio, A. Gallo, M. Fusconi, G. Pagliuca, S. Martellucci, R. Turchetta, L. Longo, and M. D. Vincentiis. "Goodpasture's syndrome: A clinical update." *Autoimmunity Reviews* 14, no. 3, 2015: 246–253.
9. Tan, E. M., A. S. Cohen, J. F. Fries, A. T. Masi, D. J. Mcshane, N. F. Rothfield, J. G. Schaller, N. Talal, and R. J. Winchester. "The 1982 revised criteria for the classification of systemic lupus erythematosus." *Arthritis & Rheumatology* 25, no. 11, 1982: 1271–1277.
10. Lublin, F. D., and S. C. Reingold. "Defining the clinical course of multiple sclerosis results of an international survey." *Neurology* 46, no. 4, 1996: 907–911.
11. Agca, R., S. C. Heslinga, S. Rollefstad, M. Heslinga, I. B. McInnes, M. J. L. Peters, T. K. Kvien et al. "EULAR recommendations for cardiovascular disease risk management in patients with rheumatoid arthritis and other forms of inflammatory joint disorders: 2015/2016 update." *Annals of the Rheumatic Diseases* 76, no. 1, 2017: 17–28.
12. Chen, C. H., H. A. Chen, H. T. Liao, C. H. Liu, C. Y. Tsai, and C. T. Chou. "Suppressors of cytokine signalling in ankylosing spondylitis and their associations with disease severity, acute-phase reactants and serum cytokines." *Clinical and Experimental Rheumatology* 34, no. 1, 2016: 100–105.
13. Gnanapriya, S., R. Suganya, G. Sumithra Devi, and M. Suresh Kumar. "Data mining concepts and techniques." *Data Mining and Knowledge Engineering* 2, no. 9, 2010: 256–263.
14. Wilson, R. A., and F. Keil. "The MIT encyclopaedia of the cognitive sciences Bradford." (1999).
15. Mitchell, M. "*An Introduction to Genetic Algorithms*", Cambridge, Massachusetts: MIT Press, ISBN: 0-262-13316-4, 1996.
16. Hastie, T., R. Tibshirani, and J. Friedman. "Overview of supervised learning." In *The Elements of Statistical Learning*, pp. 9–41. Springer, New York, 2009.
17. Kohavi, R., and J. Ross Quinlan. "Data mining tasks and methods: Classification: decision-tree discovery." In *Handbook of data mining and knowledge discovery*, pp. 267–276. Oxford University Press, Inc., 2002.
18. Aha, D. W., D. Kibler, and M. K. Albert. "Instance-based learning algorithms." *Machine Learning* 6, no. 1, 1991: 37–66.
19. Man, K.-F., K.-S. Tang, and S. Kwong. "Genetic algorithms: Concepts and applications [in engineering design]." *IEEE transactions on Industrial Electronics* 43, no. 5, 1996: 519–534.
20. Forbus, K. D. "Qualitative process theory." *Artificial Intelligence* 24, no. 1, 1984: 85–168.
21. Agrawal, R. "Fast algorithms for mining association rules in large databases." In *Proc. 20th International Conference on Very Large Data Bases*, Sept, 1994, pp. 478–499. 1994.
22. Alpaydin, E. *Introduction to Machine Learning*. Cambridge, Massachusetts London, England: MIT press, 2014.
23. Wang, L., F.-S. Wang, and M. E. Gershwin. "Human autoimmune diseases: A comprehensive update." *Journal of Internal Medicine* 278, no. 4, 2015: 369–395.
24. Wang, L., F.-S. Wang, C. Chang, and M. Eric Gershwin. "Breach of tolerance: Primary biliary cirrhosis." In *Seminars in Liver Disease*, 34, no. 3, pp. 297–317. Thieme Medical Publishers, 2014.
25. Salinas, G. F., F. Braza, S. Brouard, P.-P. Tak, and D. Baeten. "The role of B lymphocytes in the progression from autoimmunity to autoimmune disease." *Clinical Immunology* 146, no. 1, 2013: 34–45.
26. Willrich, M. A. V., D. L. Murray, and M. R. Snyder. "Tumor necrosis factor inhibitors: Clinical utility in autoimmune diseases." *Translational Research* 165, no. 2, 2015: 270–282.

27. Cui, Y., Y. Sheng, and X. Zhang. "Genetic susceptibility to SLE: Recent progress from GWAS." *Journal of Autoimmunity* 41, 2013: 25–33.
28. National Diabetes Statistic Report, 2017, Available at: https://www.cdc.gov/features/diabetes-statistic-report/index.html cited on 15-12-2017
29. Kawasaki, E. "Type 1 diabetes and autoimmunity." *Clinical Pediatric Endocrinology* 23, no. 4, 2014: 99–105.
30. Kahaly, G. J., and M. P. Hansen. "Type 1 diabetes associated autoimmunity." *Autoimmunity Reviews* 15, no. 7, 2016: 644–648.
31. Atkinson, M. A., and G. S. Eisenbarth. "Type 1 diabetes: New perspectives on disease pathogenesis and treatment." *The Lancet* 358, no. 9277, 2001: 221–229.
32. Iyer, A., S. Jeyalatha, and R. Sumbaly. "Diagnosis of diabetes using classification mining techniques." *International Journal of Data Mining & Knowledge Management Process*, 5, no. 1, 2015: 1–14.
33. Vijayan, V., and A. Ravikumar. "Study of data mining algorithms for prediction and diagnosis of diabetes mellitus." *International Journal of Computer Applications* 95, no. 17, 2014: 12–16.
34. Repalli, P. "Prediction on diabetes using data mining approach." Stillwater: Oklahoma State University 2011.
35. Breault, J. L. "Data mining diabetics databases: Are rough sets a useful addition." In *Proceedings of the 33rd Symposium on Interface, Computing Science and Statistics, Orange, CA, USA*, pp. 13–16. 2001.
36. Parthiban, G., A. Rajesh, and S. K. Srivatsa. "Diagnosis of heart disease for diabetic patients using naive bayes method." *International Journal of Computer Applications* 24, no. 3, 2011: 7–11.
37. Motka, R., V. Parmarl, B. Kumar, and A. R. Verma. "Diabetes mellitus forecast using different data mining techniques." In *Computer and Communication Technology (ICCCT), 2013 4th International Conference on*, pp. 99–103. IEEE, 2013.
38. Hosseinpour, N., S. Setayeshi, K. Ansari-asl, and M. Mosleh. "Diabetes Diagnosis by Using Computational Intelligence Algorithms." *International Journal of Advanced Research in Computer Science and Software Engineering* 2, no. 12, 2012: 71–77.
39. Disease, L. Available at: https://www.ninds.nih.gov/Disorders/All-Disorders/Neurological-Sequelae-Lupus-Information-Page, cited on Dec 15, 2017.
40. Tan, E. M., A. S. Cohen, J. F. Fries, A. T. Masi, D. J. Mcshane, N. F. Rothfield, J. G. Schaller, N. Talal, and R. J. Winchester. "The 1982 revised criteria for the classification of systemic lupus erythematosus." *Arthritis & Rheumatology* 25, no. 11, 1982: 1271–1277.
41. Gomathi, S., and V. Narayani. "Applying decision tree algorithm to predict lupus using Rapid Miner." *International Journal of Applied Engineering Research* 10, no. 9, 2015: 2015.
42. Reynolds, J. A., and I. N. Bruce, "*Overview of the Management of Systemic Lupus Erythematosus*," Arthritis Research UK Topical Reviews, Chesterfield Derbyshire: Spring 2013.
43. Myasthenia Gravis Fact Sheet, Available: https://www.ninds.nih.gov/Disorders/Patient-Caregiver-Education/Fact-Sheets/Myasthenia-Gravis-Fact-Sheet, cited on: Dec 15, 2017.
44. Lindstrom, J. M., M. E. Seybold, V. A. Lennon, S. Whittingham, and D. D. Duane. "Antibody to acetylcholine receptor in myasthenia gravis Prevalence, clinical correlates, and diagnostic value." *Neurology* 51, no. 4, 1998: 933–933.
45. Liu, F., Z. Hou, L. Li, and R. Luo. "Constructing data mining model of five viscera correlation theory of Myasthenia Gravis based on rough set and association rules." In *Bioinformatics and Biomedicine Workshops (BIBMW), 2011 IEEE International Conference on*, pp. 778–783. IEEE, 2011.

46. https://www.niddk.nih.gov/health-information/endocrine-diseases/graves-disease (Accessed on Dec 16, 2017).
47. Smith, B., and R. Hall. "Thyroid-stimulating immunoglobulins in graves'disease." *The Lancet* 304, no. 7878, 1974: 427–430.
48. Weetman, A. P. "Graves' disease." *New England Journal of Medicine* 343, no. 17, 2000: 1236–1248.
49. Ozyilmaz, L., and T. Yildirim. "Diagnosis of thyroid disease using artificial neural network methods." In *Neural Information Processing, 2002. ICONIP'02. Proceedings of the 9th International Conference on*, 4, pp. 2033–2036. IEEE, 2002.
50. Orunesu, E., M. Bagnasco, C. Salmaso, V. Altrinetti, D. Bernasconi, P. Del Monte, G. Pesce, M. Marugo, and G. S. Mela. "Use of an artificial neural network to predict graves' disease outcome within 2 years of drug withdrawal." *European Journal of Clinical Investigation* 34, no. 3, 2004: 210–217.
51. Ioniţă, I., and L. Ioniţă. "Prediction of thyroid disease using data mining techniques." *BRAIN. Broad Research in Artificial Intelligence and Neuroscience* 7, no. 3, 2016: 115–124.
52. Wen, Z., and C. Fiocchi. "Inflammatory bowel disease: Autoimmune or immune-mediated pathogenesis?" *Journal of Immunology Research* 11, no. 3–4, 2004: 195–204.
53. Snook, J. "Are the inflammatory bowel diseases autoimmune disorders?" *Gut* 31, no. 9, 1990: 961.
54. Inflammatory Bowel Disease (IBD), Available on: https://www.ncbi.nlm.nih.gov/pubmedhealth/PMHT0022800/, Cited on Dec 16, 2017.
55. Cooper, J. G., and G. P. Purcell. "Data mining for correlations between diet and Crohn's disease activity." In *AMIA Annual Symposium Proceedings*, 2006, p. 897. American Medical Informatics Association, 2006.
56. Johnson, E. *Health care Data Mining Using In-database Analytics to Predict Diagnosis of Inflammatory Bowel Disease.* Master of Science Thesis: University of Washington, 2012.
57. Abouzahra, M., K. Sartipi, D. Armstrong, and J. Tan. "Integrating Data from EHRs to Enhance Clinical Decision-making: The Inflammatory Bowel Disease Case." In *Computer-Based Medical Systems (CBMS), 2014 IEEE 27th International Symposium on*, pp. 531–532. IEEE, 2014.
58. IFNB Multiple Sclerosis Study Group. "Interferon beta-1b is effective in relapsing-remitting multiple sclerosis I. Clinical results of a multicenter, randomized, double-blind, placebo-controlled trial." *Neurology* 43, no. 4, 1993: 655–655.
59. Multiple Sclerosis, Available at: https://medlineplus.gov/multiplesclerosis.html cited: 15-12-2017.
60. Lublin, F. D., and S. C. Reingold. "Defining the clinical course of multiple sclerosis results of an international survey." *Neurology* 46, no. 4, 1996: 907–911.
61. Poser, C. M., D. W. Paty, L. Scheinberg, W. Ian McDonald, F. A. Davis, G. C. Ebers, K. P. Johnson, W. A. Sibley, D. H. Silberberg, and W. W. Tourtellotte. "New diagnostic criteria for multiple sclerosis: Guidelines for research protocols." *Annals of Neurology* 13, no. 3, 1983: 227–231.
62. Bejarano, B., V. Segura, and P. Villoslada. "Data mining in multiple sclerosis: computational classifiers. Introduction and methods (Part I)."
63. Bejarano, B., V. Segura, and P. Villoslada. "Data mining in multiple sclerosis: computational classifiers. Introduction and methods (Part II)."
64. Guo, P., Q. Zhang, Z. Zhu, Z. Huang, and K. Li. "Mining gene expression data of multiple sclerosis." *PloS One* 9, no. 6, 2014: e100052.

8 Data Mining Techniques in Imaging of Embryogenesis

Diptesh Mahajan and Gaurav K. Verma

CONTENTS

8.1 INTRODUCTION

Over the last two decades, there has been an explosive production of data. This boom in datasets has observed a growth along two dimensions simultaneously—the number of fields and the number of cases (Feyyad 1996). The voluminous amount of structured and unstructured data generated has given rise to a new term—"big data." This term is not only defined by its volume, but its quality, diversity, and reliability (Sarkar 2016). In the field of biological sciences, the emergence of high throughput technology has led to the generation of huge molecular datasets at various levels of biological systems leading to the development of the "omics" era in life sciences (Li and Chen 2014). Researchers turn more often now towards big data to explore information at diverse scales, as it gives them a quantitative approach over the traditional qualitative approaches. The technological advancements in image analysis techniques coupled with the rapid price decline in high throughput instruments have helped even the small-scale laboratories to generate big data. These developments have broadened the scope of fields like developmental biology, cancer biology, drug design, and many other subsequent fields.

In this chapter, we primarily focus on the science of embryology. Embryology deals with the study of prenatal developmental stages and congenital disorders occuring during embryonic development; it is also known as tetralogy. The classical embryology approach can detect the patterns of abnormal morphologic development but fails to recognize their cause (Carlson 2002). Contemporary molecular and imaging techniques have made it feasible to track the dynamic cellular or subcellular activities in the early stages of the embryo. This has helped us in identifying the reasons behind any inappropriate turn taken by the cell, which lead to birth defects. The advancements of optical design of microscopes have evolved from confocal microscopy and two-photon microscopy to light sheet microscopy, which can periodically generate cell-datasets of the developing embryo, one cell at a time. This exponential escalation of datasets has generated data well beyond petabytes in size (Marx 2013). With such huge datasets available, reliance on traditional analysis forms a bottleneck, hence leading to the development of the interdisciplinary field of bioinformatics. This field uses tools employing various data mining techniques to highlight new biological insights or hidden trends from the available data (Luscombe, Greenbaum, and Gerstein 2001). Data mining tools like SBML and CellML are used to elucidate the interaction of genes, proteins, and metabolites from the already existing data. This chapter critically reviews the techniques utilized to generate "big data" in embryology, and the languages used to mine the useful data with reference to some ongoing projects and platforms.

8.2 IMAGING TECHNIQUES

Contemporary imaging techniques provide an insight of the multidimensional and multiparameter activities at the cellular level. These usually capture the physical parameters including cell population, tissue property, and surface area (Eils and Athale 2003), as well as biological parameters, which include cell division, motility, and morphology changes (Pantazis and Supatto 2014).

Currently there are four techniques most widely used to track in vivo embryonic development.

8.2.1 Microscopy Techniques

i. *Confocal Fluorescence Microscopy*: The advent of fluorescence imaging has led to the emergence of 3D imaging of live embryo. This was a major leap over the conventional 2D static analysis of fixed embryos. Confocal microscopy uses point illumination, which is used to excite the biological fluorophores. Analysis of the emission spectrum of these fluorophores helps us in identifying the cell structure and its subsequent functions (Khairy and Keller 2011). This technique has been most widely used to study live morphogenesis (Pawley and Masters 2008) and to track cell lineage (Bao et al. 2006). Among vertebrates, the fluorescently labeled cells and matrix motion can be tracked during the early embryonic development. Fish embryos provide a better understanding of embryogenesis as they are relatively smaller, and can be nurtured in nearly ambient environments and remain transparent throughout the early stages of cardiogenesis (Megason and Fraser 2007). Among vertebrates of higher order, the primitive streak formation, vasculogenesis, and gastrulation can be visualized prominently through confocal microscopy. Avian embryos allow for long-term optical access, as their germ disks are sleek and flat and are thus often used to study the early stages of embryogenesis in higher vertebrates (Yalcin et al. 2010). However, the process of embryonic development is highly sensitive to photodamage, and their development occurs at a scale beyond the range of confocal microscopy. The recent developments in multiphoton and light sheet microscopy have rectified these limitations to some extent.

ii. *Multiphoton Microscopy*: The extent to which the images can be analyzed is restricted by the optical properties of the embryonic tissues. These properties are highly dependent on the species and vary in time and space during development (Supatto et al. 2009). Unlike single photon illumination in confocal microscopy, multiphoton microscopy uses two photons, which substantially suppresses the background noise. It also uses nonlinear contrast mechanisms wherein the imaging is carried out without fluorescence probes, hence cells which cannot be photobleached can also be tracked. Multiphoton microscopy has been extensively used to reconstruct cell lineages and to study cell division patterns at a deeper tissue level, predominantly within drosophila (Fowlkes et al. 2008) and zebrafish (Hsieh et al. 2014) embryos. The recent advances include improved laser sources with increased excitation wavelength, which in-turn increases optical penetration and reduces photodamage (Andresen et al. 2009). However, the effectiveness of fluorescence and multiphoton microscopy techniques is limited to the earlier stages of embryonic development. A more sophisticated approach is required to analyze the later stages of embryogenesis using deeper tissue penetration.

iii. *Light-Sheet Microscopy*: Recording the images at multidimensional and multiparameter scales often compromises the speed performance. The aforementioned approaches use point illumination, hence can record images only one pixel at a time with pixel rates less than 1 MHz. To capture the rapid ongoing processes like the beating heart development, cilia motility, and fluid flow dynamics often a higher pixel rate is required. High-speed imaging also gives us a more detailed analysis of slower processes including cell proliferation, migration, and shape changes. Light-sheet microscopy is a promising solution for fast imaging of the live embryos. A distinctive advantage of this technique is it can carry out fast imaging with a wider angle of view and minimal photodamage. The whole-embryo imaging of the first 24 hours of the zebrafish has been beautifully captured by this technique, making it possible to track thousands of cells in parallel with high pixel rates and resolution (Keller et al. 2008). However, this technique also has its shortcomings; the high speed imaging results in uneven image quality and limited imaging depth. These shortcomings result in relatively inaccurate intensity-based measurements and compromise deep tissue imaging.

The limitations of these microscopy techniques have given rise to alternate techniques like Bessel beam (Planchon et al. 2011) or multidirectional illumination (Huisken and Stainier 2007), which aim to improve either the depth of penetration or homogeneity of the fluorescence excitation and increase the spatial resolution throughout the field of view.

8.2.2 Ultrasound Imaging

This noninvasive technique sends short pulses of high frequency (ultrasonic) sound waves, which get reflected and/or transmitted when they encounter a tissue. Clinically, this technique is used for prenatal diagnosis of the embryo, which includes the development of follicles in the ovary and development of the fetus, the gestational sac, fetal parameters, and the placenta. It uses a probe containing one or more acoustic transducers. When the sound waves emitted hit a material with nonidentical acoustic impedance, a fraction of the wave gets reflected, which the probe detects as an echo. The time taken by the echo to travel to the probe is detected, allowing the depth of the tissue to be calculated. These calculated distances of the tissue from the probe are used to generate a series of 2D images in three different planes. These images are then superimposed by the computer to construct a 3D image (Jensen 2007). This technique provides real-time information and can provide visualization of the later fetal stages, which helps to detect abnormal morphology (Yu et al. 2008) and hemodynamic changes (Oosterbaan et al. 2009). Abnormal morphology may include early diagnosis of a trophoblastic disease, interruption of bladder line, genetic abnormality, and other subsequent anomalies. The hemodynamic changes are measured by Doppler ultrasound, which detects the change in blood flow and blood pressure by rebounding ultrasounds off circulating red blood cells. The rapid evolution of ultrasound technology has made it possible for us to trace information from organ to molecular level and this is primarily due to the development of microbubbles. These bubbles act as intravascular tracers and can be target specific if attached with suitable ligands (Thomasson et al. 2004). The past few years have seen the increased use of nanoparticles for extravascular tracing, as they can easily migrate to regions of vascular injury, or to regions where vascular permeability is aberrantly high. The images generated by nanoparticle tracers have a lower signal-to-noise ratio compared to microbubble tracers, and therefore have an improved image quality (Kherlopian et al. 2008).

8.2.3 Computed Tomography/Micro Computed Tomography

The basic principle behind this technique is the differential absorption of X-rays by different components like fat, bone, air, and water. Thus, depending on the tissue present, the amount of X-rays absorbed will differ. This difference in absorption is detected by the computer, which generates cross-sectional images or "slices" of the part exposed. These slices are called tomographic images and are more informative than conventional X-rays. The tomographic slices generated by the computer are then stacked digitally to generate 3D images in order to easily identify and locate the tissue structures along with any possible abnormality present, such as an intrinsic tumor (Kalender 2006). The soft tissue contrast is relatively low for tumors, but this can be enhanced with iodinated contrast agents, which can detect tumors more easily.

For the study of animal specimens, micro-CT is used and emits higher energy X-rays than CT. The increase in energy helps in improving the resolution of the tomographic slices, but the increase in radiation also leads to adverse health effects. Micro-CT systems have low cost of maintenance and a brief scan time when compared to the other imaging techniques (Kim et al. 2011). Although micro-CT imaging has the highest possible theoretical resolution, the soft tissues poorly attenuate X-rays; therefore, most of the embryonic development imaging is performed on fixed tissues stained with molecules having high atomic weight to enhance X-ray attenuation (Metscher 2009). The lethal radiations of micro-CT often lead to abnormalities in developing embryos, making it difficult for scientists to study embryonic developments afterwards. Recent developments use gold nanoparticles as contrast agents, which increase the residence time and reduces cytotoxic effects (Murphy et al. 2008).

Both CT/micro-CT have high spatial resolutions, which can allow for visualization of the fine anatomical details; however, the radiation dose is a concern and cannot be neglected. Thus, they have limited applications in human studies.

8.2.4 Magnetic Resonance Imaging (MRI)/Micro-MRI/Functional-MRI/Magnetic Resonance Spectroscopy

This scan is performed by placing the subject under the influence of a strong magnetic field and radio waves, which generate images that cannot be visualized with ultrasound or CT-scans. The magnetic field aligns the hydrogen protons present in water molecules, which have a designated spin. A radio frequency (RF) pulse is emitted simultaneously, which the hydrogen proton absorbs causing its spin to flip through a process known as precession. After the RF pulse is turned off, the protons return to their native spin and realign with the applied magnetic field. While returning to their native state, the protons emit radio signals detected and processed by the scanners. The varying spin density among different tissues helps in generating an excellent tissue contrast (Hornak 2008).

Micro-MRI also follows a similar principle but employs a much stronger magnetic field. An increase in the magnetic field provides images with a better contrast, but may lead to peripheral nerve stimulation of the subject.

Functional-MRI is used to study the brain activity in response to a predetermined stimulus. The stimulus initiates a response from certain areas of the brain, leading to an increased metabolism in that part. The increased metabolism in turn initiates an increase in blood flow. Thus, the concentration of oxygenated blood increases in that region resulting in an imbalance in oxygenated and deoxygenated blood. This imbalance leads to a change in image contrast, which can be detected by the fMRI (Parrish et al. 2000).

Magnetic resonance spectroscopy is an emerging imaging technique combining the ability of nuclear magnetic resonance (NMR)—to detect biochemical moieties—and the ability of MRI to isolate individual 3D pixels. This is mainly carried out by emitting phase modulated RF pulses, which eliminates the possibility of contaminating nearby voxels. A combination of NMR with MRI helps in generating anatomical and biochemical information in parallel (Arias-Mendoza and Brown 2004).

The key advantage of MRI over other imaging techniques is its exceptional tissue contrast due to high signal-to-noise ratio and lack of ionizing radiation.

8.3 LANGUAGES

The imaging techniques discussed above generate enormous volumes of data. It is manually impossible to draw conclusions from such huge sample sizes. Thus, we need the help of computational analysis to interpret the required biological data. Computational technologies are essential to automate the processing, quantification, and analysis of high-information content biomedical images (Raza 2012). The first step is to store these datasets in structured formats, in a way that simplifies sharing and accessibility. The next step includes integrating and interpreting these stored datasets at genomic, proteomic, metabolomic, cellular, or whole organism level (Zaki et al. 2003).

To summarize, a system biologist faces a number of problems without a domain-specific language (Hucka et al. 2004):

i. In order to work with data obtained from multiple simulation/analysis tools, they need to re-encode the models received from different sources. This makes the process both error prone and time consuming.
ii. The development of new software tools has rendered the old systems obsolete. This has resulted in the loss of a number of usable models.
iii. Models developed by authors are usually developed under different modeling environments. The definitions to obtain these models are often too complex and cannot be retrieved easily and reused.

Thus, the development of computer-oriented, domain-specific languages, have become a necessity to study the dynamic interaction of genes, proteins, or metabolites at different levels under different modeling environments.

Currently two languages, SBML and CellML, are most widely used to interpret these biochemical interactions. Both these languages are based on XML. The reason prompting system biologists to choose XML as a basis is its user-friendly nature. XML is both human and machine-readable and this feature has encouraged its use in the field of bioinformatics (Achard et al. 2001).

8.3.1 CellML

The components defined in CellML (Figure 8.1) are linked together by connections to generate a model (Cuellar et al. 2015). The groups indicate the existence of a logical/physical group and metadata provides the basis for the model.

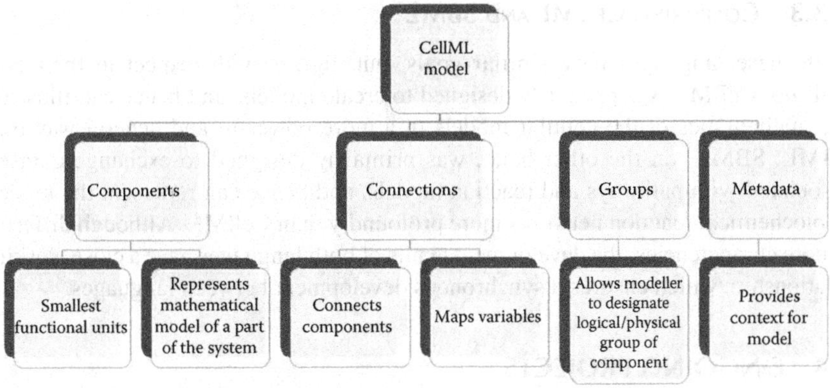

FIGURE 8.1 Components of a CellML model.

8.3.2 SBML

SBML allows models to represent arbitrary complexity. All the components constituting the model are described using a specific type of data object, which categorizes the relevant information (Hucka et al. 2015). The latest level of a SBML model definition consists of lists of components, as described in Figure 8.2.

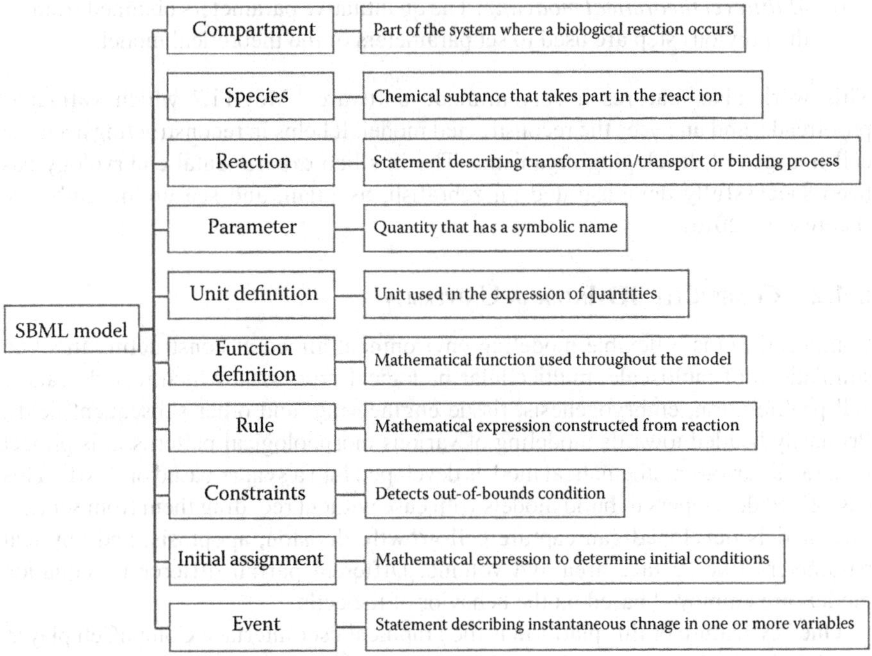

FIGURE 8.2 Components of a SBML model.

8.3.3 Comparing CellML and SBML

Both these languages have similar goals, but digress with respect to their core abilities. CellML was primarily designed to create models, and hence can illustrate the mathematics of the cellular models in a more powerful and general way than SBML. SBML, on the other hand, was primarily designed to exchange datasets associated with pathways and reaction models, and hence can represent the models of biochemical reaction networks more profoundly than CellML. Although differing in core competencies, the development teams of both languages have a close working relationship, which ensures a synchronous development between languages.

8.4 ONGOING PROJECTS

8.4.1 Bio-Emergences—Paris Saclay Institute of Neuroscience

This project aims at pioneering novel techniques and strategies to improve the multistep approach of quantitatively understanding the morphogenesis of organisms, both under normal and pathological conditions. The architecture of the workflow can be broadly categorized into three steps

i. *Multiscale-multimodal observation*: This step involves in vivo/in toto imaging by techniques such as illumination and multiphoton microscopy.
ii. *Quantification*: Also termed phenomenological reconstruction, this step uses an algorithmic method to extract the relevant quantitative information.
iii. *Multilevel theoretical modeling*: The quantitative parameters obtained from the previous step are used to set parameters of the theoretical model.

This work also features a custom-made software "Mov-IT," which validates, proofreads, and analyses the reconstructed model. It helps in reconstructing accurate cell lineages of developing organisms. The in silico experimental embryology has been successfully demonstrated on zebrafish, ascidian, and sea urchin embryos. (Faure et al. 2016).

8.4.2 CompuCell 3D-Indiana University

Compucell 3D is a flexible modeling environment aimed at constructing in silico simulations of multiscale, multicellular biological processes, which include cancer cell proliferation, embryogenesis, tissue engineering, and other subsequent fields. Primarily headed towards modeling of various morphological patterns, this project integrates various mathematical models developed into a syntax based on XML. This has helped developers to build models with ease without recoding them from scratch. The models developed can capture cell growth, division, apoptosis, and physical parameters like surface area and volume. Different partial differential equation models are employed based on the behavior of the cells.

One key feature of this platform is the graphical user interface CompuCell player, which helps in 3D visualization and supports the feature of switching to 2D cross sections in each dimension along with ability to alternate the fields. To improve performance, the player can be run in silent mode to generate a greater number of images between two steps.

In summary, this problem-solving environment has the following features (Swat et al. 2012)

i. Can run simulations in 3D.
ii. Designed for high performance and high memory consumption to analyze sizeable numbers of cells in parallel.
iii. Flexible and extensible allowing incorporation of new cell behaviors at a higher level/model.

8.5 FUTURE PROSPECTS

The reason behind a substantial number of cases showing abnormal embryonic development still remains unknown. With such huge data volumes increasing, the science of embryogenesis needs to shift its focus from in vivo to in silico experiments. The cell developmental models of various cells can now be reconstructed using computer simulations with the aid of applied mathematics. Thus, critical data mining approaches followed by in silico resynthesis of embryos will enable scientists to accurately unravel the factors causing subtle changes in cell behavior. This will eventually help them uncover the causes behind abnormal development (Knudsen et al. 2017).

REFERENCES

Achard, F., G. Vaysseix, and E. Barillot. 2001. "XML, Bioinformatics and Data Integration." *Bioinformatics* 17 (2): 115–25. doi:10.1093/bioinformatics/17.2.115.

Andresen, V., S. Alexander, W. M. Heupel, M. Hirschberg, R. M. Hoffman, and P. Friedl. 2009. "Infrared Multiphoton Microscopy: Subcellular-Resolved Deep Tissue Imaging." *Curr Opin Biotechnol* 20 (1): 54–62. doi:10.1016/j.copbio.2009.02.008.

Arias-Mendoza, F., and T. R. Brown. 2004. "In Vivo Measurement of Phosphorous Markers of Disease." *Disease Markers* 19 (2–3): 49–68.

Bao, Z., J. I. Murray, T. Boyle, S. L. Ooi, M. J. Sandel, and R. H. Waterston. 2006. "Automated Cell Lineage Tracing in Caenorhabditis Elegans." *Proceedings of the National Academy of Sciences* 103 (8): 2707–12. doi:10.1073/pnas.0511111103.

Carlson, B. M. 2002. "Embryology in the Medical Curriculum." *Anatomical Record* 269 (2): 89–98. doi:10.1002/ar.10075.

Cuellar, A., W. Hedley, M. Nelson, C. Lloyd, M. Halstead, D. Bullivant, D. Nickerson, P. Hunter, and P. Nielsen. 2015. "The CellML 1.1 Specification." *Journal of Integrative Bioinformatics* 12 (2): 259. doi:10.2390/biecoll-jib-2015-259.

Eils, R., and C. Athale. 2003. "Computational Imaging in Cell Biology." *The Journal of Cell Biology* 161 (3): 477–81. doi:10.1083/jcb.200302097.

Faure, E., T. Savy, B. Rizzi, C. Melani, O. Stašová, D. Fabrèges, R. Špir et al. 2016. "A Workflow to Process 3D+time Microscopy Images of Developing Organisms and Reconstruct Their Cell Lineage." *Nature Communications* 7. doi:10.1038/ncomms9674.

Feyyad, U. M. 1996. "Data Mining and Knowledge Discovery: Making Sense out of Data." *Ieee Expert Intelligent Systems And Their Applications* 11 (5): 20–25. doi:10.1109/64.539013.

Fowlkes, C. C., C. L. Luengo Hendriks, S. V. E. Keränen, G. H. Weber, O. Rübel, M. Y. Huang, S. Chatoor et al. 2008. "A Quantitative Spatiotemporal Atlas of Gene Expression in the Drosophila Blastoderm." *Cell* 133 (2): 364–74. doi:10.1016/j.cell.2008.01.053.

Hornak, J. P. 2008. "The Basics of MRI." *Biomedical Engineering* 24 (2003): 2–6.

Hsieh, C.-S., C.-Y. Ko, S.-Y. Chen, T.-M. Liu, J.-S. Wu, C.-H. Hu, and C.-K. Sun. 2014. "In Vivo Long-Term Continuous Observation of Gene Expression in Zebrafish Embryo Nerve Systems by Using Harmonic Generation Microscopy and Morphant Technology." *Journal of Biomedical Optics* 13 (6): 64041. doi:10.1117/1.3050423.

Hucka, M., A. Finney, B. J. Bornstein, S. M. Keating, B. E. Shapiro, J. Matthews, B. L. Kovitz et al. 2004. "Evolving a Lingua Franca and Associated Software Infrastructure for Computational Systems Biology: The Systems Biology Markup Language (SBML) Project." *Systems Biology* 1 (1): 41–53. doi:10.1049/sb.

Hucka, M., F. T. Bergmann, S. Hoops, S. M. Keating, S. Sahle, J. C. Schaff, L. P. Smith, and D. J. Wilkinson. 2015. "The Systems Biology Markup Language (SBML): Language Specification for Level 3 Version 1 Core." *Journal of Integrative Bioinformatics* 12 (2): 266. doi:10.2390/biecoll-jib-2015-266.

Huisken, J., and D. Y. R. Stainier. 2007. "Even Fluorescence Excitation by Multidirectional Selective Plane Illumination Microscopy (mSPIM)." *Optics Letters* 32 (17): 2608. doi:10.1364/OL.32.002608.

Jensen, J. A. 2007. "Medical Ultrasound Imaging." *Progress in Biophysics and Molecular Biology*. doi:10.1016/j.pbiomolbio.2006.07.025.

Kalender, W. A. 2006. "X-Ray Computed Tomography." *Physics in Medicine and Biology*. doi:10.1088/0031-9155/51/13/R03.

Keller, P. J., A. D. Schmidt, J. Wittbrodt, and E. H. K. Stelzer. 2008. "Reconstruction of Zebrafish Early Embryonic Development by Scanned Light Sheet Microscopy." *Science* 322 (5904): 1065–69. doi:10.1126/science.1162493.

Khairy, K., and P. J. Keller. 2011. "Reconstructing Embryonic Development." *Genesis*. doi:10.1002/dvg.20698.

Kherlopian, A. R., T. Song, Q. Duan, M. A. Neimark, M. J. Po, J. K. Gohagan, and A. F. Laine. 2008. "A Review of Imaging Techniques for Systems Biology." *BMC Systems Biology*. doi:10.1186/1752-0509-2-74.

Kim, J. S., J. Min, A. K. Recknagel, M. Riccio, and J. T. Butcher. 2011. "Quantitative Three-Dimensional Analysis of Embryonic Chick Morphogenesis Via Microcomputed Tomography." *Anatomical Record* 294 (1): 1–10. doi:10.1002/ar.21276.

Knudsen, T. B., B. Klieforth, and W. Slikker Jr. 2017. "Programming Microphysiological Systems for Children's Health Protection." *Experimental Biology and Medicine* 242, no. 16 (2017): 1586–92. doi:10.1177/1535370217717697.

Li, Y., and L. Chen. 2014. "Big Biological Data: Challenges and Opportunities." *Genomics, Proteomics & Bioinformatics* 12 (5). Beijing Institute of Genomics, Chinese Academy of Sciences and Genetics Society of China: 187–89. doi:10.1016/j.gpb.2014.10.001.

Luscombe, N. M., D. Greenbaum, and M. Gerstein. 2001. "Review What Is BioinformaticsAn." *Gene Expression* 40 (4): 83–100. doi:10.1053/j.ro.2009.03.010.

Marx, V. 2013. "The Big Challenges of Big Data." *Nature* 498 (7453): 255–60. doi:10.1038/498255a.

Megason, S. G., and S. E. Fraser. 2007. "Imaging in Systems Biology." *Cell*. doi:10.1016/j.cell.2007.08.031.

Metscher, B. D. 2009. "Micro CT for Comparative Morphology: Simple Staining Methods Allow High-Contrast 3D Imaging of Diverse Non-Mineralized Animal Tissues." *BMC Physiology* 9 (1). doi:10.1186/1472-6793-9-11.

Murphy, C. J., A. M. Gole, J. W. Stone, P. N. Sisco, A. M. Alkilany, E. C. Goldsmith, and S. C. Baxter. 2008. "Gold Nanoparticles in Biology: Beyond Toxicity to Cellular Imaging." *Accounts of Chemical Research* 41 (12): 1721–30. doi:10.1021/ar800035u.

Oosterbaan, A. M., N. T. C. Ursem, P. C. Struijk, J. G. Bosch, A. F. W. Van Der Steen, and E. A. P. Steegers. 2009. "Doppler Flow Velocity Waveforms in the Embryonic Chicken Heart at Developmental Stages Corresponding to 5–8 Weeks of Human Gestation." *Ultrasound in Obstetrics and Gynecology* 33 (6): 638–44. doi:10.1002/uog.6362.

Pantazis, P., and W. Supatto. 2014. "Advances in Whole-Embryo Imaging: A Quantitative Transition Is Underway." *Nature Reviews Molecular Cell Biology*. doi:10.1038/nrm3786.

Parrish, T. B., D. R. Gitelman, K. S. LaBar, and M. Marsel Mesulam. 2000. "Impact of Signal-to-Noise on Functional MRI." *Magnetic Resonance in Medicine* 44 (6): 925–32. doi:10.1002/1522-2594(200012)44:6<925::AID-MRM14>3.0.CO;2-M.

Pawley, J. B., and B. R. Masters. 2008. Handbook of Biological Confocal Microscopy. *Journal of Biomedical Optics*. Vol. 13. doi:10.1117/1.2911629.

Planchon, T. A., L. Gao, D. E. Milkie, M. W. Davidson, J. A. Galbraith, C. G. Galbraith, and E. Betzig. 2011. "Rapid Three-Dimensional Isotropic Imaging of Living Cells Using Bessel Beam Plane Illumination." *Nature Methods* 8 (5): 417–23. doi:10.1038/nmeth.1586.

Raza, K. 2012. "Application of Data Mining in Bioinformatics." *Indian Journal of Computer Science and Engineering* 1 (2): 114–18. doi:10.4010/2016.1759.

Sarkar, R. R. 2016. "The Big Data Deluge in Biology: Challenges and Solutions." *Journal of Informatics and Data Mining* 1 (2): 1–3. doi:10.21767/2472-1956.100014.

Supatto, W., A. McMahon, S. E. Fraser, and A. Stathopoulos. 2009. "Quantitative Imaging of Collective Cell Migration during Drosophila Gastrulation: Multiphoton Microscopy and Computational Analysis." *Nature Protocols* 4 (10): 1397–1412. doi:10.1038/nprot.2009.130.

Swat, M. H., G. L. Thomas, J. M. Belmonte, A. Shirinifard, D. Hmeljak, and J. A. Glazier. 2012. "Multi-Scale Modeling of Tissues Using CompuCell3D." *Methods in Cell Biology* 110: 325–66. doi:10.1016/B978-0-12-388403-9.00013-8.

Thomasson, D. M., A. Gharib, and K. C. P. Li. 2004. "A Primer on Molecular Biology for Imagers: VIII. Equipment for Imaging Molecular Processes." *Academic Radiology*. doi:10.1016/j.acra.2004.07.008.

Yalcin, H. C., A. Shekhar, A. A. Rane, and J. T. Butcher. 2010. "An ex-Ovo Chicken Embryo Culture System Suitable for Imaging and Microsurgery Applications." *Journal of Visualized Experiments*, no. 44. doi:10.3791/2154.

Yu, Q., L. Leatherbury, X. Tian, and C. W. Lo. 2008. "Cardiovascular Assessment of Fetal Mice by In Utero Echocardiography." *Ultrasound in Medicine and Biology* 34 (5): 741–52. doi:10.1016/j.ultrasmedbio.2007.11.001.

Zaki, M. J., J. T. L. Wang, and H. T. T. Toivonen. 2003. "Data Mining in Bioinformatics." *ACM SIGKDD Explorations Newsletter* 5 (2): 198. doi:10.1145/980972.981006.

9 Machine Learning Approach to Overcome the Challenges in Theranostics

A Review

Bishwambhar Mishra, Sayak Mitra, Karthikeya Srinivasa Varma Gottimukkala, and Shampa Sen

CONTENTS

9.1 INTRODUCTION TO THERANOSTICS

Modern medicine is currently undergoing a paradigm shift from conventional disease treatments based on the diagnosis of a generalized disease state to a more personalized—and customized—treatment model based on molecular-level diagnoses. This involves the use of novel biosensors able to precisely extract disease-related information from complex biological systems. Moreover, with the recent progress in chemical biology, materials science, and synthetic biology it has become possible to simultaneously conduct diagnosis and targeted therapy (theranostics/theragnosis) by directly connecting the readout of a biosensor to a therapeutic output. These advances pave the way for more advanced and better-personalized treatment for intractable diseases, while producing fewer side effects. In this review, we describe the recent advances in the development of cutting-edge theranostic agents, which contain both diagnostic and therapeutic functions in a single integrated system.

By comparing the advantages and disadvantages of each modality, we discuss the future challenges and prospects of developing ideal theranostic agents for the next generation of personalized medicine. "Right drug, right patient, right moment, right space, and right dose" is a key concept of current personalized medicine to maximize the efficacy of treatment while minimizing side effects (Hood 2003). The formulations of theranostic agents are highlighted in Figure 9.1. A robust new paradigm for advancing treatments against cancers and other diseases is provided by the integration of multiple moieties such as imaging and therapeutics into a single unit. (Shah and Shah 2012).

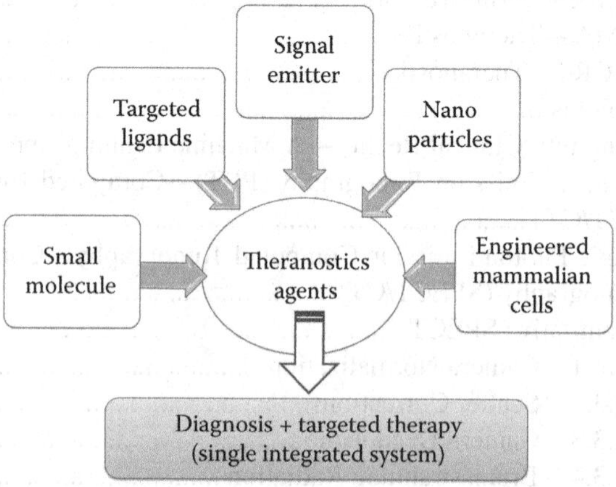

FIGURE 9.1 Construction of the theranostics in a single integrated system.

In this context, theranostics have emerged as a new concept in next-generation medicine to combine simultaneous diagnosis and targeted therapy. Cutting-edge theranostic agents are designed to contain both diagnostic and therapeutic functions in a single integrated system capable of sensing the endogenous biomarkers of a host organism, and therefore can exert therapeutic functions with a minimal delay (Crawley and Thompson 2014). For example, it has become possible to entrap imaging agents and therapeutic drugs in a single integrated molecule, or nanoparticle, using the advances in material science and chemical biology; the drugs and agents are released when and where the system senses the presence of the disease.

The basic principle of theranostics has been in practice for decades, dating back to the 1940s, where the use of radioiodine was used to image and treat thyroid cancers. More recently, advances in molecular biology, genetics, and proteomics have provided a wealth of biochemical and functional information for cancers, which also includes the identification of many cell surface receptors that are overexpressed in tumor cells (Kelkar and Reineke 2011). Theranostics take the advantages of these cells surface receptors to visualize the site of diseases. More over, it is helpful for defining presence or absence of a molecular target, and delivering the cytotoxic payloads directly to tumor sites (Jeelani et al. 2014).

Depending on the application, molecular targeting agents can be combined with a wide variety of imaging agents including radionuclides (single photon emission computed tomography [SPECT], and positron emission tomography [PET]), optical probes (fluorescence), or metal chelates (MRI). The presence of a molecular target, as well as its distribution within a patient, can be provided to the clinicians by combining molecular imaging with a theranostic approach. Furthermore, theranostics has the potential to facilitate patient screening, guide clinical trial enrollment, and monitor the efficacy of therapeutics. A well-studied example of a theranostic target is the human epidermal growth factor receptor, Her-2. Her-2 is overexpressed in aggressive breast cancers with poor prognosis. Breast cancer can be identified in patients using a diagnostic test which detects the large amounts of Her-2, that is, guiding the treatment with a therapeutic monoclonal antibody against this target (Ferber et al. 2014).

The field of theranostics is rapidly facilitating the shift from "trial and error" medicine to personalized medicine, and holds great promise for improved patient outcomes. From the clinician's perspective, theranostics provide a valuable tool for identifying and selecting patients with a particular molecular phenotype indicative of positive response to treatment. This approach has the potential to help improve drug efficacy by understanding which patients might benefit the most from treatment (Jeelani et al. 2014). The working model of theranostics in the human system is illustrated by Figure 9.2.

The precise cell specific targeting properties of a theranostic may help to improve the safety profile of a drug and minimize off-target effects to normal tissues—often seen with various chemotherapies. Finally, from an economic perspective theranostics can lead to more cost-effective and efficient drug programs, guiding preclinical drug development—or clinical trial eligibility—to help maximize the probability of successful outcomes. The demanding aspects of theranostics, in terms of research, are highlighted in the published papers shown in Figure 9.3 (https://www.ncbi.nlm.nih.gov/pubmed/).

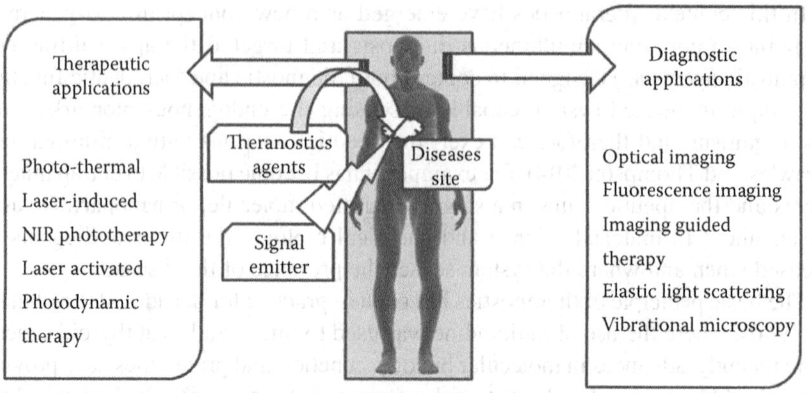

FIGURE 9.2 Schematic overview of theranostics as a working model in the targeted organ of human system.

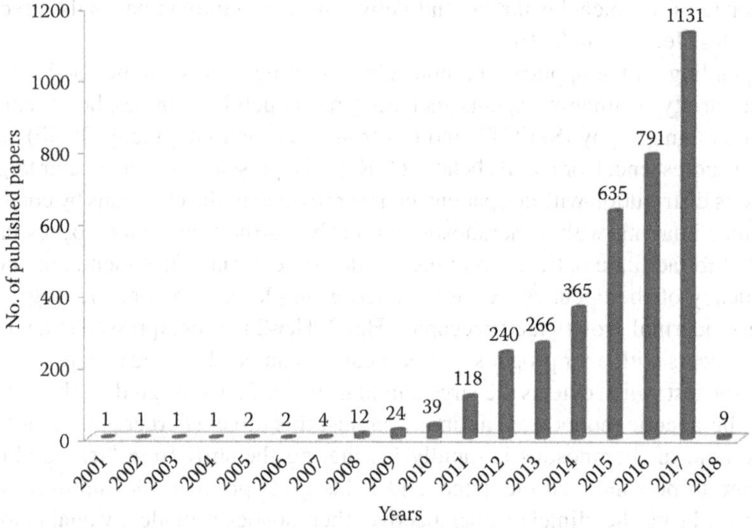

FIGURE 9.3 Number of published papers in theranostics research until the present, according to PubMed.

9.2 POTENTIAL TARGETS FOR THERANOSTIC AGENTS IN NUCLEAR MEDICINE

For nuclear imaging, either gamma emitters or positron emitters are mandatorily required for detection and visualization through a gamma camera, SPECT, or PET. These noninvasive molecular imaging techniques are used for the diagnosis and monitoring of the subsequent therapy efficiency. The radionuclides most commonly used for SPECT are Tc-99m, In-111, and I-123/125 (Nayak and Brechbiel 2011). Ga-68, F-18, I-124, Zr-89, Y-86, and Cu-64 (Table 9.1) are the most commonly used radionuclides for PET.

TABLE 9.1
Applications of Theranostic Agents in Nuclear Medicine

S.L No	Diagnostic Agents	Therapeutic Agents	Mechanism of Reaction	Application Examples	References
1	Tc-99 m(ϒ) $t_{1/2}$ = 6.06 h	Re-188(26% ß, 15% ϒ) $t_{1/2}$ = 17 h	Chelator direct labelling (sulfhydryl group)	Antibodies metal particles diphosponates	Gustafsson et al. (2012) Häfeli et al. (2001) Uematsu (2005) García-Garayoa et al. (2007) Lange et al. (2016)
2	Cu-64(19% ß+ /38% ß– /43% ϒ) $t_{1/2}$ = 12.7 h	Cu-67(ß–) $t_{1/2}$ = 61.8 h Lu-177(ß–/ ϒ) $t_{1/2}$ = 6.7 d	Chelator direct labeling (nanoformulations)	Liposomes nanoparticles antibodies	Enrique et al. (2015) Szymański et al. (2012) Stockhofe et al. (2014) Loon et al. (2017)
3	Zr-89(23% ß+) $t_{1/2}$ = 78.4 h	Y-90(>99% ß–) $t_{1/2}$ = 64 h	Chelator direct labeling (nanoparticles)	Antibodies (i.e., Zevalin) nanoparticles	Perk et al. (2006) Kramer et al. (2017)
4	Y-86(32% ß+) $t_{1/2}$ = 14.7 h	Y-90(>99% ß–) $t_{1/2}$ = 64 h	Chelator	Small molecules (e.g., DOTATOC, PSMA inhibitors)	Rösch et al. (2017) Banerjee et al. (2015)
5	F-18(97% ß+) $t_{1/2}$ = 110 min	I-131(90% ß–;10% ϒ) $t_{1/2}$ = 8 d	Nucleophilic substitution electrophilic attack	ICF-15002	Rbah-Vidal et al. (2017)
6	In-111(>99.9%EC $t_{1/2}$ = 67.9 h	In-111 (>99.9% EC; Auger electron emitter)	Chelator (DTPA)	PRRT	Gotthardt et al. (2006)
7	Ga-68(89% ß+) $t_{1/2}$ = 68 min In-111(>99.9%EC) $t_{1/2}$ = 67.9 h	Lu-177(ß–/ ϒ) $t_{1/2}$ = 6.7 d Y-90(>99% ß–) $t_{1/2}$ = 64 h	Chelator (introduced via bifunctional linker)	PSMA-617	Gotthardt et al. (2006)

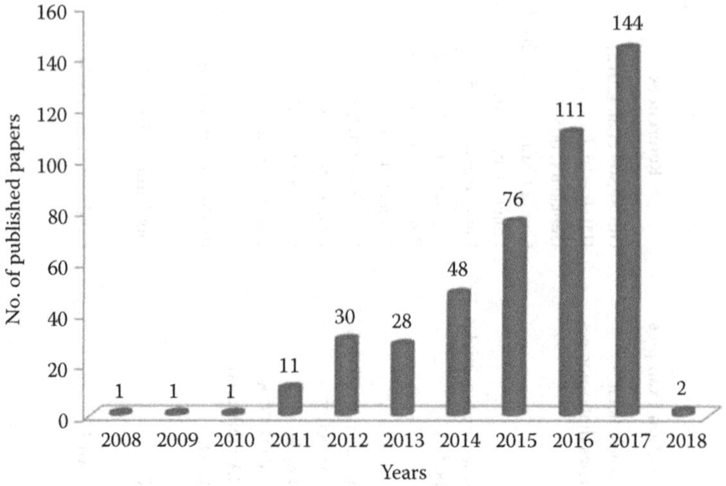

FIGURE 9.4 Number of published papers in nuclear medicine with theranostics until the present, according to Pubmed.

Therapeutic tracers are most likely radiolabeled with beta minus particles (ß–), which have a range between 1 and 10 mm with energies between 0.1 and 1 MeV. The most commonly used radionuclides are I-131, Lu-177, and Y-90. Cu-67 is acquiring more attention in preclinical studies, as it is chemically identical and exchangeable with its positron-emitting isotope Cu-64. I-131 is characterized by a 90% beta minus (ß–) emission with a mean tissue penetration of 0.4 mm and 10% gamma emission, which can be imaged with SPECT or a gamma camera. The mean tissue penetration of Y-90 in its pure beta minus emitter is 2.5 mm. The high penetration depth results in a crossfire effect resulting in damage to healthy organs depending on the tracers biodistribution (Morgenroth et al. 2017). In peptide receptor radionuclide therapy (PRRT), Y-90 was responsible for crossfire effect which could be compensated by Lu-177. Lu-177 has two advantages compared to Y-90: the mean tissue penetration of Lu-177 is 0.6 mm and it emits gamma rays, which allows for therapy monitoring and dosimetry evaluation. To apply the theranostic strategy to commonly used Tc-99m labeled probes, the generator nuclide Re-188 (W-188/Re-188 generator), which possess comparable coordination chemistry, is already being used in preclinical and early phase clinical studies for therapeutic application (Dash et al. 2013). Re-188 emits beta minus particles (25.6%) and gamma rays for imaging (15.1%). The demanding aspects is illustrated by published papers in Figure 9.4 (https://www.ncbi.nlm.nih.gov/pubmed/).

9.2.1 Metabolic Active Radiopharmaceuticals

Inorganic and organic materials are used to design nanotheranostics. Metal-based nanoparticles offer a route for direct radiolabeling. They are doped with radioisotopes and are activated through neutron reactions or proton beams. As shown for super paramagnetic iron oxide nanoparticles (SPIONs), radioactivity can be taken up either via specific trapping or via ion exchange and simple equilibrium. SPIONs are being radiolabeled with Ra-223 and show promising stability of the radioactive particles in vitro (Broda et al. 2016). As an example, for activation through neutron reactions, gold nanoparticles can be activated and intrinsically labeled with β emitting Au-198 (t 1/2 = 2.7 d; energy <1 MeV). The particles are attracted due to their optical properties, as well as their size-dependent cytotoxicity, which can be enhanced for therapeutic use through radiolabeling (Same et al. 2016). Surface modification has allowed the process for targeting ligands or additional radiolabeling via specific coordination chemistry, and for optimization of pharmacokinetics and biocompatibility (Ernsting et al. 2013). Antibodies such as cetuximab can also target ligands. The antibodies can be radiolabelled with I-131 and be grafted to gold nanoparticles for theranostic application. Despite surface modifications, the quantum dots can be doped with radiometals like Zr-89 or Cu-64 by the process of ion exchange (Cheng et al. 2016).

9.2.2 Specifically Binding Radiopharmaceuticals

The basic architecture of theranostics is as follows (Figure 9.5):

- A targeting agent, which directs the theranostic agent to a molecular target on the surface of a cell or tumor.
- A chelate in the form of an imaging agent (which enables visualization of the target), or a therapeutic drug (for delivery of treatment to the target site).
- A linker which serves as a connection between the two entities.

The various applications of nuclear theranostics have been summarized in Table 9.1.

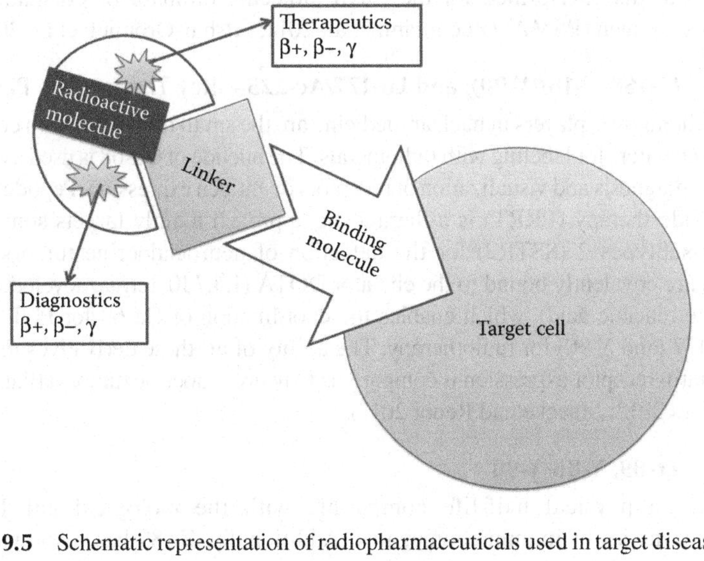

FIGURE 9.5 Schematic representation of radiopharmaceuticals used in target diseased cells.

9.2.2.1 Theranostic Application of Y-90

Y-90 is a rare-earth lanthanide, which emits high energy ß-particles and is used in radiotherapy. In PRRT, it was almost completely displaced by Lu-177, but is still commonly used in radio-immunotherapy and radio-immobilization therapy with Y-90 microspheres. While radio-immobilization therapy is usually planned with Tc-99m labeled macroaggregated albumin (MAA), the surrogate cannot predict side effects. Therefore, Y-90 has gained attention as a theranostic tool itself. Imaging of Y-90 can be performed via PET due to its internal pair production with detectable peaks of 511 keV (Goedicke et al. 2013), as well as detection of radiation with gamma cameras. Y-90 is also known to produce Cerenkov luminescence, which can be visualized with optical Cerenkov luminescence imaging (Wright et al. 2015).

9.2.2.2 Iodine—The Classic Theranostic Pair

Years before the term theranostic became viral in the medical field, the theranostic approach was already common practice in nuclear medicine. Radioiodine theranostics is a typical example regarding the concept of personalized medicine and is used extensively for the treatment of thyroid cancer. At the beginning, I-131 was used as a therapeutic and diagnostic tool, but due to its emission of ß-particles and high-radiation dose it cannot be considered as an ideal diagnostic agent. I-123, on the other hand, has a more efficient interaction with sodium iodide crystals, while showing no β-emission and a lower energy, which results in a more efficient collimation. Since I-123 provided radiation protection and superior images, it has been increasingly used as the isotope of choice for diagnostic imaging as a surrogate for I-131. Nonetheless, I-124 with its longer half-life and its suitability for PET has become an important tool in clinical and preclinical imaging (Silberstein 2012). Apart from thyroid cancer imaging, the iodine theranostic pairs are used for radio-labeling of targeted tracers found in different tumor types. Meta-iodobenzylguanidine (MIBG) is being used as a theranostic tool for endocrine tumors (Vöö et al. 2011; Sisson et al. 2012). Radiolabeling with I-124 or I-131 was also performed for the small molecule inhibitor of prostate-specific membrane antigen (PSMA) (Zechmann et al. 2014; Afshar-Oromieh et al. 2017).

9.2.2.3 Ga-68, F-18/(Y-90), and Lu-177/Ac-225—Key Theranostic Players

The key theranostic players in nuclear medicine are the small molecules with covalently attached chelators for labeling with radiometals. The nuclide of Ga-68 is used as positron emitter in diagnosis and visualization of receptors or antigen expression. Peptide receptor radionuclide therapy (PRRT) is a theranostic approach mainly targets somatostatin receptor subtypes 2 (SSTR2) for the detection of neuroendocrine tumors (NET). Peptides are covalently bound to the chelator DOTA (1,4,7,10-Tetraazacyclododecane-1,4,7,10-tetraacetic acid), which enables the coordination of Ga-68 for PET imaging and Lu-177 (and Y-90) for radiotherapy. The ability of all these derivatives to address somatostatin receptor expression is comparable to neuroendocrine tumors (Blankenberg and Strauss 2012; Maecke and Reubi 2017).

9.2.2.4 Zr-89, Y-86/Y-90

Zr-89 has a physical half-life compatible with the biological half-life and pharmacokinetics of immune conjugates. Additionally, Zr-89 has a comparatively low positron energy of 356 keV, which results in high resolution PET images, and radiolabeling can be performed in a straightforward manner using DFO as a chelator (Reiner et al. 2015). For improved therapy selection and planning, short-lived radioisotopes might not be able to provide the information needed for efficient therapy. Long-lived PET isotopes can resolve the issue and reveal detailed information of biodistribution and radiation dose on healthy organs. Similarly to the copper isotopes, the half-life of several days is ideal for in vivo imaging of macromolecules and nanosized structures. One example would be Zr-89-labeled carbon nanotubes for the treatment of tumor vasculature (Reiner et al. 2015). In case of nanoparticle labeling, it is also possible to design a dual radiolabeled nanotheranostic bearing both Y-90 for therapy and Zr-89 for high resolution simultaneous imaging. However, due to the long half-life, the radiation dose needs to be observed carefully and the dose-limiting organ needs to be defined. For most macromolecules, the liver is considered as the critical organ. In some cases, accumulation of Zr(IV) cations in the bone can render the bone marrow as the dose limiting organ.

9.3 APPLICATION OF THERANOSTICS

9.3.1 Neuroendocrine Tumors

A neuroendocrine neoplasm or neuroendocrine tumor (NET) is a heterogeneous group of epithelial neoplasms predominantly characterized by neuroendocrine differentiation (Phan et al. 2010). These tumors are characterized by a high expression of somatostatin receptors (SSTRs) and specific markers, including chromogranin A and CD53 protein (Severi et al. 2017). Currently, NET is classified based on the grading (Ki67 and mitotic index [MI]) and location of the neoplasms. Although NETs can originate from any region of the body, 60%–70% derive from the gastroenteropancreatic system, called GEP-NET. Approximately 30% of NETs have associated symptoms; therefore, they are considered as functioning tumors. These symptoms are often related to the hormonal hypersecretion, which frequently include diarrhea, flushing, hyperglycemia, or hypoglycemia. NETs of the lungs are also known as pulmonary carcinoid tumors, which account for 1%–2% of all lung tumors and approximately 30% of all NETs. Pulmonary NETs are classified according to their mitotic count rate and the presence of necrosis (Phan et al. 2010).

Theranostics of NETs include diagnosis of the disease with PET/computed tomography (CT) using Ga-68-labeled somatostatin analogs. These bind specifically to different SSTR subtypes and allow the molecular imaging and characterization of NETs with a very high diagnostic sensitivity and specificity for the early identification of metastases (Modlin et al. 2010; Phan et al. 2010). This has been shown to have a high impact on patient management (Hofman et al. 2012). In pulmonary NETs, SSTR PET/CT has been shown to have a significant impact on treatment strategy in up to 18% of the patients (Prasad et al. 2015). This subsequently allows the best management approach for each patient (Kulkarni et al. 2014a), as well as the evaluation of treatment response posttherapy (Kulkarni et al. 2014b). Following the determination of SSTR expression using PET/CT, PRRT can be instituted using therapeutic radionuclides (e.g., beta- or alpha-emitting radioisotopes) labeled with the same tracer for personalized treatment.

9.3.2 PSMA—Theranostics

Prostate cancer (PCa) is a major health problem in men. Prognosis is dependent on the stage of the disease and the five-year survival rate for advanced, disseminated PCa is only 29% (Siegel et al. 2012). Metastatic castration-resistant PCa (mCRPC) is associated with poor prognosis and a diminished quality of life. Treatment options for mCRPC patients include taxane-based therapies (docetaxel, cabazitaxel) and novel second-line hormonal therapies (enzalutamide and abiterone), which all show moderate survival benefits. Unfortunately, those therapies are only temporarily effective and development of treatment resistance has been observed (Heidenreich and Porres 2014; Heidenreich et al. 2014). The bone-seeking alpha-emitting radiopharmaceutical radium-223 chloride (XofigoTM) has recently been introduced, resulting in a reduction of bone pain and a moderately improved survival rate in mCRPC patients with bone metastases (Hoskin et al. 2014; Humm 2015). However, the use of this promising radiopharmaceutical is limited to the treatment of bone metastases. Therefore, there is an urgent need to develop more specific therapies targeting PCa visceral lesions as well.

Recent developments in PCa-targeted radiopharmaceuticals have focused on prostate-specific membrane antigen (PSMA) ligands. PSMA is a favorable target for imaging and radionuclide therapy of PCa, because it is over expressed in 90%–100% of local PCa lesions, as well as in cancerous lymph node metastases and bone lesions. Furthermore, reports indicate that PSMA expression levels are further enhanced in high-grade, metastatic, and castration-resistant PCa (Bostwick et al. 1998).

Numerous studies have shown the value of radiolabeled PSMA-targeted agents to visualize PCa lesions, including PCa metastases (Barrett et al. 2013; Pandit-Taskar et al. 2015). Among the different PSMA-targeting tracers available, several tracers have made their way to the clinic. For PET imaging of PCa, Glu-NH-CO-NH-Lys-(Ahx)-68Ga-HBED-CC, also known as 68Ga-PSMA-11 or 68Ga- PSMA-HBED-CC, is the most commonly used compound (Rowe et al. 2015; Budäus et al. 2016). Unfortunately, HBED-CC (N,N'-bis[2-hydroxy-5-(carboxyethyl)benzyl] ethylenediamine-N,N'-diacetic acid), the chelator used in this tracer, is not suitable for radiolabeling with radiometals for therapeutic applications (e.g., 177Lu, 90Y, and 213Bi). More recently, a novel 18F-labeled PSMA inhibitor, 18F-DCFPyL, was introduced in the clinic (Szabo et al. 2015). For therapy, anti-PSMA monoclonal antibodies (J591) radiolabeled with 177Lu and 90Y (Tagawa et al. 2013; Vallabhajosula et al. 2016) and the small-molecule PSMA inhibitor [131I] MIP-1095 have shown promising therapeutic efficacy, although with moderate toxicity. Moreover, PSMA ligands with DOTA-derived chelators have been developed, which can be labeled with diagnostic radionuclides, for example, 68Ga for PET and 111In for SPECT, as well as therapeutic radionuclides, for example, 177Lu, 90Y or 213Bi for radionuclide therapy. Such theranostic compounds, that is, PSMA I&T and PSMA-617 are of great interest because they combine the potential of diagnosis and radionuclide therapy using one PSMA targeting molecule (Weineisen et al. 2014; Benešová et al. 2015).

Radio-ligand properties, for example, lipophilicity, target affinity, and metabolic stability have been carefully optimized for theranostic applications. Monoclonal antibodies have shown slow blood clearance, which may result in myelotoxicity.

Small-molecule PSMA inhibitors showed the advantage to localize rapidly in tumor lesions, including soft-tissue and bone metastases. On the other hand, these small inhibitors showed high and specific uptake in PSMA-expressing kidneys and salivary glands, which is of major concern for its application for radionuclide therapy. Strategies should therefore be investigated to limit radiation-induced damage in healthy organs. Here, we report on the preclinical application of the DOTAGA-chelated urea-based PSMA inhibitor PSMA I&T for SPECT/CT imaging and radionuclide therapy of PCa.

9.3.3 CXCR4—Theranostics

CXCR4 is a member of the chemokine receptor subfamily of seven transmembrane domain, G-protein coupled receptors, whose sole known natural ligand is CXCL12/SDF-1. CXCR4 is an unusual chemokine receptor by virtue of having expanded roles beyond leukocyte recruitment, including fundamental processes such as the development of the hematopoietic, cardiovascular, and nervous systems during embryogenesis. The receptor was first discovered as one of the coreceptors for HIV, and thereafter was also found to be expressed by multiple cancers including breast, prostate, lung, colon, and multiple myeloma.

A number of recent studies have correlated high levels of CXCR4 expression in cancers with poor prognosis and with resistance to chemotherapy, in part through enhancing interactions between cancers and stroma. A possible role for CXCR4 and chemokine receptors generally in cancer and metastasis was first suggested in the studies of breast cancer showing that the receptor plays a role in directing metastatic cells to CXCL12-expressing organs. Collectively, the data on CXCR4 in cancer suggest this receptor increases tumor cell survival and/or growth and/or metastasis, making it a potentially attractive therapeutic target.

Due to the role of CXCR4 in HIV, multiple CXCR4 antagonists—although not sufficient for the treatment of HIV, are currently being evaluated and/or used for stem cell mobilization and as antitumor therapy. Some of the antagonists were also shown in animal models to be of use in evaluating CXCR4 expression in whole tumors noninvasively by molecular imaging. The research on CXCR4 has been ongoing for the last decade and has yielded more than 11,398 papers in PubMed, as of December 2017.

The primary pathology, in which CXCR4 was found to be essential, is infection by human immunodeficiency virus (HIV), and this finding initiated vast research efforts on the receptor. The role of CXCR4 by the T-trophic HIV as a coreceptor was later identified as a late phase of the disease, while the main coreceptor of the virus during the beginning of an HIV infection is another chemokine receptor, CCR5. The detailed role of CXCR4 in HIV infection, and the switch of the virus from CCR5-dependent M-trophic form to CXCR4-dependent T-trophic form has been elaborately discussed by Elisa Vicenzi et al. (2013).

9.4 MAJOR PROBLEMS

The fundamental principle of theranostics has been followed since decades, with the utilization of radioiodine to diagnose (image) and manage disease. Since its inception until now, theranostics has been fostering unremarkable tailored and targeted therapy. In the context of pharmacotherapy, the fundamental aspects of individual absorption, distribution, metabolism, and excretion are understood with regards to the complex interactions and detailed characterization, especially the functional variations. Interestingly, epigenetic variations and the impact of environmental factors, including the circadian rhythms, on an individual's response to drugs needs to be assessed specifically for each individual (Lütje 2015).

Apart from the above variations, educating the health care provider, payor, regulator, and the patient is also mandatory. Theranostics has an extended impact on patients in more than individual centric care. The patients always expect a positive outcome through the therapeutic diagnosis, which is an alarming aspects of theranostics. The existing less targeted and tailored pharmacotherapy has resulted in the loss for the overall economy of the pharmacological market accounting to $400 billion of total global pharmaceutical market, that is $825 billion.

The most common reason for failure and avoidable deaths—including expensive hospitalizations, has been found to be the side effects associated with less targeted nonspecific therapies, which are based on principles contrary to theranostics. Furthermore the age, gender, diet, and lifestyle of a patient including the intestinal microflora plays a significant role in the patient's response to a drug.

9.5 TRANSITIONING INTO CLINICAL REALITY—A MACHINE LEANING APPROACH

9.5.1 Positron Emission Tomography (PET)—Computed Tomography (PET/CT)

PET uses small amounts of radioactive materials called radiotracers, a special camera, and a computer, which evaluate your organ and tissue functions. By identifying body changes at the cellular level, PET may be used to detect the early onset of disease before it is evident on other imaging tests.

A PET scan creates pictures of organs and tissues in the body. First, a technician injects the patient with a small amount of a radioactive substance, which the organs and tissues pick up and absorb. Areas using more energy pick up more of the radioactive substance and thus cancer cells pick up a lot of it, as they tend to use more energy than healthy cells. Then a scan shows where the radioactive substance is present in the body. For this, a CT scan uses X-rays to create a 3D picture of the inner body, and it detects and shows anything abnormal, including tumors. Sometimes, a special dye, called as contrast medium is given before the scan to provide better details in the image (Rbah-Vidal 2017).

A PET-CT does not hurt but the patient needs to lie still for the entire scan. He/she might also need to keep their arms above the head, which could become uncomfortable. The technician generally asks patients to hold their breath at times because motion from breathing may cause blurry pictures. It is also advised to raise, lower, or tilt the table during the scan to allows the capture of pictures from different angles.

9.5.2 Single Photon Emission Computed Tomography—Computed Tomography (SPECT/CT)

SPECT-CT is where two different types of scans are taken and the images or pictures from each are fused or merged together to form an overlapping image. The fused scan can provide more precise information about how different parts of the body function and can clearly identify problems such as tumors (lumps) or Alzheimer's disease, etc. SPECT images are obtained following an injection of a radiopharmaceutical used for nuclear medicine scans. The injected medication sticks to specific areas in the body, depending on what radiopharmaceutical is used and the type of scan being performed. For example, it shows bones for a bone scan, and gall bladder and bile ducts for a hepatobiliary scan.

The radiopharmaceutical can be detected by a nuclear medicine gamma camera. The camera or cameras rotate over a 360° arc around the patient, allowing the reconstruction of an image in 3D. In computed tomography (CT) images are obtained while lying on a bed moving through a ring, or "donut" shaped X-ray machine. Again, the X-ray machine rotates over a 360-degree arc around the patient, allowing for image reconstruction in 3D. The X-ray machine from the CT scanner rotates much faster than the gamma camera; therefore, the CT part of the study takes less time than that of the SPECT study. The similarity between the SPECT and CT method of image processing allows the images to be combined. Combining the information from the nuclear medicine SPECT study and the CT study allows the information about function from the nuclear medicine study to be easily combined with the information about how the body structure "looks" in the CT study.

One must lie still in a ring-shaped scanner for at least 30–40 minutes. The first 3–5 minutes involves the CT scan component, while the remainder of the time is required for the SPECT study. It is very important to remain still for the entire duration of the two studies so that the SPECT and CT can be accurately combined. If not, the images from one study will not correspond exactly to the images from the other study, and may result in interpretation difficulties.

There are no risks involved in the nuclear medicine, that is, SPECT scan or the CT scan procedures. The tests involve small doses of ionizing radiation from the radiopharmaceuticals injected into the vein.

Importantly, the SPECT component of the test requires no additional injection of radiopharmaceuticals beyond what would otherwise be given for nuclear medicines tests. The CT is usually done using a low-dose radiation technique, which is around 20%–25% of the radiation exposure of a normal CT scan.

9.5.3 Quantitative SPECT

While accurate quantitation of activity distribution may be advantageous for many diagnostic imaging studies, it is absolutely essential for personalized dosimetry in radionuclide therapies. Factors affecting the measure of the quantitative SPECT are illustrated in Figure 9.6.

With the increased availability of hybrid SPECT/CT systems, fully quantitative determination of radiotracer distribution in tumors and critical organs has become both

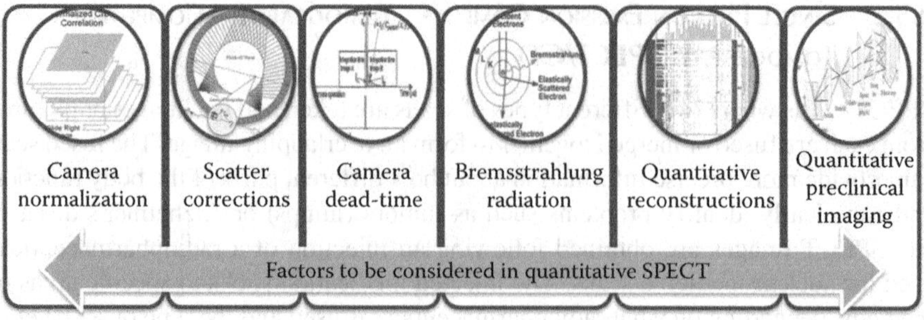

FIGURE 9.6 Factors affecting the measure of the quantitative SPECT of various organ in the human system.

feasible and practical. This is required to develop methods to accurately reconstruct activity distributions, expressed in absolute units (Bq or Ci), from conventional clinical SPECT studies performed using both diagnostic and therapeutic radiotracers.

9.5.3.1 Camera Normalization

Once quantitatively accurate images are reconstructed, the counts in each voxel must be translated into activity values. This is done using an experimentally determined camera normalization factor. This is required to investigates the accuracy of the camera normalization factor, determined using the planar scan of a point source versus that obtained from a tomographic scan of a large phantom filled with activity.

9.5.3.2 Scatter Corrections

For quantitative imaging it is essential to remove all the scattered photons from the photopeak energy window, and several methods have been developed for this purpose. This project evaluates the accuracies of various simple methods (such as triple-energy-window) and compares them with more complex but also more reliable approaches (such as Monte Carlo and analytical photon distribution calculation). Corrections for self-scatter and high-energy scatter are being investigated.

9.5.3.3 Camera Dead-Time

Activities of radiotracers used in diagnostic imaging are low, resulting in low count rates with no camera dead-time. However, in typical radionuclide therapy procedures patients are injected with high activities (of the order of GBq), emitting high fluxes of gammas. In order to accurately quantify a patient's activity distribution in these situations, various investigations and developments are carried out in order to correct the reconstructed images for these dead-time losses.

9.5.3.4 Bremsstrahlung Radiation

Targeted radionuclide therapies typically use radioisotopes, which decay through β emissions (electrons). These electrons create Bremsstrahlung photons, which may affect the quantitative accuracy of images. This is required to investigate the characteristics of bremsstrahlung emissions created by ß-emissions in tissues, and to analyze the spectra recorded by SPECT cameras with different types of collimators.

9.5.3.5 Quantitative Reconstructions

A graphical user interface (GUI) named SPECToR allows the quantitative reconstruction of images from data acquired using different SPECT/CT cameras. Parameters of the reconstruction algorithm, as well as quantitative corrections can be varied. The comparison of quantitative accuracies of images, which are reconstructed using commercial software obtained from SPECToR.

9.5.3.6 Quantitative Preclinical Imaging

Development of new pharmaceuticals and new treatments prior to their introduction into clinics often involve the use of small animal models, which often lead to the crucial analysis of pharmacokinetic, targeting, and toxicity during preclinical imaging. Such imaging requires cameras with high resolution, providing accurate quantification of tracer uptakes. This is required to investigate the quality and quantitative accuracy of images obtained from the VECTor/CT camera using different diagnostic and therapeutic radioisotopes.

9.5.3.7 Patient Specific Dosimetry/Dose Calculations

As therapeutic uses of radionuclides in nuclear medicine have increased, the use of patient-specific methods for calculation of radiation dose has become more important. Most current resources make use of standardized models of the human body representing median individuals, but the use of image-based and more realistic models will soon take their place, and will permit adjustments to represent individual patients and tailor therapy planning uniquely for each subject.

One of the important beneficial applications of the use of radiation is in the healing arts. Unlike in many other uses of radiation, the patient receives a direct benefit from the study and therefore evaluation of the risk/benefit relationship is more straightforward. An important example is the use of radiopharmaceuticals, that is, nuclear medicine in both diagnosis and therapy. Diagnostic uses of radiopharmaceuticals are well established and are employed to evaluate a broad variety of patient conditions. Radiation doses for diagnostic agents are developed by studying the biokinetics of the radiopharmaceuticals in preclinical and clinical studies. In the therapeutic approach, extrapolation methods are applied to convert values measured in animal organs for humans, and in the diagnostics case, the quantitative data has been observed in the human subjects and can be used directly for input to dose calculations (Stabin 2006).

The basic goal of all forms of radiation therapy (using external or internal radiation sources) is to deliver a lethal radiation dose to the unhealthy tissues of concern while avoiding or limiting the expression of undesired effects in other normal tissues of the patient. Radioactive iodine (131 I) has been used for many years to treat benign thyroid disease with and without patient-specific treatment planning (Canzi et al. 2006; Carlier et al. 2006). Treatment of thyroid cancer with 131I NaI is the most common application of radionuclide therapy in nuclear medicine and has been used for many decades. Patient therapy is most often based on administration of fixed levels of radiation to all subjects, rather than targeting absorbed doses, although there have been exceptions.

I-131 labeled meta-iodobenzylguanidine (mIBG) has been used for many years in the treatment of adult and pediatric neuroendocrine tumors (including phaeochromocytoma, paraganglioma, and neuroblastoma), typically with administrations of 7.4 GBq to more than 30 GBq in adults. To treat cancer, several monoclonal antibodies have been developed.

A general equation for the absorbed dose rate in an object uniformly contaminated with radioactivity (for example an organ or tissue with radiopharmaceutical uptake) may be shown as:

$$\dot{D}_T = \frac{k\,A_S \sum_i y_i E_i \phi_i}{m_T}$$

where

D_T = absorbed dose rate to a target region of interest (Gy/sec)
A_S = activity (MBq) in source region S
y_i = number of radiations with energy E_i emitted per nuclear transition
E_i = energy per radiation for the ith radiation (MeV)
ϕ_i = fraction of energy emitted in a source region, which is absorbed in a target region
m_T = mass of the target region (kg)
k = proportionality constant (Gy-kg/MBq-sec-MeV)

The proportionality constant k includes the various factors needed to obtain the dose rate in the desired units from the units employed for the other variables, and it is essential that this factor is properly calculated and applied. The time versus activity curve for theranostics agents for a specific organ is given in Figure 9.7.

9.5.3.8 Image-Based Computational Tools

Several centers have implemented the use of image fusion techniques to develop 3D maps of the dose, instead of average organ dose, which is estimated from standard

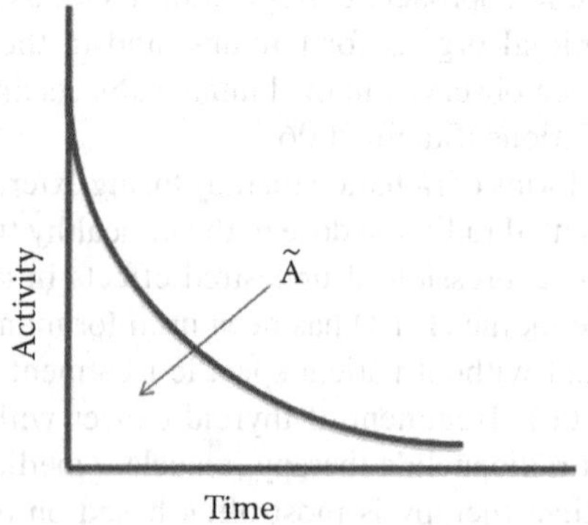

FIGURE 9.7 Time/activity curve to study the activity of an organ in the human system.

models thatare generally available. This suggests the treatment planning for internal emitters may soon be far more sophisticated and similar to that used in external beam therapy for individualized patient therapy planning. Examples include the 3D-ID code from the Memorial Sloan-Kettering Cancer Center (Beauregard et al. 2011), the SIMDOS code from the University of Lund, the RTDS code at the City of Hope Medical Center, the RMDP code from the Royal Marsden Hospital, the DOSE3D code, and the PEREGRINE code developed by various researchers (Willowson et al. 2012).

Radiation dose calculations for radiopharmaceuticals have been standardized by the implementation and dissemination of tools such as the RADAR web site and the OLINDA/EXM software (Grbic et al. 2013). Current efforts suggest a move towards more image-based and patient-specific methods in internal dose calculations for therapeutic applications in nuclear medicine (e.g., the 3D-ID code). Current evidence in the literature strongly supports the idea that patient-specific dose calculations are needed to improve patient outcomes when internal emitters are used in therapy, as is commonly accepted in external radiation therapy (Lehmkuhl et al. 2013).

9.6 CONCLUSION

The usage of theranostic agents is thought to be an effective approach because it allows one to selectively deliver the active drug to the cancer site and to simultaneously monitor the therapeutic efficacy through the use of imaging technology. We hope that this review will be beneficial for researchers aiming to improve and expand the variety of theranostic agents by overcoming the drawbacks of each agent and using the merits of each agent synergically.

REFERENCES

Afshar-Oromieh A. et al. "Repeated PSMA-Targeting radioligand therapy of metastatic prostate cancer with ^{131}I-MIP-1095." *European Journal of Nuclear Medicine and Molecular Imaging* 44, no. 6, 2017: 950–959.

Banerjee S.R. et al. "Preclinical evaluation of ^{86}Y-labeled inhibitors of prostate-specific membrane antigen for dosimetry estimates." *Journal of Nuclear Medicine: Official Publication, Society of Nuclear Medicine* 56, no. 4, 2015: 628–634.

Barrett J.A. et al. "First-in-Man evaluation of 2 high-affinity PSMA-Avid small molecules for imaging prostate cancer." *Journal of Nuclear Medicine* 54, no. 3, 2013: 380–387.

Beauregard J.M., Hofman M.S., Pereira J.M., Eu P., Hicks R.J. et al. "Quantitative 177LuSPECT (QSPECT)imaging using a commercially available SPECT/CT system." *Cancer Imaging* 11, no. 4, 2011: 56–66.

Benešová M. et al. "Preclinical evaluation of a Tailor-made DOTA-conjugated PSMA inhibitor with optimized linker moiety for imaging and endoradiotherapy of prostate cancer." *Journal of Nuclear Medicine* 56, no. 6, 2015: 914–920.

Blankenberg F.G., Strauss H.W. "Nuclear medicine applications in molecular imaging." *Journal of Magnetic Resonance Imaging* 16, no. 4, 2012: 352–61.

Bostwick D.G., Pacelli A., Blute M., Roche P., Murphy G.P. "Prostate specific membrane antigen expression in prostatic intraepithelial neoplasia and adenocarcinoma: A study of 184 cases." *Cancer* 82, no. 11, 1998: 2256–61.

Broda J. et al. "Assessing the intracellular integrity of phosphine-stabilized ultrasmall cytotoxic gold nanoparticles enabled by fluorescence labeling." *Advanced Health Care Materials* 5, 2016: 3118–3128.

Budäus L. et al. "Initial experience of ^{68}Ga-PSMA PET/CT imaging in high-risk prostate cancer patients prior to radical prostatectomy." *European Urology* 69, no. 3, 2016: 393–396.

Canzi C., Zito F., Voltini F., Reschini E., Gerundini P. "Verification of the agreement of two dosimetric methods with radioiodine therapy in hyperthyroid patients." *Medical Physics* 33, no. 8, 2006: 2860–2867.

Carlier T. et al. "Optimized radioiodine therapy for Graves' disease: Two MIRD-based models for the computation of patient-specific therapeutic I-131 activity." *Nuclear Medicine Communications* 27, no. 7, 2006: 559–566.

Cheng L. et al. "Facile preparation of multifunctional WS2/WO*x* nanodots for chelator-free 89Zr-labeling and in vivo PET imaging." *Small* 12, no. 41, 2016: 5750–5758.

Crawley N., Thompson M. "Theranostics in the growing field of personalized medicine: An analytical chemistry perspective." *Analytical Chemistry* 86, no. 1, 2014: 130–160.

Dash A., Knapp F.F., Pillai M.R.A. "Targeted radionuclide therapy—an overview." *Current Radiopharmaceuticals* 6, no. 3, 2013: 152–180.

Enrique M.-A., Mariana O.-R., Mirshojaei S.F., Ahmadi A. "Multifunctional radiolabeled nanoparticles: Strategies and novel classification of radiopharmaceuticals for cancer treatment." *Journal of Drug Targeting* 23, no. 3, 2015: 191–201.

Ernsting M.J., Murakami M., Roy A., Li S.-D. "Factors controlling the pharmacokinetics, biodistribution and intratumoral penetration of nanoparticles." *Journal of Controlled Release* 172, no. 3, 2013: 782–794.

Ferber S. et al. "Polymeric nanotheranostics for real-time non-invasiveoptical imaging of breast cancer progression and drug release." *Cancer Letter* 352, no. 1, 2014: 81–89.

García-Garayoa E., Schibli R., Schubiger P.A., "Peptides radiolabeled with Re-186/188 and Tc-99m as potential diagnostic and therapeutic agents." *Nuclear Science and Techniques* 18, no. 2, 2007: 88–100.

Goedicke A. et al. "Study-parameter impact in quantitative 90-Yttrium PET imaging for radioembolization treatment monitoring and dosimetry." *IEEE Transactions on Medical Imaging* 32, no. 3, 2013: 485–492.

Gotthardt M., Kwekkeboom D.J., Feelders R.A., Brouwers A.H., Teunissen J.J. et al. "Nuclear medicine imaging and therapy of neuroendocrine tumours." *Cancer Imaging* 6, no. Spec No A, 2006: S178–S184.

Grbic S. et al. "Image-based computational models for TAVI planning: From CT images to implant deployment." *Medical Image Computing and Computer-assisted Intervention* 16, no. 2, 2013: 395–402.

Gustafsson A.M. et al. "Comparison of therapeutic efficacy and biodistribution of ^{213}Bi- and ^{211}At-labeled monoclonal antibody MX35 in an ovarian cancer model." *Cancer* 39, no. 1, 2012: 15–22.

Häfeli U., Pauer G., Failing S., Tapolsky G. "Radiolabeling of magnetic particles with rhenium-188 for cancer therapy." *Journal of Magnetism and Magnetic Materials*, 225, no. 1–2, 2001: 73–78.

Heidenreich A. et al. "EAU guidelines on prostate cancer. Part II: Treatment of advanced, relapsing, and castration-resistant prostate cancer." *European Urology* 65, no. 2, 2014: 467–79.

Heidenreich A., Porres D. "Prostate cancer: Treatment sequencing for CRPC—what do we know?" *Nature Reviews Urology* 11, 2014: 189–190.

Hofman M.S. et al. " High management impact of Ga-68 DOTATATE (GaTate) PET/CT for imaging neuroendocrine and other somatostatin expressing tumours." *Journal of Medical Imaging and Radiation Oncology* 56, no. 1, 2012: 40–7.

Hood E. "Pharmacogenomics: The promise of personalized medicine." *Environ Health Perspect* 111, no. 11, 2003: 581–589.

Hoskin P. et al. "Efficacy and safety of radium-223 dichloride in patients with castration-resistant prostate cancer and symptomatic bone metastases, with or without previous docetaxel use: A prespecified subgroup analysis from the randomised, double-blind, phase 3 ALSYMPCA trial." *The Lancet. Oncology* 15, no. 12, 2014: 1397–406.

Humm J.L., Sartor O., Parker C., Bruland O.S., Macklis R. "Radium-223 in the treatment of osteoblastic metastases: A critical clinical review." *International Journal of Radiation Oncology, Biology, Physics* 91, no. 5, 2015: 898–906.

Jeelani S. et al. "Theranostics a treasured tailor for tomorrow." *Journal of Pharmacy & Bioallied Sciences* 6, 2014: S6-S8.

Kelkar S.S., Reineke T.M. "Theranostics: Combining imaging and therapy." *Bioconjugate Chemistry* 22, no. 10, 2011: 1879–1903.

Kramer L. et al. "Quantitative and correlative biodistribution analysis of 89Zr-labeled mesoporous silica nanoparticles intravenously injected into tumor-bearing mice." *Nanoscale* 9, no. 27, 2017: 9743–9753.

Kulkarni H.R., Baum R.P. "Patient selection for personalized peptide receptor radionuclide therapy using Ga-68 somatostatin receptor PET/CT." *PET Clinics* 9, no. 1, 2014a: 83–90.

Kulkarni H.R., Baum R.P. "Theranostics with Ga-68 somatostatin receptor PET/CT: Monitoring response to peptide receptor radionuclide therapy." *PET Clinics* 9, no. 1, 2014b: 91–97.

Lange R. et al. "Treatment of painful bone metastases in prostate and breast cancer patients with the therapeutic radiopharmaceutical rhenium-188-HEDP." *Nuklearmedizin* 55, no. 5, 2016: 177–209.

Lehmkuhl L. et al. "Role of preprocedural computed tomography in transcatheter aortic valve implantation." *Rofo* 185, no. 10, 2013: 941–949.

Loon J. et al. "PET imaging of zirconium-89 labelled cetuximab: A phase I trial in patients with head and neck and lung cancer." *Radiotherapy and Oncology* 122, no. 2, 2017: 267–273.

Lütje S. et al. "PSMA ligands for radionuclide imaging and therapy of prostate cancer: Clinical status." *Theranostics* 5, no. 12, 2015: 1388–1401.

Maecke H.R. and Reubi J.C. "Somatostatin receptors as targets for nuclear medicine imaging and radionuclide treatment." *Journal of Nuclear Medicine* 52, no. 6, 2017: 841–844.

Modlin I.M., Pavel M., Kidd M., Gustafsson B.I. "Review article: Somatostatin analogues in the treatment of gastroenteropancreatic neuroendocrine (carcinoid) tumours." *Alimentary Pharmacology & Therapeutics* 31, no. 2, 2010: 169–88.

Morgenroth A., Vogg A.T., Neumaier B., Mottaghy F.M., Zlatopolskiy B.D. "Radioiodinated indomethacin amide for molecular imaging of cyclooxygenase-2 expressing tumors." *Oncotarget* 8, no. 11, 2017: 18059–18069.

Nayak T.K., Brechbiel M.W. "^{86}Y based PET radiopharmaceuticals: Radiochemistry and biological applications." *Medicinal Chemistry* 7, no. 5, 2011: 380–388.

Pandit-Taskar N. et al. "A phase I/II study for analytic validation of ^{89}Zr-J591 ImmunoPET as a molecular imaging agent for metastatic prostate cancer." *Clinical Cancer Research: An Official Journal of the American Association for Cancer Research* 21, no. 23, 2015: 5277–5285.

Perk L.R. et al. "Preparation and evaluation of 89Zr-Zevalin for monitoring of 90Y-Zevalin biodistribution with positron emission tomography." *European Journal of Nuclear Medicine Molecular Imaging* 33, no. 11, 2006: 1337–1345.

Phan A.T. et al. "NANETS consensus guideline for the diagnosis and management of neuroendocrine tumors: Well-differentiated neuroendocrine tumors of the thorax (Includes Lung and Thymus)." *Pancreas* 39, no. 6, 2010: 784–798.

Prasad V. et al. "Somatostatin receptor PET/CT in restaging of typical and atypical lung carcinoids." *EJNMMI Research* 5, 2015: 53.

Rbah-Vidal L. et al. "Theranostic approach for metastatic pigmented melanoma using ICF15002, a multimodal radiotracer for both PET imaging and targeted radionuclide therapy." *Neoplasia* 19, no. 1, 2017: 17–27.

Reiner T., Lewis J.S., Zeglis B.M. "Harnessing the bioorthogonal inverse electron demand diels-alder cycloaddition for pretargeted PET imaging." *Journal of Visualized Experiments* 96, 2015: 52335.

Rösch F., Herzog H., Qaim S.M. "The beginning and development of the theranostic approach in nuclear medicine, as exemplified by the radionuclide pair ^{86}Y and ^{90}Y." Ed. Klaus Kopka. *Pharmaceuticals* 10, no. 2, 2017: 56.

Rowe S.P. et al. "^{18}F-DCFBC PET/CT for PSMA-based detection and characterization of primary prostate cancer." *Journal of Nuclear Medicine: Official Publication, Society of Nuclear Medicine* 56, no. 7, 2015: 1003–1010.

Same S., Aghanejad A., Akbari Nakhjavani S., Barar J., Omidi Y.. "Radiolabeled theranostics: Magnetic and gold nanoparticles." *Bioimpacts* 6, no. 3, 2016: 169–181.

Severi S. et al. "Peptide receptor radionuclide therapy in the management of gastrointestinal neuroendocrine tumors: Efficacy profile, safety, and quality of life." *OncoTargets and Therapy* 10, 2017: 551–557.

Shah R.R., Shah D.R. "Personalized medicine: Is it a pharmacogenetic mirage?" *British Journal of Clinical Pharmacology* 74, no. 4, 2012: 698–721.

Siegel R. et al. "Cancer treatment and survivorship statistics, 2012." *CA: A Cancer Journal for Clinicians* 62, no. 4, 2012: 220–41.

Silberstein E.B. "Radioiodine: The classic theranostic agent." *Seminars in Nuclear Medicine* 42, no. 3, 2012: 3164–170.

Sisson J.C., Ynik G.A. et al. "Theranostics: Evolution of the radiopharmaceutical meta-iodobenzylguanidine in endocrine tumors." *Seminars in Nuclear Medicine* 42, no. 3, 2012: 171–184.

Stabin M.G. Internal Radiation Dosimetry. *Nuclear Medicine*, 2nd Ed., Vol 1, Chapter 22, 313–331, edited by RE Henkin, D Bova, GL Dillahay, JR Halema, SM Karesh, RH Wagner, AM Zimmer. Mosby, St. Louis, MO 2006.

Stockhofe K., Postema J.M., Schieferstein H., Ross T.L. "Radiolabeling of nanoparticles and polymers for PET imaging." *Pharmaceuticals* 7, no. 4, 2014: 392–418.

Szabo Z. et al. "Initial evaluation of [^{18}F]DCFPyL for prostate-specific membrane antigen (PSMA)-targeted PET imaging of prostate cancer." *Molecular Imaging and Biology: MIB: The Official Publication of the Academy of Molecular Imaging* 17, no. 4, 2015: 565–574.

Szymański P., Frączek T., Markowicz M., Mikiciuk-Olasik E., "Development of copper based drugs, radiopharmaceuticals and medical materials." *BioMetals* 25, no. 6, 2012: 1089–1112.

Tagawa S.T. et al. "Phase II study of Lutetium-177 labeled anti-prostate-specific membrane antigen (PSMA) monoclonal antibody J591 for metastatic castration-resistant prostate cancer." *Clinical Cancer Research: An Official Journal of The American Association for Cancer Research* 19, no. 18, 2013: 5182–5191.

Uematsu T. et al. "Comparison of FDG PET and SPECT for detection of bone metastases in breast cancer." *American Journal of Roentgenology* 184, no. 4, 2005: 1266–1273.

Vallabhajosula S. et al. "Radioimmunotherapy of metastatic prostate cancer with ^{177}Lu-DOTAhuJ591 anti prostate specific membrane antigen specific monoclonal antibody." *Current Radiopharmaceuticals* 9, no. 1, 2016: 44–53.

Vicenzi E., Liò P., Poli G. "The puzzling role of CXCR4 in human immunodeficiency virus infection." *Theranostics* 3, no. 1, 2013: 18–25.

Vöö S., Bucerius J., Mottaghy F.M. "I-131-MIBG therapies." *Methods* 55, no. 3, 2011: 238–245.

Weineisen M., Simecek J., Schottelius M., Schwaiger M., Wester H.-J. " Synthesis and preclinical evaluation of DOTAGA-conjugated PSMA ligands for functional imaging and endoradiotherapy of prostate cancer." *EJNMMI Research* 4, 2014: 63.

Willowson K. et al. "CT-based quantitative SPECT for theradionuclide 201Tl:experimental validation and a standardized uptake value for brain tumour patients." *Cancer Imaging* 12, 2012: 31–40.

Wright C.L., Zhang J., Tweedle M.F., Knopp M.V., Hall N.C. "Theranostic imaging of Yttrium-90." *Biomed Research International* 2015 Article ID 481279 doi:10.1155/2015/481279.

Zechmann C.M. et al. "Radiation dosimetry and first therapy results with a $^{124}I/^{131}I$-Labeled small molecule (MIP-1095) targeting PSMA for prostate cancer therapy." *European Journal of Nuclear Medicine and Molecular Imaging* 41, no. 7, 2014: 1280–1292.

10 Emotion Detection System

Adrish Bhattacharya, Vibhash Chandra, and Leonid Datta

CONTENTS

10.1 INTRODUCTION

10.1.1 Background

The recent past has witnessed enormous advancements pertaining to artificial intelligence, which has become the buzzword of scientists, technocrats, and students alike. We have created machines able to not only understand what is said to them, but which can also laugh, joke, or make sarcastic comments. Indeed, we have created smart machines able to respond to almost anything we want. First phones, after that, watches and now TVs, homes, and cars—every imaginable utilitarian device has become smart based on the exponential growth in artificial intelligence. However, as we delve deeper into the solutions these devices provide, a basic constraint emerges limiting the practical applications of these systems. This limiting factor is due to its inability to process paralinguistic features of commands. To simplify, our machines can't read between the lines. For instance, if you are tired from working late and place an order for a hot cup of coffee to a smart brewing machine, it wouldn't brew a nice cup of espresso. Instead, it would brew your favorite cappuccino, which is the most frequently ordered variety. But in the same situation, a close associate will definitely bring you a hot brew of black coffee. Similarly, there are numerous situations in day-to-day life the latest smart machines fail to address appropriately. A close friend is always better at recommending books and movies than even the most advanced recommendation engines powered by artificial intelligence. This is especially true when we are not using our regular devices. To summarize, even with our terrific developments in artificial intelligence and big data, these devices are nowhere as smart as we need them to be. It's quite ironic that with huge IQs, our systems hardly have any common sense! Let alone, any degree of emotional quotient.

10.1.2 Emotion Recognition—Putting the Smart in Smart Devices

From our introduction, we understand that to make machines truly smart, it is imperative to design systems able to understand human emotions without the users having to express them explicitly. Reflecting on our everyday experiences, it is easy to note that there is hardly any moment when we feel unemotional. Thus, as humans we generate abundant volumes of emotion related data that is open for exploitation.

However, the study of affective science is hardly a new field. Even ancient Romans and Greeks studied expressive speech and later on their manuals formed the basis for western philosophical work on the subject of the emotional appeal. Active research on recognition of emotions can be traced to the early 1960s when phoneticians, linguists, emotion psychologists, and engineers began exploring affective science.

In the twenty first century, it is the need of the hour to create systems that converge the results of the earlier generations while giving them the ability to efficiently utilize newer and emerging data types. Later in the chapter, we shall discuss various methods of speech analysis and explore the challenges they present. Initially, it's important to understand what constitutes emotion, that is, how to determine and contrast one emotion from other. Without a detailed understanding of emotions, it is not possible to teach machines to identify them. Therefore, let us go deeper into the classification of emotions in the next section.

10.2 TYPES OF EMOTION

It is seemingly a simple task for us to list the various emotions we experience—joy, melancholy, angst, fear, excitement, pleasure, anxiety, boredom, dismay, ecstasy, disappointment, delight, and so on. However, the boundaries between two different emotions are often fuzzy and not discrete. For instance, dismay and disappointment can be closely related, with one of them being the cause and other the effect. Therefore, when we venture into identifying emotions, the task presents newfound challenges not generally encountered in everyday life. In fact, the debate on classification of emotions has riddled experts till today. As a result, there is no unanimous consensus as how to classify emotions. However, two fundamental viewpoints of classification of emotions exist:

1. Differentiating emotions based on their constructs.
2. Dimensional categorization of emotions based on similarity.

10.2.1 The Basic Emotions

When we discuss basic emotions, we assume the discrete nature of emotions. The discrete emotion theory states that every human possesses a cultural-independent and discrete set of emotions. It is argued these basic emotions can be distinguished by an individual's biological processes (Colombetti 2009; Ekman 1992).

In 2017, Cowen and Keltner (2017) quoted:

> Across self-report methods, we find that the [2185] videos [selected and shown to volunteer subjects] reliably elicit 27 distinct varieties of reported emotional experience. Further analyses revealed that categorical labels such as amusement better capture reports of subjective experience than commonly measured affective dimensions (e.g., valence and arousal). Although reported emotional experiences are represented within a semantic space best captured by categorical labels, the boundaries between categories of emotion are fuzzy rather than discrete. By analyzing the distribution of reported emotional states we uncover gradients of emotion—from anxiety to fear to horror to disgust, calmness to aesthetic appreciation to awe, and others—that correspond to smooth variation in affective dimensions such as valence and dominance. Reported emotional states occupy a complex, high-dimensional categorical space.

The categories suggested are:

- A—admiration, adoration, aesthetic, appreciation, amusement, anger, anxiety, awe, awkwardness
- B—boredom
- C—calmness, confusion, craving
- D—disgust
- E—empathic pain, entrancement, excitement
- F—fear
- H—horror
- I—interest
- J—joy

- N—nostalgia
- R—relief, romance
- S—sadness, satisfaction, sexual desire, surprise

Smith's list of emotions—Watt Smith listed 154 different emotions, including foreign ones. This is one of the most comprehensive lists of emotions available.

Richard and Bernice Lazarus (Lazarus and Lazarus 1996), in their book *Passion and Reason,* listed 15 separate types of emotions including shame, anger, love, pride, relief, sadness, fright, gratitude, aesthetic experience, happiness, jealousy, anxiety, compassion, depression, envy, guilt, and hope.

William James proposed four basic emotions based on the bodily involvements of individuals—fear, grief, love, and rage.

Paul Ekman, Wallace Friesen, and *Phoebe Ellsworth* suggested six basic emotions—anger, disgust, fear, happiness, sadness, and surprise.

10.2.2 Emotion Categories

For a long time, researchers have attributed emotions according to multiple dimensions. The Wilhelm Max Wund of modern psychology, described emotions in three dimensions, namely "pleasurable or unpleasurable," "arousing or subduing," and "strain or relaxation" (Wundt 2009). Harold Schlosberg, in 1954, dubbed these three distinct dimensions of emotions as "pleasantness–unpleasantness," "attention–rejection," and "level of activation" (Schlosberg 1954). Unlike models of basic emotion theories, which propose different emotions arising from separate neural systems, the dimensional models argue for the presence of an interconnected neurophysiological system. This idea is central to the different affective states (Posner, Russell, and Petersona 2005).

These models aim to define human emotions by defining where they lie in these proposed dimensions. Although several models have been proposed by researchers, only a few remain accepted by the mainstream scientific community. Among them, the positive activation–negative activation (PANA) model and the vector model are the ones most widely accepted (Rubin and Talarico 2010). A short description of the Plutchick and PANA model has been given below.

10.2.3 Plutchik's Model

Robert Plutchik, who first proposed the model, argued for a mixed three-dimensional model fusing the discrete and dimensional theories. Plutchick had arranged the affective states within concentric circles, with the basic emotions inside the inner circles and the complex emotions in circles of larger radii (Plutchick 2001).

10.2.4 PANA Model

The PANA, or "consensual" model, was devised by Watson and Tellegen (Watson and Tellegen 1985). PANA model argues that the basic differentiating factor between emotional classes is due to positive (Figure 10.1) and negative (Figure 10.2) affect (Watson and Tellegen 1985).

Negative and forceful	Negative and uncontrollable	Negative thoughts	Negative and passive
• Anger	• Anxiety	• Pride	• Boredom
• Annoyance	• Embarrassment	• Doubt	• Despair
• Contempt	• Fear	• Envy	• Disappointment
• Disgust	• Helplessness	• Frustration	• Hurt
• Irritation	• Powerlessness	• Guilt	• Sadness
	• Worry	• Shame	• Agitation
			• Stress
			• Shock
			• Tension

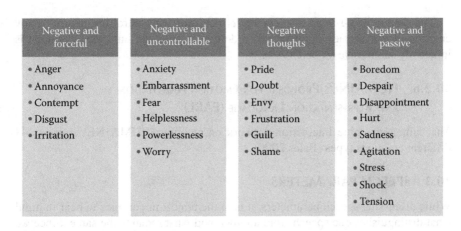

FIGURE 10.1 Negative emotions.

Positive and lively	Positive thoughts	Quiet positive	Reactive
• Amusement	• Courage	• Calmness	• Interest
• Delight	• Hope	• Contentment	• Politeness
• Elation	• Humility	• Relaxation	• Surprise
• Excitement	• Satisfaction	• Relief	
• Happiness	• Trust	• Serenity	
• Joy			
• Pleasure			
• Affection			
• Empathy			
• Friendliness			
• Love			

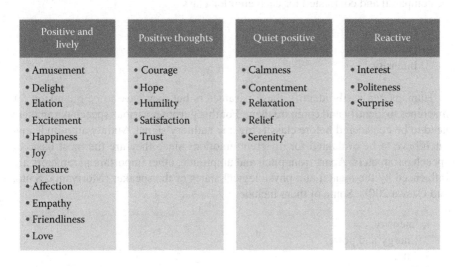

FIGURE 10.2 Positive emotions.

10.2.5 Pad Emotional State Model

Albert Mehrabian and James Russel proposed the pad emotional state model that is basically a psychological model where 3 numerical dimensions are employed to categorize different emotions. Here three numerical dimensions are employed to categorize different. The pad model puts forward the following dimensions—*pleasure, arousal,* and *dominance.*

The *arousal–nonarousal scale* measures the emotion intensity. For example, worry and anxiety are both unpleasant emotions, however, anxiety has greater intensity compared to worry.

The *pleasure–displeasure scale* measures the pleasantness of any emotion. For example, worry and anxiety are both highly unpleasant emotions but happiness is a pleasant one (Albert 1980).

The *dominance–submissiveness scale* measures the controlling nature in any emotion. For example, annoyance is a dominant emotion, but embarrassment is inherently submissive (Albert 1980).

10.2.6 HUMAINE's Proposal for Emotion Annotation and Representation Language (EARL)

The human–machine interaction network on emotion (HUMAINE) suggests 48 different emotions types (Bales 2000).

10.3 SPEECH PARAMETERS

While discussing speech parameters, it is of the utmost importance to bear in mind what distinguishes one speech from another, and what remains the same. Since we are interested in the paralinguistic analysis of speech, the following features should be compared and contrasted for each emotion class.

1. Intonation
2. Duration
3. Intensity

Humans can easily identify these parameters but they have to be quantized for machines to identify and compare them. For this purpose, various speech parameters need to be considered before classifying any auditory signal. Mainly amplitude and pitch have to be evaluated for different emotions since they are the most basic of speech parameters. Apart from pitch and amplitude, other important parameters are influenced by the mental and physiological states of the speaker (Moriyama, Saito, and Ozawa 2001). Some of them include:

1. Intensity
2. Energy and power
3. Jitter
4. Pauses
5. Speaking rate
6. Mean
7. Variance
8. Mel-frequency cepstrum coefficient (MFCC) and its derivatives

10.3.1 Amplitude and Pitch

As noted earlier, to distinguish between different types of emotions from speech we need to consider a number of parameters; however, for most purposes, a binary classification of the positive and negative emotion state meets the required purpose. The positive emotions include happiness, surprise, excitement, etc., and negative emotions include sadness, fear, and angst among others.

While considering these two classes of emotions, we observe that amplitude and pitch can appreciably distinguish between them. In general, positive emotions are characterized by an increased amplitude and pitch, while the negative emotions show relatively less amplitude and pitch. Another point of difference is that in positive emotions, the range of values of amplitude and pitch are both larger than those observed in negative emotions. Correspondingly, we can conclude that positive emotions are associated with high energy and vice-versa for negative emotions.

10.3.2 MFCC Parameters

The MFCC represents the short-term power spectrum on a nonlinear scale of frequency called the mel scale. The various terms in the mathematical representation are determined using a linear cosine transform from the logarithmic power-spectrum.

MFCCs collectively make up an MFC. In MFC, the frequency bands are equally spaced on the mel-scale, allowing it to correctly emulate the response of a human auditory system.

MFCCs is determined through the mentioned steps (Sahidullah and May 2012) (Figure 10.3):

1. Apply Fourier transforms on any windowed part of the required signal
2. Match the powers of the newly obtained spectrum onto the mel-scale
3. Take the logs of the powers at each of the mel frequencies
4. Determine discrete cosine transform of all the mel log powers
5. The amplitudes of the resulting spectrum give the MFCCs coefficient.

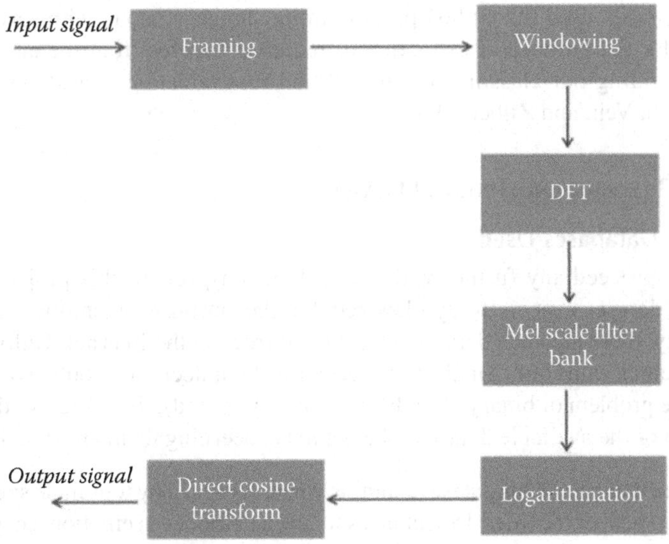

FIGURE 10.3 Schematic of MFCC process.

10.4 CHOICE OF PARAMETERS

10.4.1 Benefits of MFCC: Why MFCC is Used

As with any machine-learning endeavor, the choice of comparative parameters is of prime importance. Therefore, a clear understanding of the key differences between the various emotions is essential. As previously noted, differentiating among emotions is perhaps one of the murkiest and highly debatable fields, although there exist a few areas of mutual agreement—however limited. But one aspect of emotion recognition, which stands despite the challenges, is that every sample of the audio signal is processed by humans, and humans themselves hold the primary authority in categorizing them. This seemingly obvious fact holds great importance once we realize our ears are not equally sensitive to vibration in every frequency. Human ears have a limited capacity and detect frequencies only in a specific range. Additionally, not all frequencies in the audible range are perceived similarly. The higher and lower frequencies are subject to different sensitivities by the human ear.

Therefore, a device emulating the human brain analyzing audio signals for recognition of emotions must perceive the incoming signals as if received by the ear. Therefore, the MFCC parameters are employed for this purpose. Detailed understanding of the MFCC parameters is not required, but a primary knowledge is helpful (Kwon et al. 2003).

10.5 PROGRAMMING TOOLS USED

We are now aware of the various facets of emotions and their recognition (especially through speech) such as different types of emotion, several speech parameters and the benefit of using MFCC over any other parameter. So, let us put this theoretical knowledge towards a practical purpose in order to build a predictive model for emotion recognition using machine learning.

We primarily used the python programming language, along with several tools/modules, for the purpose of extraction of required features from the audio signals and for building our classification model and its evaluation (Casale et al. 2008; Mordkovich, Veit, and Zilber 2011).

10.5.1 Training and Testing Dataset

10.5.1.1 Databases Used

Before we proceed any further with the technical aspects of this project, we first require a dataset especially synthesized for the purpose of emotion detection. Fortunately, there are such datasets present (for free) on the Internet. Unfortunately, most of them became quite small in size because of our decision to only work towards solving the problem of binary classification (happy vs. sad); therefore, we decided to merge two of the available datasets; the details concerning them are stated below:

1. Savee database: This dataset contains voices recorded by four male speakers. Each speaker recorded 15 sentences for each of the seven emotion categories; namely, "anger," "disgust," "fear," "happiness," "neutral," "sadness," and "surprise" (Haq and Jackson 2010).

2. Emo-db corpus (berlin database of emotional speech): This dataset contains about 500 utterances spoken by actors conveying happy, angry, anxious, fearful, bored, disgusted, and neutral emotions. We had the freedom to choose utterances from 10 different actors and 10 different texts (Burkhardt et al. 2005).

Our final dataset contained 250 instances of speech audio, out of which 130 instances conveyed a "happy" emotion and the rest—120, were "sad." Training and testing sets were created from this merged database according to the methodology explained in the following sections.

10.5.1.2 Features and Target Labels

Every machine-learning algorithm requires a dataset for training. Another dataset—called testing dataset, may be used for subsequent evaluation of the model. Both of these sets contain features and target labels (if it is a supervised learning problem). Let us look into the components of the training and testing sets in detail, with respect to our classification problem.

Features: As already discussed, we exclusively used features derived from MFCCs. Using only the first 13 MFCCs for each audio signal, we calculated the mean, maximum, minimum, and variance for each coefficient, as well as its derivative across all frames. We also calculated the mean, variance, maximum, and minimum of the mean for each coefficient and its derivative. Therefore, 112-dimensional feature vector for each audio signal was created accordingly.

Target labels: We encoded our two classes of emotions such that "1" signified the emotion conveyed by the audio signal was happy, and "0" if sad. Therefore, each audio signal in the dataset had a "1" or "0" value associated to it.

The following Python modules were used for preparing this dataset: Librosa (Mcfee et al. 2015), numpy (Oliphant 2007), scikit-learn (Pedregosa et al. 2012), and matplotlib (Hunter 2007) and their dependencies.

10.5.1.3 Train-Test Split

Although it is possible, we cannot use the entire dataset for training our classification model since we won't be able to estimate its performance afterwards. Therefore, a common practice is to divide the dataset into two parts; one is used for training the model and the other for testing it, hence their names. Due to the small size of our database, we decided the split would be 80% and 20% for our training and testing datasets, respectively. The split was stratified, that is, the relative percentage of each class remained the same in both datasets.

10.5.2 Building the Classification Model

After our training and testing datasets were prepared, we needed an algorithm/s using with which we could build our models. The implementation of a large number of machine-learning algorithms are present on the Internet, and since we cannot train and evaluate all of them, we decided to work with the four most prominent ones based on previous research done in this area. These are LR(logistic regression) (Hosmer, Lemeshow, and Sturdivant 2013), SVM (support-vector machines) (Pan, Shen, and

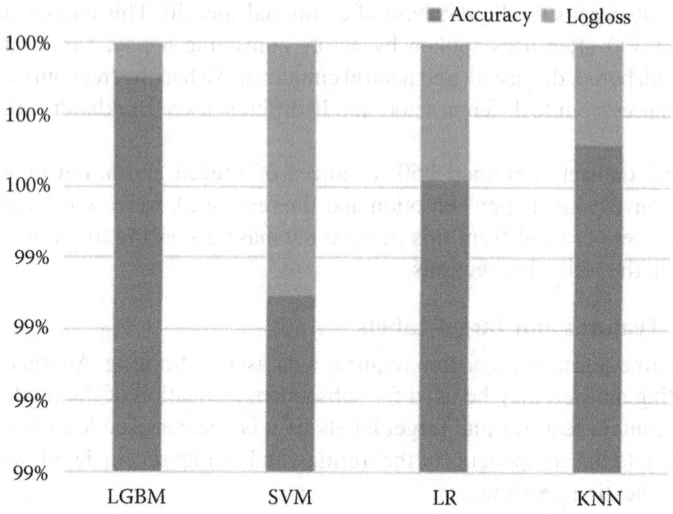

FIGURE 10.4 Comparison of accuracy scores and logloss.

Shen 2012), k-NN (k-nearest neighbor) (Mucherino et al. 2009), and LGBM (light gradient boosted model) (Ke et al. 2017). For the evaluation and optimization of each of these models, a cross-validation technique was used, and then their accuracy and log-loss score (Vovk 2015) was calculated using the test set (refer Figure 10.4).

Results:

Accuracy score:

LGBM ~ 95%
LR ~ 83%

SVM ~ 91%
KNN ~ 85%

Logloss:

LGBM ~ 0.09
LR ~ 0.59

SVM ~ 0.35
KNN ~ 0.24

10.6 APPLICATIONS

10.6.1 HCI: HCI Applications of the System

Emotion recognition systems can be widely employed for increasing the depth of human–computer interaction (HCI). HCI has taken a central stage today, as far as smart and responsive systems are concerned. Without efficient HCI, a smarter world would remain a dream. Although the advances in the field of smart systems are appreciable, there lies a massive opportunity for improvement mainly due the inability of our systems to understand true human needs. Our machines are only capable of understanding human wants and delivering upon them. We must grasp the fundamental difference between the two. After all, identifying the problem is central to solving it! (Cowie et al. 2001)

It is precisely this avenue that the emotion recognizing systems hope to tap into. Broadly speaking, they can be employed in almost any field of work to replace humans (at some point in the future perhaps), but six key areas have been identified and can be exploited using the present technology. They are:

1. Medicine
2. E-learning
3. Monitoring
4. Entertainment
5. Law
6. Marketing

10.6.2 Medical Applications

There exist numerous medical applications for emotion recognition systems. From direct delivery of treatment to patients to collecting valuable data for studying behavioral sciences, they can find uses in every realm imaginable (Lalitha et al. 2014).

Some of the main applications include:

1. Patient monitoring systems—they can be employed for keeping a 24/7 vigilance of patients, thus aiding in easier rehabilitation.
2. Companion bots—although JARVIS, from the popular sci-fi feature *Iron Man,* is still a distant dream, the emotion recognizing systems are a definite step towards it. Companion bots can be upgraded to understand human emotions making them capable of taking smarter and more important decisions. They can prove to be invaluable to the user, not only in performing certain tasks but also talking to them intelligently, thereby enhancing realism.
3. Counseling—perhaps the need of the hour, a smart counseling system can democratize access to a psychiatrist in the remotest corners of the world. Long-term analysis of a person's mental state can be valuable data for doctors who often need such information to make a better diagnosis.
4. Autism support—millions suffer from autism across the world, and despite sincere attempts some key challenges remain unsolved. Among them is the inability or difficulty autistic persons have to interpret emotions. This is a case efficiently tackled by these systems, which not only identify emotions for the user, but also help them with therapy.
5. Music therapy—this form of therapy has been clinically proven to induce appreciable degrees of positivity. It can help patients with the anxiety, depression, and stress of dealing with their conditions. Also, music therapy is advised for Alzheimer's patients.

10.6.3 Responsive Systems

A responsive device is often used to deploy an emotion recognition system, especially for HCI applications. There exist a variety of application of such systems including:

1. E-learning—this is a growing field but it is nothing revolutionary per se; however, smarter HCI can change that. Automated learning systems able to track the mood of the learner can deliver more interesting content to the student. Such systems, taking individual feedback from the students, can change their pace or method of teaching. These measures are impossible in a regular class. Even with personal tutors, human errors are bound to creep in making content curation a tough nut to crack (Shen, Wang, and Shen 2009).
2. Monitoring—a quick response, based on the anomalous behavior of a user, can be easily provided with the use of systems able to identify emotional states such fear or anger. These kinds of solutions find extensive applications in call centers and feedback systems, where angry customers can be dealt with more sensitively.
3. Entertainment systems—the true depth of scientific advancement is felt only when the results trickle down to the common person through everyday applications. Undoubtedly, entertainment devices including music players, TVs, video games, etc. are an indispensable part of our lives. Therefore, any disruptive technology must be able to bring changes in the way we interact with entertainment. Interestingly, all these entertainment devices can tune to the users' liking by using emotion recognition software. For instance, music players can suggest songs based on the mood of the user; the scriptwriter of a television series can study how the audience reacts to the plot, and modify it accordingly. Even recommendation engines can detect the mood of the users to suggest books or movies based on the current state of mind of the user. This not only improves the suggestions, increasing their relevance, but also eliminates the cold start problem of today's recommendation systems.
4. Marketing—just as in the case of television content writers, advertisement makers can readily use the data pertaining to the mood shifts of their audience to make better advertisements, which suit the users. The brands can focus on what's relevant to their customers. Today, the companies are mainly dependent on large-scale surveys to determine the likings of customers. This method is not only time consuming but also often ineffective because people can't always tell what they like most about a product—even if they want to. Our choices depend on various conscious and subconscious markers, which can be effectively determined by noticing our moods.

10.7 CONCLUSION

Emotion recognition systems have the ability to humanize digital interactions by creating artificial emotional intelligence in machines. Since our interactions with technology are becoming increasingly conversational, artificial emotional intelligence has become a critical component of today's technology. Although a lot of work is going in this field of study and some amazing breakthroughs have been made, we have only scratched the tip of the iceberg. A lot of questions still remain unanswered and newer challenges crop up as we delve deeper into this field of study. However, these systems have the potential of disrupting industries from health care to education, and anything in between. Emotion recognition systems will not only revolutionize human interactions with machines but also allow machines to be more human.

REFERENCES

Bales, R.F. 2000. "Social Interaction Systems: Theory and Measurement: Book Review." *Group Dynamics: Theory, Research, and Practice* 4 (2):199–208. https://doi.org/10.1037/1089-2699.4.2.199.

Burkhardt, F., A. Paeschke, M. Rolfes, W.F. Sendlmeier, and B. Weiss. 2005. "A database of german emotional speech." In *Interspeech*, 5:1517–1520.

Casale, S., A. Russo, G. Scebba, and S. Serrano. 2008. "Speech Emotion Classification Using Machine Learning Algorithms." *2008 IEEE International Conference on Semantic Computing* 118 (13):167–74. https://doi.org/10.1109/ICSC.2008.43.

Colombetti, G. 2009. "From Affect Programs to Dynamical Discrete Emotions." *Philosophical Psychology* 22 (4):407–25. https://doi.org/10.1080/09515080903153600.

Cowen, A.S., and D. Keltner. 2017. "Self-Report Captures 27 Distinct Categories of Emotion Bridged by Continuous Gradients." *Proceedings of the National Academy of Sciences*, 201702247. https://doi.org/10.1073/pnas.1702247114.

Cowie, R., E. Douglas-Cowie, N. Tsapatsoulis, G. Votsis, S. Kollias, W. Fellenz, and J.G. Taylor. 2001. "Emotion Recognition in Human-Computer Interaction." *Signal Processing Magazine, IEEE* 18 (1):32–80. https://doi.org/10.1109/79.911197.

Ekman, P. 1992. "An Argument for Basic Emotions." *Cognition and Emotion* 6 (3–4):169–200. https://doi.org/10.1080/02699939208411068.

Haq, S., and Jackson, P.J.B. 2010. *Machine Audition: Principles, Algorithms and Systems*, 532. https://doi.org/10.4018/978-1-61520-919-4.

Hosmer, D.W., S. Lemeshow, and R.X. Sturdivant. 2013. *Applied Logistic Regression*, Third Edition. Wiley Series in Probability and Statistics. https://doi.org/10.1002/0471722146.

Hunter, J.D. 2007. "Matplotlib: A 2D Graphics Environment." *Computing in Science and Engineering* 9 (3):99–104. https://doi.org/10.1109/MCSE.2007.55.

Ke, G., Q. Meng, T. Wang, W. Chen, W. Ma, and T.-Y. Liu. 2017. "A Highly Efficient Gradient Boosting Decision Tree." *Advances in Neural Information Processing Systems* 30 (Nips):3148–56. http://papers.nips.cc/paper/6907-a-highly-efficient-gradient-boosting-decision-tree.pdf.

Kwon, O.-w., K. Chan, J. Hao, and T.-w. Lee. 2003. "Emotion Recognition by Speech Signals." *Eighth European Conference on Speech Communication and Technology*, 125–28. http://ergo.ucsd.edu/~leelab/pdfs/ES030151.pdf.

Lalitha, S., A. Madhavan, B. Bhushan, and S. Saketh. 2014. "Speech Emotion Recognition." *2014 International Conference on Advances in Electronics, Computers and Communications (ICAECC)* 7 (Table I):1–5. https://doi.org/10.1109/ICAECC.2014.7002390.

Lazarus, R.S., and B.N. Lazarus. 1996. *Passion and Reason*. Oxford University Press, Washington, DC.

Mcfee, B., C. Raffel, D. Liang, D.P.W. Ellis, M. Mcvicar, E. Battenberg, and O. Nieto. 2015. "Librosa: Audio and Music Signal Analysis in Python." *Proc. of The 14th Python in Science Conf*, no. Scipy:1–7.

Mordkovich, A., K. Veit, and D. Zilber. 2011. "Detecting Emotion in Human Speech." http://cs229.stanford.edu/proj2011/VeitZilberMordkovich-DetectingEmotionInHumanSpeech.pdf.

Moriyama, T., H. Saito, and S. Ozawa. 2001. "Evaluation of the Relation between Emotional Concepts and Emotional Parameters in Speech." *Systems and Computers in Japan* 32 (3):56–64. https://doi.org/10.1002/1520-684X(200103)32:3<56::AID-SCJ5>3.0.CO;2-B.

Mucherino, A., P.J. Papajorgji, and P.M. Pardalos. 2009. "K-nearest neighbor classification." In *Data Mining in Agriculture*: 83–106, Springer, New York, NY.

Oliphant, T.E. 2007. "Python for Scientific Computing." *Computing in Science and Engineering* 9 (3):10–20. https://doi.org/10.1109/MCSE.2007.58.

Pan, Y., P. Shen, and L. Shen. 2012. "Speech Emotion Recognition Using Support Vector Machine." *International Journal of Smart Home* 6 (2):101–8. https://doi.org/10.5120/431-636.

Pedregosa, F., G. Varoquaux, A. Gramfort, V. Michel, B. Thirion, O. Grisel, M. Blondel, et al. 2012. "Scikit-Learn: Machine Learning in Python." *Journal of Machine Learning Research* 12:2825–30. https://doi.org/10.1007/s13398-014-0173-7.2.

Plutchik, R. The Nature of Emotions. American Scientist. Archived from the original on July 16, 2001.

Posner, J., J.A. Russell, and B.S. Petersona. 2005. "The Circumplex Model of Affect: An Integrative Approach to Affective Neuroscience, Cognitive Development, and Psychopathology." *Dev Psychopathol* 17 (3):715–34. https://doi.org/10.1007/s10955-011-0269-9.Quantifying.

Rubin, D.C., and J.M. Talarico. 2010. "NIH Public Access" 17 (8):802–8. https://doi.org/10.1080/09658210903130764.A.

Sahidullah, M., and G. Saha. 2012. "Design, Analysis and Experimental Evaluation of Block Based Transformation in MFCC Computation for Speaker Recognition." *Speech Communication* 54 (4):543–65. https://doi.org/10.1016/j.specom.2011.11.004.

Schlosberg, H. 1954. "Three Dimensions of Emotion." *Psychological Review* 61 (2):81–88. https://doi.org/10.1037/h0054570.

Shen, L., M. Wang, and R. Shen. 2009. "Affective E-Learning: Using 'emotional' Data to Improve Learning in Pervasive Learning Environment Related Work and the Pervasive E-Learning Platform." *Educational Technology & Society* 12:176–89. https://doi.org/citeulike-article-id:7412147.

Vovk, V. 2015. "The Fundamental Nature of the Log Loss Function." *Lecture Notes in Computer Science (Including Subseries Lecture Notes in Artificial Intelligence and Lecture Notes in Bioinformatics)*, 9300:307–18. https://doi.org/10.1007/978-3-319-23534-9_20.

Watson, D., and A. Tellegen. 1985. "Toward a Consensual Structure of Mood." *Psychological Bulletin*. https://doi.org/10.1037/0033-2909.98.2.219.

Wundt, W. 2009. "Outlines of Psychology (1897)." *Found. Psychol. Thought A Hist. Psychol.* 36–44. https://doi.org/10.1007/978-1-4684-8340-6_7.

11 Segmentation and Clinical Outcome Prediction in Brain Lesions

Sharmila Nageswaran, S. Vidhya, and Deepa Madathil

CONTENTS

11.1 INTRODUCTION

The diagnosis and management of brain lesions have been most challenging for many decades. Understanding of the brain is another domain, or challenge which has multiple focuses. To treat any brain lesions, the disease or condition needs to be identified precisely. It is not possible to confirm the diagnosis by an invasive procedure when it happens to be a neural system and so is the brain. Noninvasive and precise diagnostic procedures are directly needed and prediction of treatment's outcome is what is prediction of prognosis is based on the estimation of pathology done using these non invasive procedures. Hence, diagnostic imaging plays a vital role in deciding the treatment and its outcome. Segmentation and clinical outcome prediction has been a topic of interest for clinicians, health care practitioners, patients, and researchers for many decades. However, the only outcome of importance to the patient is the understanding about the disease and its progression. It will help them plan their treatment and future.

This chapter provides an insight into the segmentation of brain images from different perspectives. It is aimed to bring together specialized researchers from different disciplines who can deal with problems from an entirely different perspective but share a common aim towards this book. The chapter is divided into the following five sections:

1. Introduction
2. Background
3. Segmentation of brain lesion
4. Outcome prediction of brain lesion
5. Case studies

11.2 BACKGROUND

A lesion can be defined as a region in a tissue, or an organ, which has been damaged through disease or injury. Diagnosis of lesions in real time, using reliable algorithms, has been the main focus of the latest developments in medical image processing. In this area, the major focus of research has been the detection of lesions using MR (magnetic resonance) and CT (computed tomography) images. Usually CT and MRI are used to examine the anatomy of brain lesions. Magnetic resonance imaging (MRI) is a technology used in biomedical applications to detect and visualize minute details of the internal body structure. This technique detects differences in the tissues, and is a far better technique compared to CT in terms of image properties and quality. Cancer imaging and lesion detection are major applications related to MRI [1]. Since MRI provides more accurate results and does not involve any radiation, it has an advantage over CT scans. MRI is a technology using magnetic fields and radio waves.

The ability to separate the nuclei and the cells from the rest of the image content is one of the most important abilities in medical imaging and diagnostic systems. This process is called "segmentation." It is the most important factor in the design of an effective and robust diagnostic system. It is more efficient to segment an image with lesion than the one without a lesion. In medical image systems, segmentation of 3D images play a vital role which occurs before the implementation of algorithms for its segmentation based on the recognition of a type of lesion. on recognition of the lesion. Many statistical parameters like the size, volume, density, etc. can be inferred with accuracy from segmented 3D images. The accuracy in measurement of the size, shape, and appearance of the lesion is an important parameter in the diagnosis of brain lesions.

11.2.1 Brain Lesion

Brain lesions do not have a predefined shape often present a quite abnormal structure, making them difficult to detect and measure. It is therefore very important, from a researcher's point of view, to have an automated software package having the capability to detect brain lesions of any shape, volume, and size . There are a few important types of brain lesions, which need to be understood before processing MR or CT images. The types of brain lesion are:

1. Traumatic
2. Malignant
3. Benign
4. Vascular
5. Genetic
6. Immune
7. Plaques
8. Brain cell malfunction or death and
9. Ionizing radiation

The most common types are the traumatic, benign, and malignant lesions. The traumatic lesion is a lesion occurring due to an injury such as a road accidents or a gunshot wound on the brain, etc. The benign lesion typically does not expand in an abrupt way and does not further affect the healthy neighboring tissues, also it does not expand to the nonadjacent tissues. A malignant lesion grows with the passage of time and ultimately results in the death of the patient. Severe progressing disease is the term used for malignant lesions by medical practitioners. It is typically used for the description of a cancer.

11.2.2 Classification Methods

Accurate detection of lesion in the brain is important for humans. Additionally, the lesion has to be classified precisely. Automated detection and prediction of a lesion is something expected from every clinical setup. In medical imaging, after the image is captured by the imaging system it is digitized and sent to a computer. It is then segmented and processed further to obtain features from the segmented image. In brain images, the fundamental task is to obtain useful features from the white matter, gray matter, and cerebrospinal fluid. Manual segmentation is a traditional method used for segmenting a MR image, which is more time-consuming. Segmentation approaches vary from among experts [2]. Therefore, there is need to have an efficient and strong computer-based system to accurately examine the brain lesion statistics. The success of classification of a lesion completely depends on the accuracy of segmentation. Classification can be parametric or nonparametric.

11.2.2.1 Parametric Classification

When the estimated output of segmentation has more similarity to the data available from a known class, or set of well-classified lesions, then the approach becomes parametric. The goal is fixed in most of the parametric approaches and the learning of the classifier network is usually supervised, requiring an abundant data set available for training and testing. Most benign lesions fall into this category.

11.2.2.2 Nonparametric Classification

When the estimated output is new, and if multiple conditions of a particular pathology are to be listed with a similar output, then a new class is formed and added to the available set of parametric classes. However, when there is no existence of another similar estimation, then it paves the way for a nonparametric approach. This kind of classification is based upon the fact that whatever is deviating from the available models will all be collectively grouped under nonclassified models. Based on the responses from experts (neurosurgeon or neuroradiologist), they will be classified as a particular type of lesion. Certain malignant tumors and disease conditions expected to have unpredicted courses of progression will fall into this category.

11.2.3 Regression Methods

Regression methods or analysis is a form of predictive modeling, which brings out a relationship or correlation between a dependent and an independent variable. Autoregression or AR modeling is the most commonly used regression method in biological studies. AR is a stochastic approach used in statistical calculations. Here, the future values are estimated based on a weighted sum of past values. An autoregressive process operates under the premise that past values have an effect on current values, and current values will have an effect on future values. A series of images taken during the course of a lesion will help in predicting the outcome. This prediction method may be based on a linear, stepwise, or parabolic regression analysis. All this would completely be determined by the set of predictions following the segmentations. These regression analyses are thus used in predicting the outcome and recovery from a stroke. Therefore, it helps the clinicians in the proper planning and execution of the treatment.

11.2.4 Neural Network

Neural networks have met with a considerable amount of interest in predictive methods like regression, classification, and artificial neural networks (ANNs) for outcome prediction. This is because they offer a number of theoretical advantages. ANNs have a much wider range of flexibility to fit models to the available data when the patterns are not very obvious [3]. Apart from this, the major advantages of using ANNs in modeling are that nonlinear and random relationships can be seen between independent and dependent datasets. Also, they do not require any kind of distributional assumption. Furthermore, they also allow interactions between dependent variables. A large number of clinicians have aware of the power of ANNs as a medical analysis tool. The main issue called the "black box" has been stated by a number of statistical framework developers for ANNs [4–7]. The success of any new technology—for its integration into clinical practice—depends mostly on the number of trails in the published literature. Apart from having a large number of publications reporting on the use of ANNs in medicine, the number of clinical trials remains to be less [8–10].

ANNs have found application in nonlinear signal processing, classification, and optimization. In many of these applications their performance has been considered as superior to the classical and traditional linear approaches. In processing of biomedical signals like ECG and EEG, the most common networks used are multilayer perceptron (MLP), learning vector quantization networks, and radial basis function (RBF). The perceptron consists of neurons interconnected to different layers in which each neuron is a represented as a function given as:

$$y = f\left(w_0 + \sum_{i=1}^{N} w_i x_i\right)$$

where w_i is the assigned weight to the input x_i, and f is a function, which can be linear or nonlinear. In the case of nonlinearity, f is defined frequently as the logistic function, which is given by

$$f(u) = 1/(1+e-u) \quad \text{or} \quad f(u) = \tanh(u)$$

In a similar way, networks with RBF are implementations given by a function as,

$$y(n) = \sum_{i=1}^{N} w_i \exp\left(-\frac{x(n)-c_i}{\sigma_i}\right)$$

where *x(n)* denotes an input data vector given to the network. *N* is the number of neurons, and w_i is the coefficients, c_i is the center vector, and σ_i are the standard deviations, which denote the network parameters. The exponential can be replaced by functions like wavelets. In Figure 11.1, an MLP based neural network consisting of three layers is shown. Eight to ten linear neurons are present at the input layer, which are signals delayed by time and are samples of inputs. Also, there are three to five logistic nonlinear neurons in the hidden layer and the output layer contains a linear neuron.

In some methods, where the network is trained before the detection is done using selected samples while in other networks, online training is done. This would be the reason for the ability of neural networks to adapt to signal statistics. RBF networks are very closely related to fuzzy logic methods. The RBF networks have advantages over the MLP networks in their ability in predicting the network parameters. This helps to produce reliable and more predictable results.

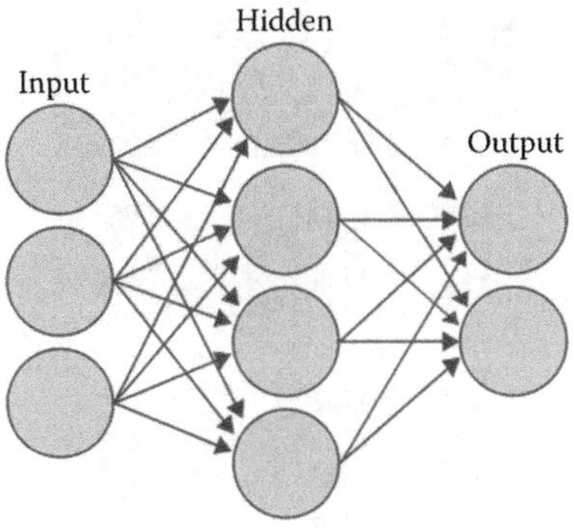

FIGURE 11.1 Typical MLP network.

11.3 SEGMENTATION OF BRAIN LESION

It is the most challenging task to segment the lesion from its background, especially when the lesion is in multiple locations. Following a good technique during the data acquisition will ease the complexity involved with the processing. However, in medical imaging technology, the dosage of exposure is strictly controlled, as to avoid the ill effects of overexposure to radiation. Hence, more sophistication is expected in processing. The stages of image processing are given in Figure 11.2.

Segmentation of a hemorrhagic stroke image is not very challenging compared to segmentation of the image of an ischemic stroke. The former has a wider intensity variation. The clot or hemorrhage, which is rich in hydroxyl ions, responds well to the MRI and is captured as white region whereas the rest of the brain appears to be grey. Because of this demarcation segmentation is comparatively simpler. However, based on the strength of the magnetic field used in the acquisition machine, contrast enhancement will be done in preprocessing. A sobel operator and a median filter are

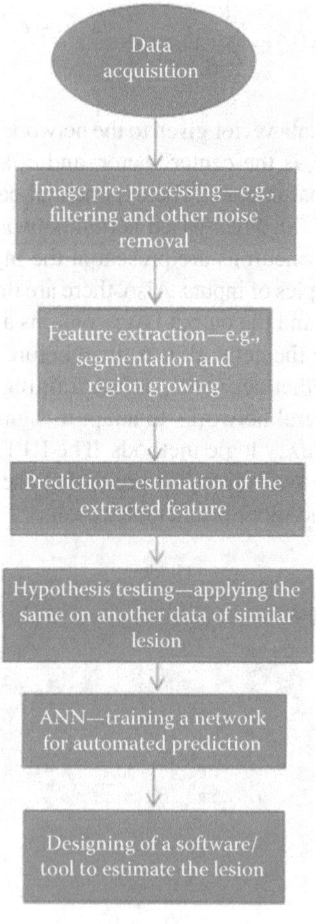

FIGURE 11.2 Stages of the brain image processing.

commonly used in preprocessing, and logical operators are widely used for feature extraction. By varying the threshold of these operators, different regions and different contrasts can be segmented.

Segmentation of an ischemic stroke image is more challenging, as there is no well-defined boundary of lesion or an absence of wider variation between the intensities in the regions of interest and those surrounding them. Because of atrophy the cell lacks hydroxyl ions and will show up as darker regions, while the rest will show up as grey. Furthermore, there is no fixed intensity level for the other regions, hence it is difficult to understand if the darker shades in other regions are because of illness or due to lower hydroxyl ion content (within physiological limits as a result of aging or any other disease). This problem can be minimized to a good extent by a suitable preprocessing method. Therefore, the precision in segmentation now depends on the effectiveness achieved in contrast enhancement. Usage of an appropriate threshold will further reduce the complexity of segmentation.

Classification of a lesion is purely based on segmentation. Based on the accuracy of segmentation, the classification would be ideal. As discussed earlier, it could either be parametric or nonparametric.

11.4 OUTCOME PREDICTION OF BRAIN LESION

The accuracy of prediction of clinical outcome is again based on the methods adapted for a lesion segmentation and classification. Other than the type of lesion, it is usually the area, volume, depth information, and its location, which are predicted by image processing methods. This would help the clinicians to plan for the treatment, surgery, follow up, and expected prognosis of a management protocol. The knowledge base linked with outcome prediction will have to undergo frequent upgrading so as to include newer types of lesions or illnesses. Sometimes in the clinical setup doctors feel that every lesion is unique. These unique lesions are the most vulnerable for false prediction, as there will not be any relevant data to confirm its prediction theory. Among the various brain lesions, the least challenging ones are tumors, while the most challenging ones are congenital and degenerative disorders combined with ischemic disorders.

11.5 CASE STUDIES

Any research topic is incomplete without its supporting documents, and case studies are the most significant reports with regards to any writing on image processing. Clinically, strokes can present as (1) hemorrhagic strokes—due to trauma or cardio vascular accidents following hypertension, and (2) ischemic strokes—due to vasoconstriction, extreme hypotension, hypoglycemia, and myocardial infarction leading to emboli lodged in cerebral vessels. Stroke or hemiplegia is the partial paralysis of the contralateral side of the body with respect to its controlling cerebral hemisphere. Irrespective of the type or cause the presentation of the condition will be paralysis, and hence they are collectively described as "stroke." Although the result is paralysis, the management of the stroke and prevention of its reoccurrence is based on the type of stroke. Hence, it is very significant to identify the type of stroke very precisely. The classification of the lesion is the crucial step, which determines the success rate of the treatment.

11.5.1 Ischemic Stroke Segmentation Challenge

The MRI image of an ischemic stroke is shown in Figure 11.3. The image shows certain dark regions marked within a circle in Figure 11.3a, which is identified as the region with the lesion, or the region of interest. This is often misread as atrophy of the brain, which could either be pathological or physiological as in degeneration due to aging. However, in the case of ischemic strokes there exist mild difference in the contrast between the region of interest and its surrounding. The contrast variation is used as a basis for further processing in image segmentation. However, this difference is well appreciated in case of acute lesions. In the case of a chronic lesion, the demarcation eventually reduces. The reason could be that the surrounding region is also becoming involved in the lesion, as the cause is not treated until diagnosed. As with any segmentation process, the first step is to reduce the noise and enhance the contrast. The same is done in this case too!

Preprocessing is done by region growing followed by contrast stretching, as the original image has normal regions, as well as regions with lesions in similar grey levels. Logical operations were performed with zero padded image, and image compliment was done using thresholding. Finally, image cropping was achieved

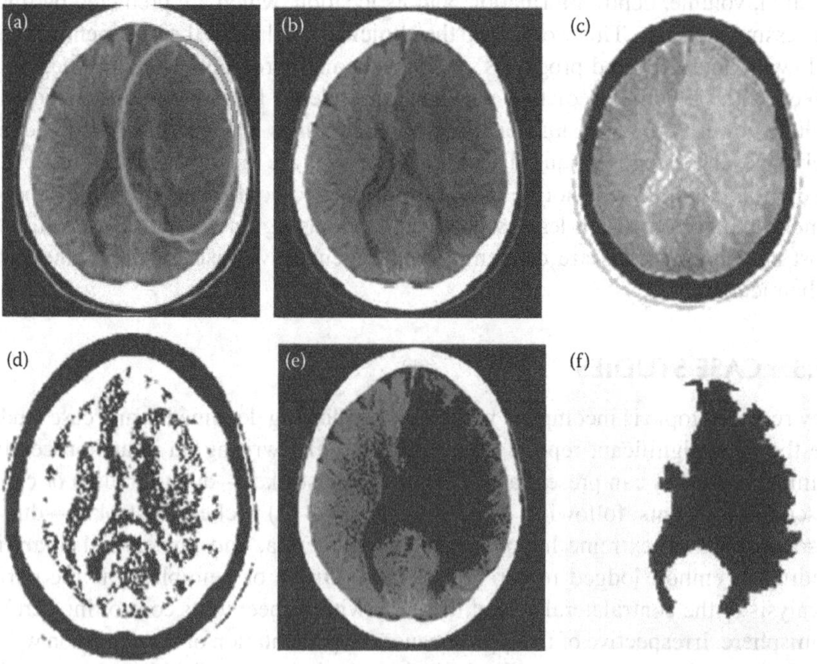

FIGURE 11.3 (a) Original image, (b) region growing, (c) contrast stretching, (d) logical operation and binary conversion, (e) image compliment and thresholding, and (f) segmented image.

with boundary conditions. This will segment the region of interest. All the images produced during this process are shown in Figure 11.3.

11.5.2 mTOP (Mild Traumatic Brain Injury Outcome Prediction) Challenge

Traumatic brain injuries are not very challenging in terms of lesion prediction but when the lesion is mild it is seldom challenging. The challenges can be overcome by appropriate thresholding and processing.

Figure 11.4a shows a hemorrhagic stroke image used for segmentation. As there is a variation in the image contrast, stretching is done as a part of preprocessing. The outcome of preprocessing yields an image having good demarcation between the region of interest and other regions. Therefore, image compliment is done to get an image showing further variation in contrast. As a part of the segmentation process, binary conversion with suitable thresholding is carried out to obtain appropriate segmentation. All the images rendered during this process are shown in Figure 11.4 [11].

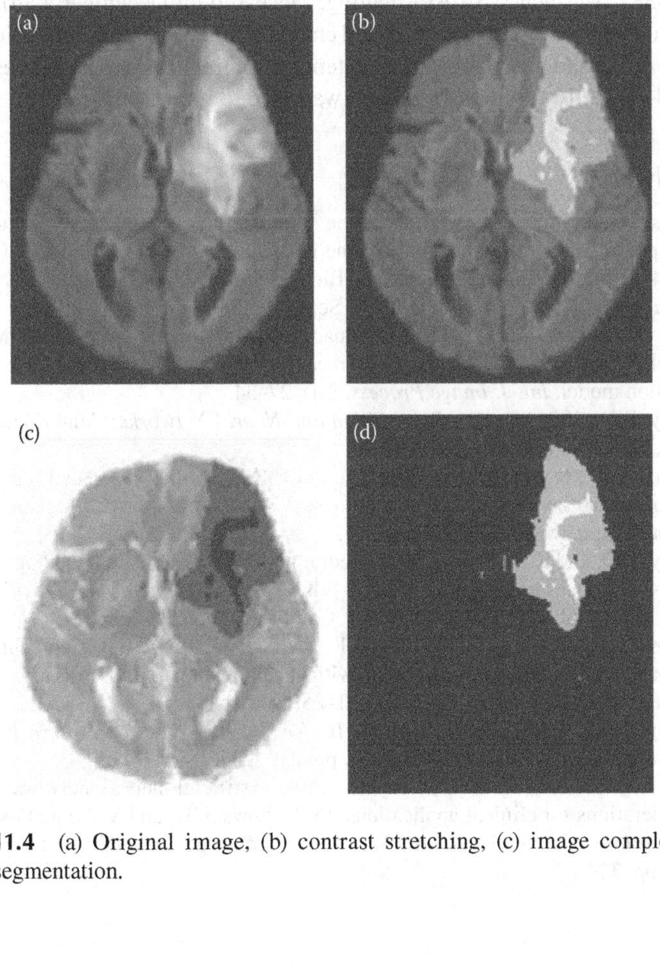

FIGURE 11.4 (a) Original image, (b) contrast stretching, (c) image complement, and (d) image segmentation.

11.6 FURTHER READING

Various researchers have worked on different aspects of image processing. Nevertheless, every image is unique and challenging, and thus processing has to be performed discretely for every image. Based on the requirement, analyzing the contrast level, histogram of the image to be processed, a researcher can read the relevant work published by an author in the same domain. A few authors have done segmentation based on image texture [11,12] while few others have done it based on contour. It all depends on the specific image.

ACKNOWLEDGMENT

The authors would like to thank Kondaveti Revanth and Abhishek Patil, who have contributed to the topics on case study and MLP network, respectively.

DISCLOSURE

All the original Images were taken from https://emedicine.medscape.com/article/338385-overview; "Stroke Imaging," accessed on December 4, 2017. This is an open source article and the authors and chief editor have nothing to disclose. Terms and Policy of the site had no objection statement relating to the reuse of these images. The MATLAB 2016 version of software was used to process the images.

REFERENCES

1. Devos, A. and L. Lukas, 2004, Does the combination of magnetic resonance imaging and spectroscopic imaging improve the classification of brain tumors?, On Page(s): 407–410, Engineering in Medicine and Biology Society, 2004. *IEMBS '04. 26th Annual International Conference of the IEEE*, Sept 1–5, 2004.
2. Masroor Ahmed, M. and D.B. Mohamad, 2008, Segmentation of brain MR images for tumor extraction by combining Kmeans clustering and PeronaMalik Anistropic diffusion model, *Int. J. Image Process.* 2(1), 27–34.
3. Ripley, B.D., 1996, *Pattern Recognition and Neural Networks.* Cambridge University Press, Cambridge, U.K.
4. Biganzoli, E., P. Boracchi, L. Mariani and E. Marubini, 1998, Feed forward neural networks for the analysis of censored survival data: A partial logistic regression approach. *Stat. Med.* 17(10), 1169–1186.
5. Ripley, B.D. and R.M. Ripley, 2001, Neural networks as statistical methods in survival analysis. In: Dybowski, R. and V. Gant (Eds.), *Clinical Applications of Artificial Neural Networks.* Cambridge University Press, Cambridge, pp. 237–255.
6. Lisbosa, P.J., H. Wong, P. Harris and R. Swindell, 2003, A Bayesian neural network approach for modelling censored data with an application to prognosis after surgery for breast cancer. *Artif. Intell. Med.* 28(1), 1–25.
7. Bishop, C.M., 2004, Error functions. In: *Neural Networks for Pattern Recognition.* Oxford University Press, Oxford, U.K., pp. 230–236.
8. Gant, V., S. Rodway and J. Wyatt, 2001, Artificial neural networks: practical considerations for clinical applications. In: Dybowski, R. and V. Gant (Eds.), *Clinical Applications of Artificial Neural Networks.* Cambridge University Press, Cambridge, U.K., pp. 329–356.

9. Lisboa, P.J., 2002, A review of evidence of health benefit from artificial neural networks in medical intervention. *Neural Netw.* 15(1), 11–39.
10. Lisboa, P.J. and A.F.G. Taktak, 2006, The use of artificial neural networks in decision support in cancer: A Systematic Review. *Neural Netw.* 19, 408–415.
11. Ali, A.H., S.I. Abdulsalam and I.S. Nema, 2015, Detection and segmentation of hemorrhage stroke using textural analysis on brain CT images. *Int. J. Soft Comput. Eng. (IJSCE)* ISSN: 2231-2307, 5(1), 11–14.
12. Ali, A.H., S.I. Abdulsalam and I.S. Nema, 2015, Detection and segmentation of ischemic stroke using textural analysis on brain CT Images. *Int. J. Sci. Eng. Res.*, ISSN 2229-5518, 6(2), 396–400.

12 Machine Learning Based Hospital-Acquired Infection Control System

Sehaj Sharma, Prajit Kumar Datta, and Gaurav Bansal

CONTENTS

12.1 INTRODUCTION

The goal of every health care institution is to discharge you healthier than you came in. However, this goal is often not accomplished because of HAI, also known as NI. HAI or NI is a category of infections a patient acquires upon visiting or being admitted to a health care facility. These NI's happen due to several reasons like lack of sanitation in hospitals or clinics, health care facilities becoming a congregation of pathogens due to several patients with a variety of illnesses, and in general suppressed immunities of patients making them more susceptible to NI. A World Health Organization (WHO) study—across developed and developing countries, reported that approximately 8.5% of patients admitted suffered some form of NI.

ML is a very popular buzzword in scientific circles. The applications of it are ubiquitous. ML is an application of Artificial Intelligence (AI) ,which provides the system the ability to learn patterns from the data provided to it, and then to make decisions and forecasts based on it. This chapter examines the various attempts by several researchers to manage NI better.

The chapter aims at educating researchers in reasonable depth about NI and the various ML attempts by several researchers to manage and control NI. This should help the reader develop further NI control programs. The types of NI are discussed with their specific causes. The chapter will then lucidly discuss the various NI control programs and their successes and failures. The final part of the chapter discusses the various ML attempts on NI-related data, and the implication of their results. It is hoped this chapter educates and inspires the reader to conduct novel work on solving the problem of NI.

Please note that the terms HAI and NI will be interchangeably used throughout this chapter, but they mean and imply the same thing.

12.2 TYPES OF NI

This section of the chapter describes the various types of NI. Discussing this is especially important since it will help the reader gain a comprehensive perspective of the problem. Discussing the problem in depth is aimed to help the reader develop techniques and protocols that are unique and generalised depending on the demographic he/she is targeting.

Section 12.2.1 will discuss four major categories of NIs in reasonable depth.

12.2.1 Central Line-Associated Bloodstream Infections (CLABSI)

CLABSI is a category of NI reporting the highest mortality rates [1–3,8]. Approximately 18% of the patients infected with CLABSI succumb to it [8]. Central lines are tubes, which provide a patient with nutrition, fluids, and medicines in the case the patient isn't able to consume them orally. Catheters are used for hemodynamic monitoring, plasmapheresis, apheresis, hemodialysis, tissue and organ transplantation, administration of liquids, blood products, chemotherapy, antibiotics, and parenteral nutrition [3]. Technically, a central line is a catheter typically attached to a patient's pulmonary artery or vein. Central lines are also called central venous catheters (CVCs). There are five types of central lines or CVCs:

i. Nontunneled CVC [10]: This catheter is placed into a large vein near a patient's neck, chest, or groin. They are usually used for long-term intravenous therapy [3]. They are the most common CLABSI causing CVC [3].

ii. Pulmonary artery catheters (PAC) [3,11]: A PAC is inserted through a central vein into the right side of the heart and into the pulmonary artery. PAC's are customarily used to quantify stroke volume, intracardiac pressures, etc. Unlike other CVCs, which are therapeutic devices, PAC's are diagnostic tools. Fortunately enough, the studies have observed PACs do not increase the chances of NI or mortality [11].

iii. Peripherally inserted CVC (PICC) [3,12,13]: PICCs are a fairly recent type of CVC. Unlike other CVCs inserted from the neck, chest, or groin, which were a source of great discomfort, PICCs are inserted via the veins in the forearm with the help of ultrasound or fluoroscopy [12]. They are believed to have a lower incidence of CLABSI [3]. However, this claim is disputed by Johansson et al. [12].

iv. Tunneled CVC [3,14]: Unlike nontunneled CVC, the cuff of a tunneled CVC is under the skin. This helps secure the catheter in place and prevents infection [14]. A tunneled CVC is reported to have lower incidence of CLABSI [3].

v. Totally implantable venous access devices (TIVAD) [15]: This CVC in particular was reported to have the lowest incidence of CLABSI and also gives the patient a greater degree of freedom. It is the technique of choice for cancer patients requiring chemotherapy [15].

vi. Umbilical catheters: An umbilical catheter is inserted in one of the two main arteries of the umbilical cord. It is used mostly in newborn babies to provide them with the necessary fluids, nutrition, and medicines.

A mass scale European study suggested that nearly 55% of the patients admitted were prescribed a central line [9]. CLABSI occurs when microorganisms move to the CVC tip [3]. They reach the CVC tip in the following ways:

i. From the microbes on patients own skin [3].
ii. An infected infusate [3].
iii. A pathogen already present in the patient's blood, which infects the patient when his immunity is low during a disease or postsurgery.
iv. Contamination by health care workers during preparation of fluids [3].
v. Due to inadequate sanitization while administering the CVC [3].

The most common microorganisms causing CLABSI are coagulase-negative staphylococci, Staphylococcus aureus, Enterococci, Enterobacteriaceae, Pseudomonas, and Candida [3].

The spread of CLABSI can be controlled by:

i. Catheter material chosen, coated with antiseptic, etc. [16].
ii. Catheter insertion site. The subclavian site is preferred [16].
iii. Ultrasound guided placement [16].
iv. Sterilization of the person inserting the catheter. Wear masks, caps, disinfect hands, etc. [16].

12.2.2 Catheter Associated Urinary Tract Infections (CAUTI)

This is the most commonly contracted NI [4]. Urinary transmitted infections (UTI) account for up to 13% or more of all reported NI's. CAUTIs are usually caused by bacteria native to the patient's urinal tract. In normal circumstances, this bacteria is flushed out by the flow of urine. However, catheters have imperfect drainage of urine and create the perfect environment for bacteria to flourish. This can develop into several diseases like orchitis, epididymitis, and prostatitis in males, and pyelonephritis, cystitis, and meningitis in all patients [1]. The two types of CAUTI are [17]:

i. Asymptomatic bacteriuria (ASB): Asymptomatic bacteriuria refers to bacteria in the urine at levels often regarded as clinically consequential (>100 000 colony composing units per milliliter of urine) in patients with no symptoms suggestive of urinary tract infection. It becomes more prevalent with age. This situation is almost universal with catheterized patients [18].
ii. Symptomatic urinary tract infection [17]: Alongside asymptomatic bacteriuria, a patient suffering from symptomatic UTI will also suffer from high fever.
iii. Bacterimia [17]: Less than 3% of patients suffering from ASB progress into bacterimia. It is a bloodstream infection and can cause mortality on a few occasions.

The most common organisms causing CAUTI are Escherichia coli, Enterococci spp, Staphylococcus, Pseudomonas aeruginosa, Candida, and Proteus mirabilis. The spread of CAUTI can be controlled by:

i. Selection of catheter: Catheters with the smallest gauge should be preferred. Interestingly, disinfected catheters didn't show a reduction in CAUTI [17].
ii. Catheter insertion must be done with clean hands, sterilized equipment, barrier precautions, and other general hygiene practices [17].
iii. Catheter maintenance carried out with appropriate hand hygiene, closed drainage system, replace system if it breaks during cleaning, and secure the catheter properly [17].

12.2.3 Surgical Site Infections (SSI)

SSI is the second-most common NI [1,5]. Within the United States alone 300,000 to 500,000 SSI's occur annually [5]. Between 5% and 10% of patients develop an SSI in the United Kingdom [18]. SSIs cause a financial setback of up to GBP 1 billion annually on the National Health Services [18]. Despite the hygiene of hospitals and procedures being impeccable in the developed countries, the incidence of SSI is high in those countries. We leave it to the reader's imagination what this plight is in developing and underdeveloped nations where hygiene isn't prioritized. Nearly 77% of the deaths associated to surgery are causes by SSIs [18]. The types of SSI are:

i. Superficial incisional: These SSIs happen at the level of the skin and subcutaneous tissue. They are indicated by redness, pain, heat, or swelling at the site of the incision, or by the presence of pus [18].
ii. Deep incisional: These SSIs occur at fascial and muscle layers. They are designated by presence of pus or an abscess, pyrexia with tenderness of the wound, or a disseverment of the edges of the incision exposing the deeper tissues [18].
iii. Organ or space infection: These SSIs occur at any component of the anatomy other than the incision opened or manipulated during the surgical procedure, for example the joint or peritoneum. They are designated by a loss of function of a joint, abscess in an organ, localized peritonitis or accumulation. Ultrasound or CT scans substantiate the infection [18].

The causes of SSI are:

i. Patients' own bacterial microflora of skin penetrates the surgical incision [18].
ii. Infected surgical equipment [18].

SSIs are primarily caused by a group of microorganisms called the Gram-positive cocci, mainly Staphylococcus aureus. Gram-negative aerobic bacteria can cause SSIs after intestinal surgery. Other bacteria can also become a reason for SSI due to the weakened immunity of the patient [18]. The strategies used to control and treat SSI are:

i. Optimization of patient health prior to surgery like stopping smoking, improving nutrition to prevent anemia, etc [18].
ii. Sanitization of surgical site and equipment [18].
iii. Extensive surveillance of NI's [18].
iv. Once infected with SSI, the medical team initiates antibiotic therapy [18].
v. Drainage of pus upon SSI [18].
vi. Cleaning of waste and dead tissue around the surgery wound [18].
vii. Specialist wound care services upon infection [18].

12.2.4 Ventilator Associated Pneumonia (VAP)

VAP occurs with patients put on mechanical ventilators through endotracheal tubes or tracheotomy tube for at least 48 hours [1,6]. There are two major categories of VAP:

i. VAP because of non-Multi Drug Resistant (MDR) Pathogens: These pathogens are fairly manageable to treat. The pathogens that fall under this category are Streptococcus pneumoniae, other streptococcus SPP, Hemophilius influenza, MSSA, Antibiotic—sensitive Enterobacteriaceae, Escherichia Coli, Klebsiella pneumoniae, Proteus SPP, Enterobacter SPP, Serratiamarcescens [6].
ii. VAP because of MDR Pathogens: As evidenced by the category, treating this type of VAP is fairly challenging. The unfortunate fact is that VAP is increasingly associated with MDR. These pathogens aren't recommended to be treated via broad-range antibiotics. The pathogens that fall under this category are Pseudomonas aeruginosa, MRSA, Acinetobacter SPP, antibiotic-resistant Enterobacteriaceae, Enterobacter SPP, ESBL—positive strains, Klebsiella SPP, Legionella, pneumophila, Burkholderiacepacia, and Aspergilius SPP [6].

The strategies to manage and treat VAP are slightly more complex than the rest. To be able to develop protocols to treat VAP, one must understand and further study the currently used strategies, which are [6]:

i. Allowing the patient to breathe through nasal or facemasks, which are not noninvasive.
ii. Short antibiotic courses.
iii. Feed the patient through a catheter, postpyloric feeding.
iv. Avoiding accumulation of gastric fluids.
v. Decontamination of the gastric tract with nonabsorbable antibiotics.
vi. Avoiding gastrointestinal feeding.
vii. Disinfecting the medical team.
viii. Disinfecting equipment.
ix. Elevating the head by 30–40 degrees.
x. Avoiding heavy sedation.

Up to 50%–70% of patients infected with VAP succumb. Other than a high probability of mortality, there are several other complications affecting patients suffering from VAP like:

i. Prolonged stay on mechanical ventilation and ICU.
ii. On a few occasions, necrotizing pneumonia and pulmonary hemorrhage occur due to pseudomonas aeruginosa [6].
iii. Bronchiectasis and pulmonary scarring leading to recurrent pneumonias, catabolic state in malnourished patients.

Like SSI, VAP too inflicts high mortality rates and leads to higher treatment costs [7].

This section has discussed the different types of NIs and their management strategies. Section 12.3 covers various programs setup by hospitals and governments to control NI.

12.3 NI PROGRAMS

This particular section discusses the strategies currently in place to manage and mitigate NIs [19]. In bits and pieces, this section will also cover some of their history, which led to the methods currently used today.

Some of the most insightful information on how the NI program has been setup in Canada was covered by Paton et al. Based on U.S. data, it had been estimated that 5%–10% (180,540 to 361,082 patients) of patients admitted to a hospital contracted some form of NI. This problem was compounded by the fact that the US public health care establishment spent around US$4,532,000,000 per annum [19]. These are numbers that could stagger any health care establishment.

Out of this need, the Canadian public health care system setup the Bureau of Communicable Disease Epidemiology (BCDE) in 1980. Over its 37-years history, BCDE has initiated several programs to control infections. One of their key programs, initiated on June 1984, was the Canadian NI Surveillance Program (CNISP). Revisions and updates were continuously made to the CNISP. One of the significant ones was the recommendation from an ad hoc committee in 1993 to implement a national strategy to control NI and antimicrobial resistance [19].

12.3.1 THE CANADIAN NOSOCOMIAL SURVEILLANCE PROGRAM (CNISP)

The BCDE setup a network of "sentinel hospitals" to become data collection centers in the CNISP. The first step of CNISP was to have sentinel hospitals acquire data to study the nationwide and regional NI trends, identify the magnitude of the problem and pathogens spreading NI. In 1995, there was widespread survey and study on methicillin-resistant *Staphylococcus aureus* and *Clostridium difficile* [19].

The next step of the CNISP was to survey the effectiveness of the currently employed techniques. This step is key to assure the effectiveness of the program. The third step was to implement the necessary health care facilities and observe their effect on the patient population [19].

We must take note of the in-built self-improvement mechanism of the CNISP. Such programs have demonstrated steadily decreasing rates of NI.

Several countries have formulated such NI programs. However, the success of these programs depends on adherence to the NI prevention protocols. The lack of adherence is discussed by Lizandra Ferrera et al. [32] in great depth. The study has been analyzed and summarized below.

12.3.2 The Failure of a NI Program in a Hospital

This section discusses and demonstrates how lack of compliance to a NI program can give a frighteningly high incidence of NI. Handwashing is one of the cheapest and easiest methods of controlling NI. This is because hands of a health care worker, or patient, are the primary ways of interacting with the environment. This makes it really easy for NI's to travel via touch [20].

In this Brazilian study, two observers were trained to collect data on the prevalence of NI and the adherence to hand sanitization from four different departments, that is, clinical, surgical, pediatric, and intensive care unit (ICU). The health care workers were surveyed twice a day. The standard of hand hygiene compliance was defined by washing hands after any contact with the patient or any inanimate object in the patient's room [20].

This compliance data was further classified into shift, sex, designation of health care worker, and the risk of cross-risk infection: highest risk being physically contacting the patient, or patient fluids, and lowest risk being deemed an indirect contact with a patient during hospital maintenance [20].

The handwashing initiative was marketed to several categories of health care workers. They were exposed to several presentations and illustrated posters to educate them about NI and handwashing. Even bottles of alcohol hand rub were distributed to the workers to encourage sanitizing hands [20].

After the hand-sanitizing program was introduced in this hospital, the average compliance increased from 21% to 24%. A 3% increase is not noteworthy. However, the compliance in nurses was up to 83.3%. Given that this initiative wasn't successful across all categories of health care workers, the decrease in NI was only 4.1% [20].

The lack of compliance to hand washing has not been discussed by this study. However, this resulted in high rates of NI and an increased spending on treating them. This also elongated the stay of patients in the hospital.

12.3.3 The Success of a NI Program in a Hospital

The Brazilian study mentioned above was inspired by the success of another Geneva-based study [21]. This study implemented a very similar hand hygiene routine to that of the Brazilian study.

Over a period of three years, handwashing compliance steadily improved from 48% to 66%. It was interesting to note that compliance was very high among nurses, but very poor among doctors. The reason for this disparity has not been explored in the Geneva-based study [21].

It was uplifting to note that the incidence of NI dropped from 16.9% to 9.9% in three years. The consumption of alcohol-based handrub solution increased from 3.5 to 15.4 L per 1000 patient-days in those three years [21].

The studies discussed in Sections 12.3.2 and 12.3.3 clearly show how effective a NI program can be if implemented correctly. Hopefully, in the future we will have sufficient studies to evaluate the success or failures of NI programs, and the motivations behind their success of failure.

After highlighting the importance of an effective NI program, we will now describe the different methods used to regulate the spread of NI.

12.3.4 NI Control Methods

A NI control program is categorized and subcategorized in the following manner:

12.3.4.1 Standard Precautions

i. Hand hygiene: It is the simplest and most effective way to control NI. The repercussions of its failure and success have been discussed in Sections 12.3.2 and 12.3.3 [1,22].
ii. Respiratory hygiene: Health care workers must wear masks as and when required, to prevent the workers from transmitting infections to the patient and vice versa [1,22].
iii. Personal protective equipment: Equipment like rubber gloves, hazardous material suits, etc., must be worn when dealing with contagious diseases [1,22].
iv. Injection safety: Only fresh syringes must be used and must be disposed of after use. Catheters must be disinfected before inserting. The prick site must be disinfected before being pricked [1,22].
v. Medication storage and handling: Medicines must be stored in areas with the least chances of contamination [1,22].
vi. Cleaning and disinfection: The hospital environment and equipment in general should be regularly cleaned and disinfected. This also extends to proper waste disposal [1,22].

12.3.4.2 Transmission-Based Precaution

i. Contact precautions: Patients must be avoided from being touched directly. Rubber gloves and adequate sanitization must be implemented before contact is made [1,22].
ii. Droplet precautions: In diseases like pneumonia, etc., this takes special priority. Masks and hazmat suits must be worn as the case demands [1,22].
iii. Airborne precautions: In diseases like swine flu, common cold, etc., the usage of masks is recommended. Adequate ventilation is also recommended [1,22].

12.3.4.3 Immunization and Vaccination Programs

These programs, when conducted at the national level, help the general population to have immunity against deadly pathogens. The effect of this immunity also protects them from NI when admitted to hospitals [1].

12.3.4.4 Education and Training of Health Care Staff

As seen in Section 12.3.3, when done right it can make a significant difference to NI statistics [1,22].

Even if the above control guidelines are adhered to, there is a lot which remains to be tackled in order to solve this challenge. Ever since the discovery of penicillin in 1928, humankind seemed to be winning the war against diseases. However, the tables were soon turned back as the microbes acquired resistance to these drugs. Section 12.3.5 examines antimicrobial use and resistance.

12.3.5 Antimicrobial Use and Resistance

Given the size and reproduction rates of microbes, they are ubiquitous [1]. They are found almost everywhere including the South Pole. While some microbes have helped forward our civilization, there are several others which are an existential bane to mankind.

Since the discovery of penicillin, antibiotics have been used extensively. They have perhaps been the biggest revolution to the pharmaceutical and health care industry. Diseases previously causing certain mortality (e.g., typhoid) are now easily manageable because of antibiotics.

Antimicrobials are a group of medicines which kill microbes. Disinfectants and antiseptics kill a wide range of microbes. Antibiotics kill or restrict growth of bacterial infections. Antifungals kill or restrict the growth of fungi. Similar descriptions could be assigned to antivirals and antiparasitics. While some antimicrobials are available to public, others can only be taken when prescribed by a licensed doctor upon diagnosis of the patient's symptoms.

Over time, with the abuse of antimicrobials, several strains of microbes have developed resistance to antimicrobials. The following fact reminds us of the gravity of this calamity. Antibiotic resistance is responsible for the death of a child every five minutes in Southeast Asia [23].

Self-medication with antibiotics, incorrect dosage, prolonged use, lack of standards for health care workers, and misuse in animal husbandry are the main factors responsible for increase in resistance. This resistance threatens the effective control against bacteria causing UTI, pneumonia, and bloodstream infections. Highly resistant bacteria such as MRSA, or multidrug-resistant Gram-negative bacteria, are the cause of high incidence rates of NIs worldwide [1,24].

Several policies have been issued by many countries and international bodies (like the WHO) to control antibiotic resistance. Most of these policies emphasize better civic hygiene, access to clean potable water, and vaccinations to prevent infections in the first place. They further elaborate on guidelines to instruct physicians on the administration of antibiotics. Pharmacists are regularly issued notices to not sell antibiotics to the general public without a doctor's prescription. The adherence to these guidelines varies from region to region.

Due to the more pertinent challenges of NI, certain advanced control strategies have been developed. These advanced strategies, however, do not undermine the earlier-mentioned strategies in any manner. The advanced infection control strategies only serve to augment the effect of basic infection control strategies and methods. The following sections cover advanced NI control programs.

12.3.6 Antibiograms to Control Antibiotic Resistance

Patients who acquire NI's usually have suppressed immunities because of their current ailment. The growing threat of antibiotic/antimicrobial resistance compounds the problem further. To control this situation, antibiograms are used.

Simply defined, antibiograms is a summary of the effectiveness of antimicrobial drugs towards specific microbial strains. It evaluates the local and wide area effects of different drugs. This helps clinicians track antimicrobial resistance and prescribe specific drugs for specific microbes [25,26].

There are several automated tools to help clinicians and researchers generate and analyze antibiograms. One such software is WHONET [25]. It is a free downloadable software developed by WHO's Collaborating Centre for Surveillance of Antimicrobial Resistance.

Using such software, researchers generate antibiograms with a biyearly or yearly frequency. These are analyzed and the results provided to doctors to help them prescribe the correct antimicrobial. Antibiograms in conjunction with the regular NI control practices make for a formidable combination to control NI.

12.3.7 Cross-Infection Programs

Simply defined, cross infection is the spread of microbes from one person to another. This not only leads to the spread of infection, but also cross colonization and cross breeding to create new strains of microbes [22]. The addition of this effect with the already existing causes of NI compounds the problem of controlling NI. However, this can be controlled by implementing basic NI control strategies.

To develop the most effective NI program, health care establishments mix and match a variety of these methods and factors. While developing any of these programs, the regional peculiarities must be kept in mind. Regional peculiarities include the attitude of people's culture, economic conditions of the region, and population and technical variances like microbial susceptibility in different regions using antibiograms. Also, methods should be figured out to ensure compliance with NI control programs.

Although a lot has been achieved by the medical fraternity to handle NI during hospitalization, the advancement of computer software technology in the last two to three decades has significantly contributed in this field. In Section 12.4, we will discuss some of the ML studies on predicting, controlling, and mitigating NI.

12.4 ML AND NIs

Hospital information systems are available in most hospitals today. The digital data available from these systems has pioneered the way for automatic surveillance systems. It has also added to the research for the automatic detection of NI and their trends. A lot of efforts around the globe have been based on this. Most of the systems, implement rule-based systems, but the evolution of ML in last decade has made it possible to use ML for NI. It can be used in mainly following areas regarding NI:

i. HAI prone score of the facility, which will eventually calculate magnitude of the problem and identify the HAI proactively.
ii. Identifying the trends in HAI across a range of hospitals or geographical territories.
iii. Proactive treatment for HAI.
iv. Understand the cost implication of HAI and its saving.
v. Evaluation of HAI prevention program.

12.4.1 Type of Data

As for any ML project, it is significant to have a substantial amount of data to develop any model. For an NI ML model preparation, a wide variety of data can be found, which can be primarily divided into structured and nonstructured categories.

12.4.1.1 Structured Data

12.4.1.1.1 Hospital Records

These days most of the hospitals have computer programs for patients' record keeping. These systems are called Hospital information systems. This data is mainly considered as structured data because it has some predefined format. For example,

i. A record for Patient Detail may have data looking like:
    1. Patient ID, Patient Name, Patient Address, Patient Phone No, Patient Date of Birth, Age, Sex, and Patient Emergency Contact Number.
ii. A record for Patients' past treatment and hospitalization may look like:
    1. Patient ID, Date of Checkup, Doctor ID, Time of Checkup, Patient Compliant, Prescription Type, and Prescription Details.
iii. A record of Patient's hospitalization may look like:
    1. Patient ID, Date of Admission, Doctor ID, Time of Admission, Disease ID, Patient Record, Cause of Admission, Cause of Discharge, admission diagnosis, previous surgery, previous intensive care unit stay, exposure to antibiotics, antacid, and drugs.

It should be noted that all of these records are structured because they have a defined format, as some fields can only hold numbers (Patient Phone No) and some fields can only hold characters (patient names). We also define their size or length; for example, a phone number can't be greater than 10 characters. The structured fields are easily managed with any database management system (DBMS), which have been the backbone of almost all hospital information systems.

12.4.2 Unstructured Data

This is the highly untapped category until a couple of years ago, but the new computer science techniques like ML and AI supported by humongous computational power and storage capacity has made it possible to understand and work upon unstructured data. For medical records, the unstructured data may look like:

i. Doctor prescription
ii. Doctor diagnosis
iii. Radiologist diagnosis
iv. X-ray reports
v. Patient discharge summary
vi. Interview with hospital staff

A very paramount technique for understanding medical prescriptions, notes, diagnoses, and discharge summaries is called natural language processing. It is the technique by which a computer program can understand and process the naturally indited notes. It is highly unstructured because there is no fine-tuned type or size of these notes. The NLP applications are different from conventional computer applications because computers traditionally require humans to interact with them in a precise programming language, which is unequivocal and highly structured. On the other hand, human verbalization is not always same. For example, we can pass on the same message in different ways while verbalizing or inditing. Hence, the interaction is equivocal and the linguistic structure can depend on many involute variables, including slang, regional dialects, and gregarious context. The purport of an NLP application is to understand all of this and engender sentiment, entity, context, and meaning out of it. All this should be automated and precise.

Advanced neural networks interpret physician notes, X-rays reports etc., fairly accurately. It is quite an advanced subject in medical research, and work is currently in progress.

12.4.3 Data Gathering Activity

It is also important to understand the data collation activity required to create a better, robust, accurate, and acceptable ML model. In case of NI we need to have a variety of data from all possible sources:

i. Hospitals
ii. Clinics
iii. Pharmacy
iv. Diagnostic labs

12.4.4 Data Sources

It should be noted that most of the research published on NI revolves around studies carried out in very small facilities, and hence it may inhibit the ability to make better

ML models. For example, most of the study are done in university hospitals and may not include all aspects in their studies. The more randomness in gathered data, the better the model would be. But there are many constraints to gather the data from various sources. For example, the study [27] was done on 688 patients. There are some studies focusing on both internal and external hospitals. For example, in a study conducted in Taiwan [30], the internal test set was for 461 hospitalizations whereas the external test set was for 2500 hospitalizations.

12.4.5 Variety of Data

It is of significant importance to understand that one model prepared by using only one type of data may not be suitable in all situations. For example, if the available information is primarily based on a study in an underdeveloped country then it is highly possible to have a higher tendency towards hygiene-related risks, causing CLABSIs. Similarly, on the flip side, if the data is collated only from a clinical facility which doesn't provide a tenured hospitalization facility then it will lack any SSI type of NI. There are some studies limited to only ICU records, while some covers entire hospitalizations. Which variety of data to use entirely depends on the problem at hand and the accessibility to data.

12.4.6 Data Collection Period

Another important angle for the data collation is that it should be collated over a period of time, and should not be considered as a onetime activity. This will remove any kind of bias being introduced in the data. It is quite critical that the data should be collected over a continuous period of time for better results. In most of the research conducted, studies were conducted over six days of hospitalizations for a period of 6 months.

12.4.7 Data Collection Locations

It is of paramount importance that the data should be collected from geographically spread hospital facilities. There are high chances that a particular region suffers through significant poverty, lack of education and modern facilities, which makes it more prone for NI. Any model prepared on that data only may have some bias towards showing NI in the patients. The opposite of this also holds true. It may be possible that an ultramodern hospital has high standards, and hence the data collected there may have minimum NI instances available in it. In order to generate a trained model it should have a sufficient number of NI infection cases.

12.4.8 Data for All Output Types

Another part of the data collection is that it should cover all the types of NIs. If the ML model is based on only one type of NI then it can't be used to identify other types of NI, as the model will not have sufficient data to train itself to identify those NI cases.

12.4.9 Data Exclusions

While we are gathering all of the data from all possible sources, it is also important to note that we may have to skip some data. For example, if all the data for hospitalizations is gathered then the patient record stays for less than 48 hours can be removed from the data of this study. It will result in a significant chunk of data being removed, but that is not useful for studying NI. Hospitalizations of less than 48 hours are not involved in the final dataset, as they contain too little information. That data may be useful for some other study, but it is not required for NI since 48 hours is too short for any NI infection.

As you will read later in this chapter in Section 12.4.13.4, the researchers drop some data, which are not useful to them. We need to understand that all data is not useful data.

12.4.10 Data From Interview

A lot of data about NI can be generated by interviewing nurses, patients, and their doctors.

12.4.11 The Imbalanced Data Problem

The main difficulty in this type of data (as in several medical diagnostic applications) is its highly skewed class distribution. As found in study [1], out of 683 patients only 11% (75 in number of the total) were infected and the rest were fine. The quandary of imbalanced datasets is very critical for the applications where the goal is to increase the apperception of the minority class (the infected patient class in the mentioned study) to its best. Here, we try to find other ways of biasing the modeling to support sensitivity (i.e., capacity to recognize positives), as predicated on a one-class Parzen density estimator and a one-class SVM.

12.4.12 Data Preparation Activity

As we have gathered data now, it's important to first prepare the data for the ML model. This comprises various activities depending on the type of data. For example, structured data will have a different set of preparation activities as compared to unstructured data. The data preparation activities for structured data are described below, whereas those for unstructured data are described in the Text Classification Section (12.4.13.4).

12.4.12.1 Handling Missing Values

As described in study [28], there were several variables having missing values during the data gathering activity. The missing values can be due to several reasons and are quite common in daily life. It may be due to erroneous or missing measurements. Help from medical experts would be required to identify the importance of these missing values. There are lots of ways to replace missing values with the most approximate values. In study [28], the missing values were replaced with the class-conditional mean for continuous variables, and the class-conditional mode for nominal ones.

12.4.12.2 Removing Redundant Variable

When the data is being collected, there is a normal tendency to collect as much data as possible for all the fields required. But all the fields (this can also be understood as variables, columns, attributes, and features) are not equally important from an ML perspective. Redundant variables should be removed. For example, if the two variables are highly correlated (in simple terms—have the same impact on output) then only one of them can be used, thus making the other redundant. As described in study [28], initially there were 83 variables at the start of the study, however, after careful examination they were able to remove 34 redundant variables and thus used only 49 variables for the study.

12.4.13 Implementation

In this section, we will mainly focus on four methods for discovering NI from patient data. Each of the methods is unique, and generates new results. The methods are:

i. One class classifier: SVM
ii. One class classifier: Parzen density estimator
iii. Text classification
iv. Linear regression and artificial neural network

12.4.13.1 One Class Classification Implementation

Normally, a classification method is two-class classification, but the one-class approach is particularly attractive in situations where cases from one class are difficult to obtain for model construction, which is the case of NI. The one-class classification is the ability to separate between new cases similar to members of the training set and all other cases that can occur. While it looks similar to conventional classification problems (i.e., two-class), one-class classification has a different way to train the classifier. The classifier of one-class classification is trained only by cases from the majority class, and it is never trained from the cases of the minority class. The purpose for the classifier is to develop the boundary separating those two classes based only on data that lies on one side of it (on which the classifier is trained). It must therefore estimate the boundary separating those two classes based only on data which lie on one side. The objective is to identify patients with one or more Nis, based on clinical and other data collected during the survey.

As we will see, the results received by one-class classifiers are better than the standard 2-class SVMs, with some trade-off between specificity and sensitivity.

12.4.13.2 One Class Classifier: SVM

The major arduousness intrinsic in the data (as in many medical diagnostic applications) is the highly skewed class distribution. As observed in study [1], out of 683 patients only 75 (11% of the total) were infected and 608 were not. The quandary of imbalanced datasets is concretely crucial in applications where the goal is to maximize apperception of the minority class (which is the infected patient class in our case). In this chapter, we investigate another way of biasing the modeling process to boost sensitivity (i.e., capacity to apperceive positives) predicated on one-class Parzen density estimator and one-class SVM [1], which can be trained to differentiate between two classes on the substructure of the examples from a single class (in this case, only "normal" or noninfected patients). The infected ones are identified as "abnormal" cases or outliers, which deviate significantly from the mundane profile.

The one-class approach engendered a sensitivity of 92.6%, whereas a standard two-class SVMs scored a baseline sensitivity of 50.6%. Clearly, the results are in favor of one-class SVM.

To train one-class SVM classifiers researchers used an RBF kernel. Generalization error was estimated using fivefold cross-validation. In these type of cases (as seen in Parzen density estimator also), a tenfold cross validation is not used because the number of infected cases in every fold will be much less and will not help. Hence, the complete dataset was randomly partitioned into five subsets. The error rates estimated on the test sets were then averaged over the five iterations.

As described in research [27], a widely used performance metric in classification is accuracy. As the name suggest, accuracy means the fraction of correctly classified data points in the test set. But it is important to note that in highly unbalanced class, such metrics might be misleading. Let's take an example, on a dataset with a 95%–5% class distribution (assume 95% are normal patients not infected by NI, whereas 5% are infected by NI), it is straightforward to attain 95% accuracy by simply assigning each new case to the majority class. But if you think more about it, the impressive accuracy of 95% for such a solution is inacceptable as a medical diagnosis, as the classifier would have failed to recognize the infected cases, which are a minority of 5%.

To discuss alternative performance criteria we adopt the standard definitions used in binary classification. These performance criteria are industry standards to evaluate and validate various ML models.

TP and TN stand for the number of true positives and true negatives, respectively. Or the positive/negative cases recognized as such by the classifier.

FP and FN represent, respectively, the number of misclassified positive and negative cases.

Sensitivity: In two-class problems, the accuracy rate on the positives is defined as: TP/(TP + FN). It is important to understand that here FN represents those cases which are actually positive, but our model has falsely termed them as negative.

Specificity: The accuracy rate on the negative class, also known as specificity, is: TN/(TN + FP). It is important to understand that here FP represents those cases which are actually negative, but our developed ML model has falsely termed them as positive.

Classification accuracy is simply (TP + TN)/TP + TN + FP + FN.

Findings: Table 12.1 lists the results of summarized performance for the one-class SVMs. This shows the best results obtained through training classifiers by utilizing various parameter configurations on noninfected cases only. So it is clear the highest sensitivity is achieved when both ν and σ are minuscule. ν is termed as the upper limit on the fraction of outliers which can be ignored. The values of sensitivity and specificity have an intriguing relationship. Let's endeavor to understand it. The sensitivity of 92.6% designates that the model is very good at identifying the authentic

TABLE 12.1
Performance of One-Class SVMs for Different Parameter Settings Utilizing an RBF Gaussian Kernel

Parameters		Accuracy	Sensitivity	Specificity
ν	σ	%	%	%
0.05	**10^{-4}**	**74.56**	**92.60**	**43.73**
	0.1	75.49	80.60	65.60
	0.15	72.51	70.39	74.40
	0.17	71.69	66.94	77.87
0.2	10^{-4}	75.69	79.28	68.27
	0.06	74.97	77.14	69.87
	0.07	74.97	76.82	70.40
	0.1	74.36	74.67	72.27

positive cases as positives; however, at the same time the specificity of 43.73% denotes that the model is also calling negative cases withal as positive cases. It clearly shows that the model is partial towards relegating most of the cases as positives.

But as you move across Table 12.1, there is a tradeoff, as the value of sensitivity decreases and the value of specificity increases. It signifies that model is now relegating more cases into a negative category, which it was not doing earlier.

12.4.13.3 One Class Classifier: Parzen Density Estimator

As performed in the study at University Hospital of Geneva (HUG), a one-class Parzen density estimator [28] was developed. As suggested above, the one-class classifier is always trained on only one class and in this study the classifier was trained on only noninfected patient data, however, the test set contained data for both infected and noninfected patients. A fivefold cross validation method was utilized for model validation where the consummate data set was divided into five subsets. On every iteration, one subset (containing one fifth of the samples) was kept aside as a test data set and the remaining four were put into a training set. The detailed explication of the Parzen density estimator is outside the scope of this book.

The one-class Parzen density estimator showed the best results were achieved by training the classifiers through utilization of different parameter configurations on noninfected cases only. The one-class approach, like this one, leads to consequential amendments in sensitivity over classical symmetrical SVMs. But a major constraint of one-class Parzen density estimator lies in the requisite of high computational systems during the testing phase.

Experimental results reported in the paper [28] are emboldening. From the perspective of sensitivity, one-class Parzen density estimators procure the highest caliber (88.6%) observed by the authors throughout a series of studies on the quandary. However, the price paid in terms of loss in specificity is quite exorbitant, and domain experts should decide about the high apperception rate.

12.4.13.4 Text Classification Method

As mentioned in [29], this study focuses on the application of text classification using SVM and gradient tree boosting. This study is quite different from the above two because it is based on unstructured data. This method uses some of the text mining processes required only for text classification and not for structured data (Figure 12.1).

Let's try to understand what the researchers have done in [29] and understand the flowchart above.

DR stands for daily patient record whereas HR is hospitalization record. All DRs of a patient together amount to the patient's HR.

This research is based on 120 inpatients at one hospital in Sweden. Not all information stored in the patients' EHRs was considered valuable by the physicians for detecting HAI.

Thus, a subset of information was retrieved:

- Journalanteckning (Engl.: record notes),
- Läkemedelsmodul (Engl.: drug module),
- MikrobiologiskaSvar (Engl.: microbiological result) and
- Kroppstemperatur (Engl.: body temperature).

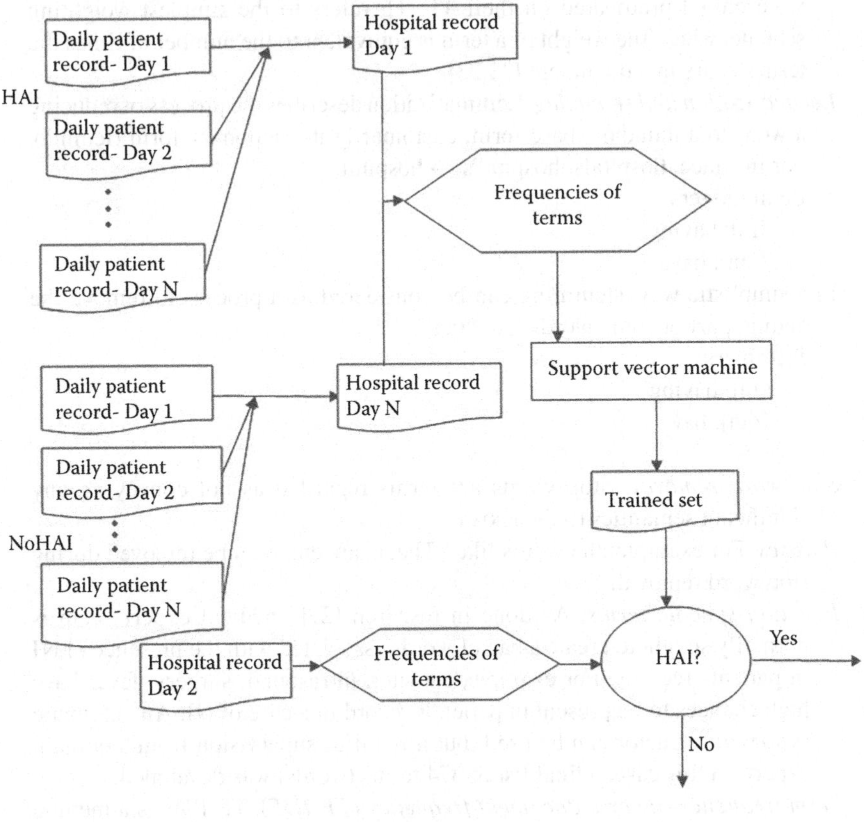

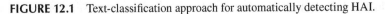

FIGURE 12.1 Text-classification approach for automatically detecting HAI.

The observation extracted from these modules contains few structured and few unstructured data. As we have already expounded on earlier, the structured data refers to the stored data in predefined fields, such as International Relegation of Diseases–10th Revision (ICD-10) diagnosis codes, medication, or body temperature. But the most intriguing part of this research [29] is the unstructured data, which refers to textual notes indited by medicos, such as daily notes or microbiological results. Since some of the 120 patients were hospitalized multiple times during the five-month period of records we received, our dataset comprises 213 HRs.

The analysis done by researchers [29] revealed that only 128 of 213 HRs contained HAI diagnoses (positive examples). Those records represent the HAI class. According to the assessment, the remaining 85 HRs contained no HAI diagnoses (negative examples), thus representing the NoHAI class. The dataset was not balanced, but instead, it was skewed toward the infected class, which is incipient to this relegation. In our earlier relegation, the data was skewed towards the NonNI class. We only utilized the class containing HAI for presage and not the class without HAI.

Since the data is unstructured (medico notes), we have to introduce some incipient terms here:

Term frequency (TF). In this method (TF 1000), the 1000 most frequent terms, were culled predicated on their TF. TF refers to the simplest weighting scheme, where the weight of a term is equivalent to the number of times the term occurs in a document [22,23].

Lemmatization and stemming. Lemmatization describes the process of reducing a word to a mundane base form, customarily its dictionary form (lemma). For instance, hospitals, hospital's → hospital.

Lemmatizers

[in]: having

[out]: have

In a simplistic way, stemming can be considered as a process to remove the ending part of some words like "ing."

Stemmers

[in]: having

[out]: hav

Stop word removal. Stop words are terms regarded as not conveying any significant semantics to the texts or

Phrases. For example, the words like "The, a, an, etc." can be removed during stop word removal.

Infection-specific terms. As done in research [29], medical experts help is normally sought to create a bag of words associated with the presence of NI in patients records. For example, catheter, ultrasound, surgery, fever have high chances to be present in patient's record in a case of NI. An automatic synonym extractor can be used, but it requires supervision from a domain expert. In this case, a final list of 374 terms (words) was developed.

Term frequency–inverse document frequency (TF-IDF). TF-IDF, is a method to understand the importance of a word in a document against a corpus.

The TF-IDF value increases proportionally with the number of times a word appears in a document, but is often offset by the frequency of the word in the corpus. This is extremely important to understand that some words appear very frequently in written or spoken communication.

12.4.13.5 Linear Regression and Artificial Neural Network

As referred to in article [29], In a study conducted in Taiwan, [31] linear regression (LR) and ANNs were used to predict HAI. The system used structured data, which comprises 16 features extracted from patient records, ranging from demographic, procedural, and therapeutic features to features concerning the general health status of the patient.

ANNs are trained using back-propagation and conjugate gradient descent. In order to evaluate the system, internal (same hospital) and external (another hospital) data sets were used. For the internal test set of 461 hospitalizations, the best result was produced using the ANN approach, reaching a recall of 96.64 percent and a specificity of 85.96 percent. For the external test set consisting of 2500 hospitalizations from different hospitals, LR gave the best result with a 82.76 percent recall and 80.90 percent specificity.

12.5 CONCLUSION

It must be understood that the different set of classifiers give different results and there is a lot of trade-off involved. For example, in the study conducted at the University Hospital of Geneva, a one-class Parzen density estimator [28] produced the highest level (88.6%) sensitivity observed by the authors among a series of studies on the problem. However, they lost specificity, thus domain experts must decide if the high recognition rate is worth the cost of treating those patients who are not infected by NI. It should also be understood that high false positives means the hospital will be considered as a breeding zone of NI in various reports, which may not be beneficial to the hospital especially when it is reported, since it is false.

A survey conducted by WHO in 2002 found that for 55 hospitals in 14 countries, an average of 8.5 percent of all hospital patients suffer from HAI.

To sum up, this chapter focuses on deploying four ML algorithms to hospitalization records:

- One-class classifier: SVM
- One-class classifier: Parzen density estimator
- Text classification
- LR and ANNs

There is always a tradeoff between sensitivity and specificity of the ML models, which requires very careful manual examination by medical experts to select the best model. But as described in the chapter, there are various means to increase the ML model accuracy, for example:

- Efficient and elaborative data gathering
- Data processing/preparation/feature selection and reduction

- Fine Tuning ML model parameters
- Validation of models by fivefold or tenfold
- Comparison of models and selecting the best one

Though in most of the studies, the data set of patient records was minute in size (thousands only), it nevertheless demonstrated the potential of applying ML techniques to patient records, including the structured and unstructured components to identify NI. The major hurdle, typical in medical diagnosis, is the quandary of rare positives.

Algorithms with high recall are especially felicitous for the screening of infections. The overall goal will always be obtaining high recall (approaching 100%) with the highest precision possible for HRs. This will enable us to implement a system able to can screen all hospitalization records and filter out all HSRs containing HAI. This would reduce hospital staff workload tremendously, as they would only need to analyze those HSRs preselected by the system.

REFERENCES

1. Khan HA, Baig FK, Mehboob R. Nosocomial infections: Epidemiology, prevention, control and surveillance. *Asian Pac J Trop Biomed.* 2017;7(5):478–482, ISSN 2221-1691, https://doi.org/10.1016/j.apjtb.2017.01.019.
2. Rabindran, Gedam DS. Central line associated bloodstream infections. *Int J Med Res Rev* 2016;4(8):1290–1291. doi:10.17511/ijmrr.2016.i08.37.
3. Nelson AD. Central Line Associated Bloodstream Infections, *Fellows Conference*, University of Utah School of Medicine. 06.27.2012.
4. CDC. Urinary tract infection (catheter-associated urinary tract infection [CAUTI] and noncatheter associated urinary tract infection [UTI] and other urinary system infection [USI]) events. Atlanta, Georgia: CDC; 2016
5. Anderson DJ. Surgical site infections. *Infect Dis Clin North Am.* 2011;25(1):135–153, ISSN 0891–5520, https://doi.org/10.1016/j.idc.2010.11.004. (http://www.sciencedirect.com/science/article/pii/S0891552010000917) Keywords: Surgical site infection; Health care–acquired infection; Outcome.
6. Bansode BR. Ventilator—Associated Pneumonia (VAP), Medicine Update 2011, The Associations of Physicians of India.
7. Morillo-García Á, Aldana-Espinal JM, de Labry-Lima AO et al. Hospital costs associated with nosocomial infections in a pediatric intensive care unit. *Gac Sanit.* 2015;29(4):282–287, ISSN 0213-9111, https://doi.org/10.1016/j.gaceta.2015.02.008. (http://www.sciencedirect.com/science/article/pii/S021391111500031X) Keywords: Intensive Care Units, Pediatric; Nosocomial infection; Health care costs; Hospital costs; Cohort studies; Regression analysis; Unidad de CuidadosIntensivosPediátricos; Infección nosocomial; Costes; Costeshospitalarios; Cohortes; Regresión.
8. Stevens V, Geiger K, Concannon C et al. Inpatient costs, mortality and 30-day re-admission in patients with central-line-associated bloodstream infections. *ClinMicrobiol Infect.* 2014;20:O318–O324. DOI: http://dx.doi.org/10.1111/1469-0691.12407.
9. Zarb P, Coignard B, Griskeviciene J et al. The European Centre for Disease Prevention and Control (ECDC) pilot point prevalence survey of health care-associated infections and antimicrobial use. Collective, *Euro Surveill.* 2012;17:20316. https://doi.org/10.2807/ese.17.46.20316-en.

10. Taslakian B, Chokr J. Placement of nontunneled central venous catheter. In: Taslakian B, Al-Kutoubi A, Hoballah JJ, editors. *Procedural dictations in image-guided intervention: nonvascular, vascular and neuro interventions*. New York: Springer; 2016. p. 379–81.
11. Rajaram SS, Desai NK, Kalra A et al. Pulmonary artery catheters for adult patients in intensive care. *Cochrane Database Syst Rev*. 2013;2:CD003408.
12. Johansson E, Hammarskjold F, Lundberg D, Arnlind MH. Advantages and disadvantages of peripherally inserted central venous catheters (PICC) compared to other central venous lines: A systematic review of the literature. *Acta Oncol*. 2013;52:886–92.
13. Chopra V, Flanders SA, Saint S. The problem with peripherally inserted central catheters. *JAMA*. 2012;308(15):1527–1528. doi:10.1001/jama.2012.12704.
14. Tunneled Central Venous Catheter (CVC) Placement, Wexner Medical Center, The Ohio State University, patienteducation.osumc.edu.
15. Di Carlo I, Cordio S, La Greca G et al. Totally implantable venous access devices implanted surgically a retrospective study on early and late complications. *Arch Surg*. 2001;136(9):1050–1053. doi:10.1001/archsurg.136.9.1050.
16. Frasca D, Dahyot-Fizelier C, Mimoz O. Prevention of central venous catheter-related infection in the intensive care unit. *Crit Care* 2010;14(2):212.
17. Nicolle LE. Catheter associated urinary tract infections. *Antimicrobial Resistance and Infection Control*. 2014;3:23. doi:10.1186/2047-2994-3-23.
18. Bagnall NM, Vig S, Trivedi P. Surgical-site infection. *Surgery* (Oxford). 2009;27(10):426–430, ISSN 0263-9319, https://doi.org/10.1016/j.mpsur.2009.08.007. (http://www.sciencedirect.com/science/article/pii/S0263931909001744) Keywords: cost; NICE guidelines; prevention; surgical-site infections.
19. Paton S. Nosocomial infection program. *Can J Infect Dis*. 1995;6(2):73–75.
20. de Almeida e Borges LF, Rocha LA, Nunes MJ, Filho PPG. Low compliance to handwashing program and high nosocomial infection in a Brazilian hospital. *Interdiscip Perspect Infect Dis*. 2012;2012:5, Article ID 579681. doi:10.1155/2012/579681.
21. Pittet D, Hugonnet S, Harbarth S et al. Effectiveness of a hospital-wide programme to improve compliance with hand hygiene. *The Lancet*. 2000;356(9238):1307–1312.
22. Hota B. Contamination, disinfection, and cross-colonization: Are hospital surfaces reservoirs for nosocomial infection? *Clin Infect Dis*. 2004;39(8):1182–1189. https://doi.org/10.1086/424667.
23. Singh PK. *Antibiotics, handle with care*. Geneva: WHO; 2016. [Online] Available from: http://www.searo.who.int/mediacentre/releases/2015/antibiotics-awareness-week-2015/en/.
24. WHO. *Antimicrobial resistance*. Geneva: WHO; 2014. [Online] Available from: http://www.searo.who.int/thailand/factsheets/fs0023/en/.
25. Joshi S. Hospital antibiogram: A necessity. *Indian J Med Microbiol*. 2010;28:277–80.
26. Saranraj P, Stella D. Antibiogram of nosocomial infections and its antimicrobial drug resistance. *Int J Pharm Biol Arch*. 2011;2(6):1598–1610.
27. Cohen G, Hilario M, Sax H, Hugonnet S, Pellegrini C, Geissbuhler A. An application of one-class support vector machine to nosocomial infection detection. *Studies in Health Technology and Informatics*. 2004;107:716–20. doi: 10.3233/978-1-60750-949-3-716.
28. Cohen G, Sax H, Geissbuhler A. Novelty detection using one-class Parzen density estimator. An application to surveillance of nosocomial infections. *Studies in Health Technology and Informatics*. 2008;136:21-6. doi: 10.3233/978-1-58603-864-9-21.
29. Ehrentraut C, Ekholm M, Tanushi H, Tiedemann J, Dalianis H. Detecting hospital-acquired infections: A document classification approach using support vector machines and gradient tree boosting. *Health Informatics Journal*. 2018;24(1):24–42. http://doi.org/10.1177/1460458216656471.

30. Chang YJ, Yeh ML, Li YC et al. Predicting hospital-acquired infections by scoring system with simpleparameters. *PLoS ONE*. 2011;6(8):e23137.
31. Dalal MK, Zaveri MA. Automatic text classification: a technical review. *Int J Comput Appl*. 2011;28(2):37–40.
32. de Almeida e Borges LF, Rocha LA, Nunes MJ, Filho PPG. Low compliance to handwashing program and high nosocomial infection in a Brazilian Hospital. *Interdisciplinary Perspectives on Infectious Diseases*. 2012; Article ID 579681, 5 pages. doi:10.1155/2012/579681.

13 No Human Doctor
Learning of the Machine

Leonid Datta, Emilee Datta, and Shampa Sen

CONTENTS

13.1 INTRODUCTION

The human mind has a unique quality for learning. This learning starts from the very first day and continues throughout life. The learnt knowledge is used for future work, which is commonly termed experience. While comparing a machine with the human mind, scientists found an important gap between them, as the machine had no learning capability or no capacity to analyze previously performed tasks (Solomonoff 1957; Case and Smith 1983). The machines were specifically coded for a task, but threw exceptions for other tasks assigned to them. This explicit gap was analyzed and learning was considered as an important and new task to be performed by the machine. Thus, the learning factor of a machine was introduced, which was a big step towards automation and ML. Initially the learning was knowledge-based, which later turned into a data-driven knowledge acquired system. The system continuously collects data either through sensors (for embedded systems) or through the input data (for software-based systems). The data set is analyzed and through different steps, knowledge is acquired (Yusuff et al. 2012). This knowledge is used as a learning factor.

The first step in approaching the concept of a "no human doctor" is learning, and "learning" initiates the availability of the data sets. From a specific data set, the behavior prediction becomes the unique capacity of the proposed system. For instance, classification is used as the basic system for predicting the stage of cancer diagnosis (André et al. 2009), while in case of observations in hospitals, the data are recorded at every moment and the patients are continuously supervised. If any kind of outliers are found in the data, or the values are found to be beyond a specific limit, the system alerts the doctor. Using the ML algorithms available at a basic level, the first step towards the automation can accomplished (Aha 1997).

13.2 ML: THE APPROACH TOWARDS AUTOMATION

ML deals with the algorithms able to automate a task. Towards the approach of no-human-doctor, the available set of data is used to make the prediction and different sensors are used to sense or input the required data. Learning algorithms can be classified into two types—supervised and unsupervised learning. In the case of supervised learning, a separate data set is available, which is taken from the previous records to train the new system. The data may also have uncleaned data or outliers. The approach of data analysis differs depending on the size of data (Jacobs et al. 1991; Jordan and Jacobs 1994), whereas unsupervised learning deals with problems where no training data set is available. Supervised learning mainly deals with problems regarding prediction of values, but unsupervised learning deals with classification problems.

Normally, before the data analysis a data set is cleaned using algorithms. This cleaning includes steps like removing null values, removing outliers, etc. After the data set is cleaned, it is used as a training data set or knowledge base for algorithms (Abe 1997; Angluin 1988).

13.3 DIFFERENT CATEGORIES OF ALGORITHMS

13.3.1 Regression

As mentioned earlier, classification techniques can be used to predict whether a tumor is malignant or benign. In other words, the probability of the tumor belonging to either type is calculated (Specht 1991).

But in the case of regression, we take prediction as the value found through the equation of a graph. The values available are the training data set. If we consider the case study of cancer prediction using different variables, regression can be the way. Yusuff et al. (2012) used a logistic regression technique in their experiments.

The question may arise as to how accurate the prediction is or the specific values taken may also vary in parameters. To measure this, a term, cost-function, has been defined to measure the extent to which the prediction is not accurate. If the cost-function value is large, then the prediction is less accurate.

If we consider a hypothesis $y = h(x)$ where x is the independent variable and y is the value to be predicted, then for a set of values $\times$ [represented as matrix X containing n values] and a given set of $(n-1)$ values of y, and the predicted value [represented as matrix Y containing n values], then the cost-function is defined as $J = (1/2n)\ \mathrm{sum}(h(x) - y)^2$.

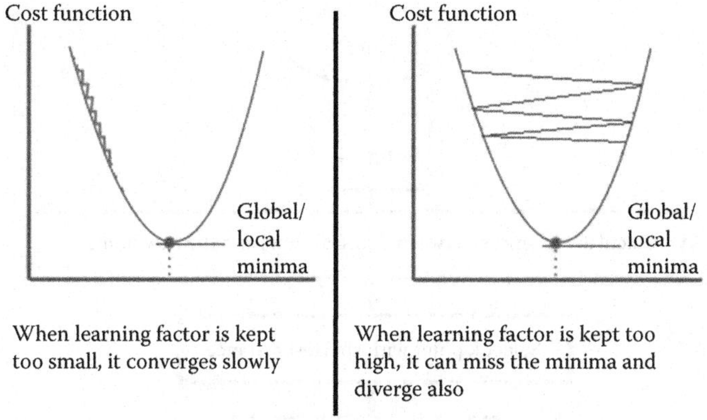

FIGURE 13.1 Variation of cost-function with learning factor.

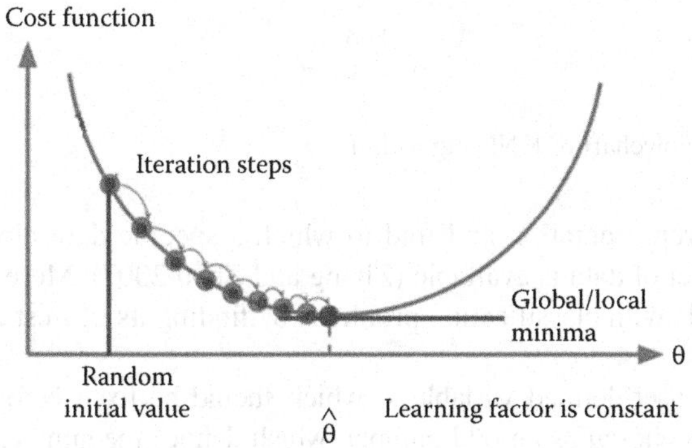

FIGURE 13.2 Cost-function with n-iterations.

Since it is a quadratic equation, the graph plotted must be parabolic in shape and open upwards, and thus one global minima must exist. So, the parameters must be chosen in such a way that we reach the point of global minima with minimum no of iterations.

If α is considered as the learning factor for the algorithm and J as the cost function, then considering a linear relation we predict a linear equation $\theta(j) := \theta(j) - (*(d/d\,\theta)(J))$. Since for continuous iterations the value of $d/d\theta$ (j) becomes gradually less, the learning factor α is kept constant. Figure 13.1 shows two cases where the learning factors are chosen either too big or too small. At a certain number of iterations the global minima is met. At that point the $\tan(\theta)$ value of the tangent becomes zero and continuously two iterations produce the same value. At this point the global minima is reached. Figure 13.2 describes this condition.

13.3.2 KNN Classification

Among the different basic algorithms available, K-nearest neighbor (KNN) is the most important one because of its simplicity. When KNN algorithm mainly deals

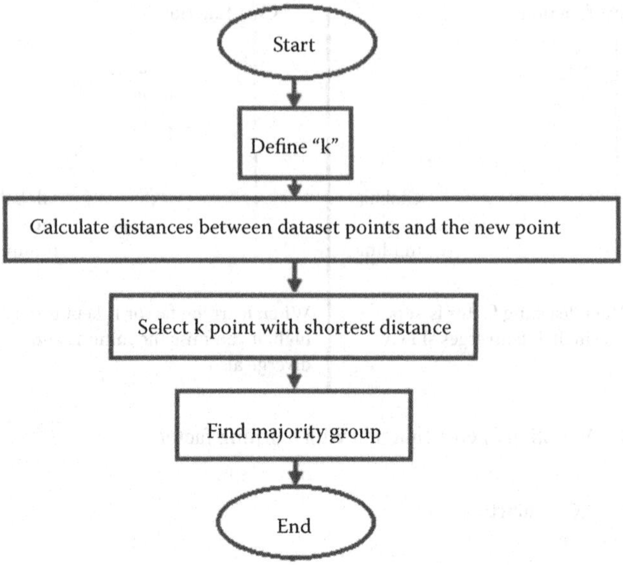

FIGURE 13.3 Flowchart of KNN algorithm.

with graphical representation and find to which a specific data plot belongs when more than one set of data is available (Zhang and Zhou 2007). More straightforward to define, it deals with classification problems by finding its nearest neighbor (Cover and Hart 1967).

KNN uses a user-defined variable k, which should be fixed before the algorithm is applied. "k" is chosen as an odd number, which defines the number of final nearest neighbors to be found. After the k has been defined, the graphical distance between the points of the knowledge data set and the new defined point are calculated. These distances are observed and the k number of points having shortest distance value are considered. Since the k is odd, the set containing k nearest neighbors will have one majority group. The new element will be considered as the member of the majority group. This is initially done for classification between two sets but it can be extended to n classification problems. Figure 13.3 shows a flowchart containing the steps of a KNN algorithm.

Towards the approach of no-human-doctor, this KNN can be used as a very basic technique to examine the patient behavior recorded through the sensors and also will help in immediate prediction. The training data set can be dynamically allocated while implementing, and the data set can be updated continuously along with the process. This approach will boost the diagnosis problem. While the initial implementation was with two variables only, the concept can be extended to a n-variable concept. Figure 13.4 describes an example with two data set and two cases with $k = 5$ and $k = 9$.

13.3.3 Decision Tree (DT)

Based on the data collected, we need a decision-making tool to predict the final result. For predicting a result, we observe the behavior of the patient termed as symptoms. Based

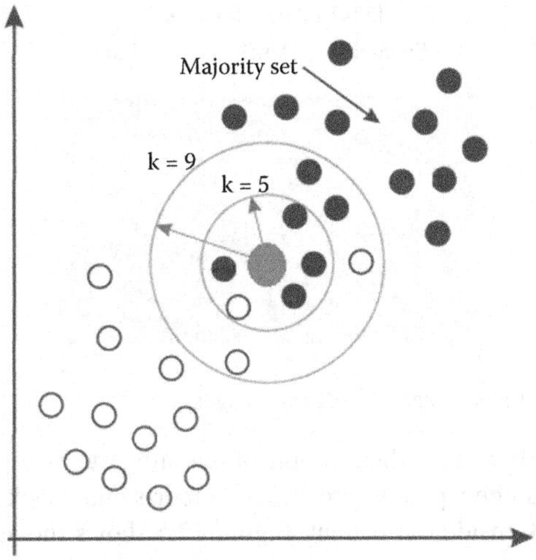

FIGURE 13.4 Example of KNN with two k values.

on the symptoms, the nature of the disease can be predicted (Tompson et al. 2014). For predicting the decision, a tree-like structure is formed, which helps to make it easy for the system—a DT (Iba and Langley 1992). It contains several nodes and these are divided into three types. One is the root node—from this node we start our iteration. Another is the leaf node where the iteration finishes. The remaining ones are called just nodes.

The DT is a valuable tool for object recognition. This object recognition can be applied for detecting an affected place (e.g., ulcer) and also for automatic analysis of X-ray or other scan reports (Haussler 1992; Freund and Mason 1999).

In the case of X-ray analysis, object recognition helps in finding if any object is stuck inside the body or even in the case any cracks are found. The image is taken and the gray scale values are plotted in a matrix for detection. The image is transferred through different filters and then the DT algorithm is applied (Clare and King 2001; Comite et al. 2010).

13.4 DEEP LEARNING: ASSOCIATION WITH THE NO-HUMAN-DOCTOR CONCEPT

Deep learning is a broader arm of ML technology, which emphasizes more detailed learning using different categories of algorithms. The main algorithms currently being researched are image restoration, speech recognition, visual processing, and more importantly natural language processing (Collobert et al. 2011). Natural language processing broadly contains emotion/sentiment analysis, information retrieval, spoken language understanding, and context analysis. This can be applied in the no-human-doctor approach for understanding the critical level of a patient's condition, to process and detect the initial symptoms of the patients. Collected data from this section can be used as the training data set for speech recognition, and initial symptoms can be verified through an image recognition system. As in the case of cancer patients, the initial scan reports can be verified through software to

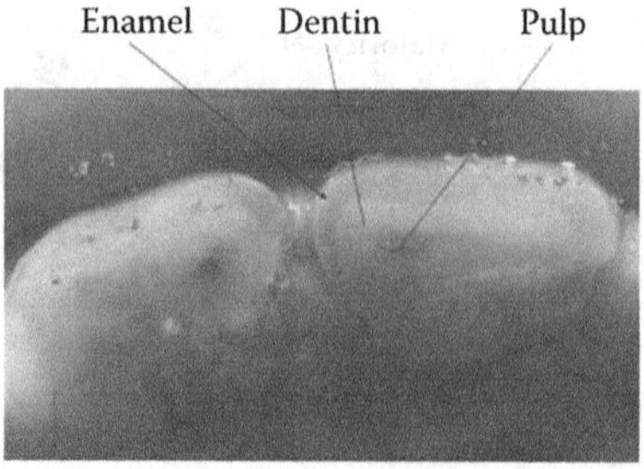

FIGURE 13.5 Object identification in X-ray images.

judge the condition. If we take the example of scanning the X-ray reports, then visual image processing can be a path to process the reports and check for cracks in bones or find objects stuck inside the patient. Figure 13.5 shows the application of object recognition in a scanned report of teeth.

For treatment of psychiatric patients, AI has shown great success in detecting the types of depression. Using learning algorithms on the observation data captured while the brain is at rest, researchers have categorized various subtypes of depression as mixtures of anxiety and lack of pleasure. On the other hand, Alzheimer's prevention should be started at a very early stage; the disease can be detected through data analysis of multicenter DTI Data (Dyrba et al. 2013). On the other hand, issue-regulated splicing code can be a good example of the use of deep learning in automation (Leung et al. 2014).

For long-term treatment, as in cases of schizophrenia, AI can be used to detect the initial symptoms and to predict the likelihood of a previously unseen patient having schizophrenia. Even the severity of symptoms, such as inattentiveness and bizarre behavior, could be predicted from brain scans through image recognition and object detection (Krizhevsky, Sutskever, and Hinton 2012).

13.5 METHODOLOGIES

For implementation of a simple regression technique, one of the simplest tools is GNU octave. GNU octave deals with data sets as matrix elements and using matrix multiplication commands.

On the other hand, GNU octave provides an interface to plot a 3D graph with a broader view, making it convenient to visualize the plotted data for finding the required prediction.

As seen in Figure 13.6, the typed command is

```
[x,y] = meshgrid(−2:.2:2);
g = x.* exp(−x − y);
surf(x, y, g)
print -deps graph.eps
```

For the equation of a random prediction of independent variable g, with dependent variables x and y as $g = x^*(e^{-x-y})$.

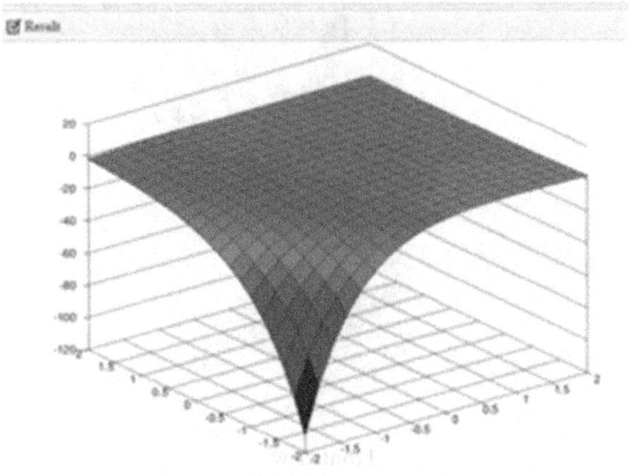

FIGURE 13.6 Graph of g = x*(exp(−x−y)) plotted through GNU octave.

Similarly, in case of a KNN algorithm, different categories of distances are considered depending on the problem. To name a few of the distance measurement methods, we include Euclidean, Mahalanobis, cosine, correlation, spearman, etc. Among these, Euclidian distance measures the graphical distance between the point and the data set, while Mahalanobis distance measures the number of standard deviations the point covers from the mean.

13.6 CASE STUDY

13.6.1 Da Vinci Robotic Machine

The da Vinci robotic machine, more popularly known as the da Vinci Surgical System, helps doctors to perform surgeries which are critical to perform with hands, especially those dealing with nerves (Gettman et al. 2002).

A machine is needed in order to handle surgeries dealing with sensitive organs and body parts. The da Vinci system typically has four hands to perform the job as a peripheral device. The hands are controlled by a doctor sitting at the back of the machine, where he/she has a 3D view of the surgery being performed. It is superior to working with a microscopic glass because the latter typically reduces the 3D nature of the picture and the sense of depth is less accurate. Along with a 3D view, it has a speaker and a microphone to communicate with the other team members who are handling the system. Beneath this, the machine contains two small input devices which are free to move 360°, and the sensors are attached to the thumb and index fingers of both hands. It typically works as if the surgery was being performed by those hands only. The doctor's legs control the movement of the camera. The system contains a large output screen where the preset functions are displayed continuously. Figures 13.7 and 13.8 illustrate the arrangement of the surgery system. This machine helps to perform the surgery with a minimum of discomfort to the patient and a minimal chance of error by the doctor.

From the case study of the da Vinci surgical system, it is clear a surgery can be performed by a peripheral device through a control unit. The next generation system

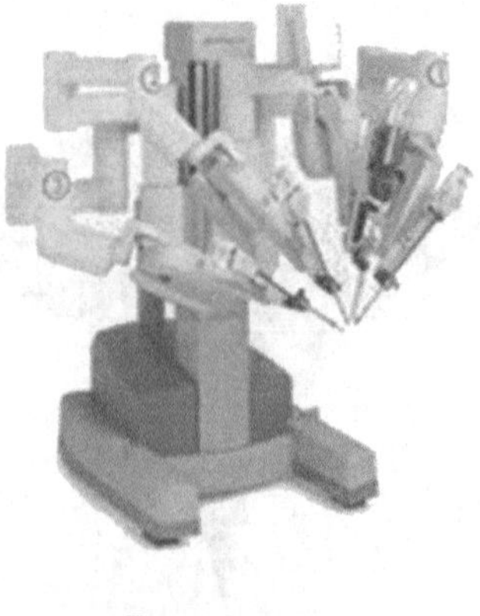

Front view

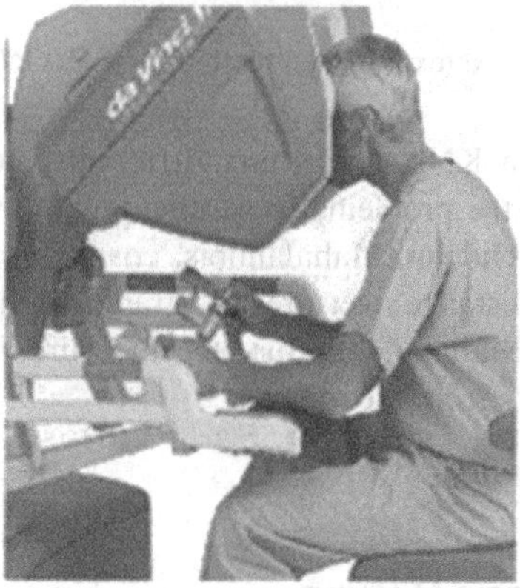

Doctor controlling movement

FIGURE 13.7 Front view of the da Vinci Robotic Machine.

can thus be designed to perform surgeries without the direct control of a human. The camera output can be directly sent for analysis by image object detection and the pathway must be preset to the data. The data collected through the graphical analysis of the image can be used to set the next path. The movement of the handle can be decided through the DT system.

On the other hand, for balloon surgery the movement of the balloon can be fixed using object detection on the image collected through the camera of the balloon head.

13.6.2 Other Case Studies

The smart tissue autonomous robot can be considered an important milestone for automated surgery systems. It uses a 3D-sensed imaging method to guide the robot for performing surgeries. The robotic machine performed open surgery, laparoscopic and robot-assisted surgery demonstrating a performance measure deemed good enough.

Orthopilot (Kowalski and Górecki 2004) is another example, although designed in the 1990s. It had many disadvantages including that of cost, but it showed the

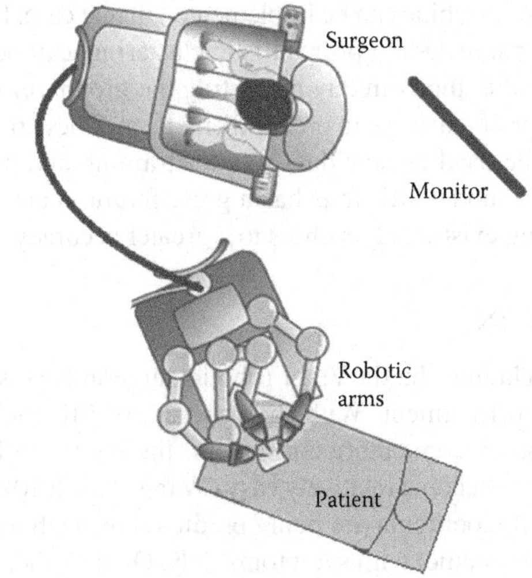

FIGURE 13.8 Arrangement of the da Vinci Robotic Machine.

path towards the automation of surgeries. This was especially the case for total knee arthoplasty, unicondylar knee arthoplasty, etc. where it performed well. Orthopilot was implemented with a good navigation system, which opened the way for later robotic designs (Figure 13.9).

The data collected from different sources concerning performance of surgeries using robotic machines were analyzed using ML algorithms for performance evaluation. The LikeSmart Tissue Autonomous Robot was estimated to perform with 70%–80% accuracy range. This analysis helps in improving surgical skills of the machines.

13.7 FUTURE RESEARCH

In the proposed system, future work may include more regression techniques where nonlinear regression follows more complex approaches, including n-dimensional cases.

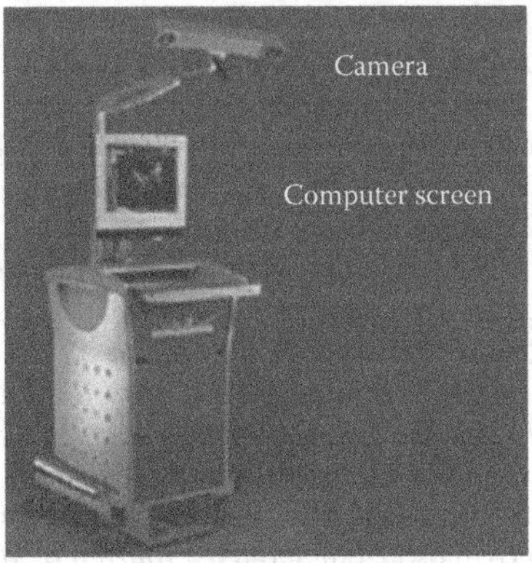

FIGURE 13.9 Front view of Orthopilot.

Even the support vector machine can be implemented in that case. Different categories of clustering algorithms, including probabilistic clustering, can be a convenient way for grouping tasks. Rather than directly predicting the group, fuzzy logic is one way to classify the behavior of a disease in probabilistic form. These probabilistic models, if implemented, can be used as new data sets for training and predicting trends or unknown patterns in a model. ML thus has a great future in automation of medical treatment by enhancing existing algorithms to a greater accuracy level.

13.8 CONCLUSION

Robotic machines including the da Vinci robotic surgeon have shown the path for automation of medical treatment. With the progress of ML tools and algorithms, the analysis of data has become more convenient; finding the unknown and hidden patterns in a data set has become relatively easy (Wang et al. 2010). Complex diseases like Alzheimer's or schizophrenia are being predicted through data analysis, which was initially difficult to predict with symptoms only. On the other hand, the progress of language processing and object recognition is boosting the automation field for symptom analysis, X-ray report checking, etc. The system can thus be implemented using more advanced algorithms of deep-learning technology towards the no-human-doctor concept, which will benefit society and science alike.

DISCLAIMER

The image of teeth for object recognition was taken from https://emedicine.medscape.com; accessed on Dec 31, 2017. This is an open source article and the authors and chief editor have nothing to disclose. Terms and Policy of the site had no objection statement relating to the reuse of the images.

REFERENCES

Abe, N. 1997. "Towards Realistic Theories of Learning." *New Generation Computing* 15 (1): 3–25. doi:10.1007/BF03037558.

Aha, D. W. 1997. "Lazy Learning." *Artificial Intelligence Review* 11: 7–10. doi:10.1007/978-94-017-2053-3.

André, F., S. Michiels, P. Dessen, V. Scott, V. Suciu, C. Uzan, V. Lazar et al. 2009. "Exonic Expression Profiling of Breast Cancer and Benign Lesions: A Retrospective Analysis." *The Lancet Oncology* 10 (4). Elsevier Ltd: 381–90. doi:10.1016/S1470-2045(09)70024-5.

Angluin, D. 1988. "Queries and Concept Learning." *Machine Learning* 2 (4): 319–42. doi:10.1023/A:1022821128753.

Case, J., and C. Smith. 1983. "Comparison of Identification Criteria for Machine Inductive Inference." *Theoretical Computer Science* 25 (2): 193–220. doi:10.1016/0304-3975(83)90061-0.

Clare, A., and R. D. King. 2001. "Knowledge Discovery in Multi-Label Phenotype Data," *In: Proceedings of the 5th European Conference on Principles of Data Mining and Knowledge Discovery,* pp. 42–53. doi:10.1007/3-540-44794-6_4.

Collobert, R., J. Weston, L. Bottou, M. Karlen, K. Kavukcuoglu, and P. Kuksa. 2011. "Natural Language Processing (Almost) from Scratch." *Journal of Machine Learning Research* 12: 2493–537. doi:10.1.1.231.4614.

Comite, F. D., R. Gilleron, M. Tommasi, F. D. Comite, R. Gilleron, M. Tommasi, and Learning Multi-label Alternating, and Decision Trees. 2010. "Learning Multi-Label Alternating Decision Trees from Texts and Data To Cite This Version: HAL Id: Inria-00536733."

Cover, T., and P. Hart. 1967. "Nearest Neighbor Pattern Classification." *IEEE Transactions on Information Theory* 13 (1): 21–27. doi:10.1109/TIT.1967.1053964.

Dyrba, M., M. Ewers, M. Wegrzyn, I. Kilimann, C. Plant, A. Oswald, T. Meindl et al. 2013. "Robust Automated Detection of Microstructural White Matter Degeneration in Alzheimer's Disease Using Machine Learning Classification of Multicenter DTI Data." 8 (5). doi:10.1371/journal.pone.0064925.

Freund, Y., and L. Mason. 1999. "The Alternating Decision Tree Learning Algorithm." *Proceeding of the Sixteenth International Conference on Machine Learning*, 10 str. doi:10.1007/s13398-014-0173-7.2.

Gettman, M. T., R. Neururer, G. Bartsch, and R. Peschel. 2002. "Anderson-Hynes Dismembered Pyeloplasty Performed Using the Da Vinci Robotic System." *Urology* 60 (3): 509–13. doi:10.1016/S0090-4295(02)01761-2.

Haussler, D. 1992. "Decision Theoretic Generalizations of the PAC Model for Neural Net and Other Learning Applications." *Information and Computation* 100 (1): 78–150. doi:10.1016/0890-5401(92)90010-D.

Iba, W., and P. Langley. 1992. "Induction of One-Level Decision Trees (Decision Stump)." *ML92: Proceedings of the Ninth International Conference on Machine Learning*, Aberdeen, Scotland, July 1–3, 1992.

Jacobs, R. A., M. I. Jordan, S. J. Nowlan, and G. E. Hinton. 1991. "Adaptive Mixtures of Local Experts." *Neural Computation* 3: 79–87.

Jordan, M. I., and R. A. Jacobs. 1994. "Hierarchical Mixtures of Experts and the EM Algorithm." *Neural Computation* 6 (2): 181–214. doi:10.1162/neco.1994.6.2.181.

Kowalski, M., and A. Górecki. 2004. "Total Knee Arthroplasty Using the OrthoPilot Computer-Assisted Surgical Navigation System." *Ortop Traumatol Rehabil* 6 (4): 456–60.

Krizhevsky, A., I. Sutskever, and G. E. Hinton. 2012. "ImageNet Classification with Deep Convolutional Neural Networks." *Advances In Neural Information Processing Systems*, 1–9. http://dx.doi.org/10.1016/j.protcy.2014.09.007.

Leung, M. K. K., H. Y. Xiong, L. J. Lee, and B. J. Frey. 2014. "Deep Learning of the Tissue-Regulated Splicing Code." *Bioinformatics* 30 (12): 121–29. doi:10.1093/bioinformatics/btu277.

Solomonoff, R. J. 1957. "An Inductive Inference Machine." *IRE Convention Record, Section on Information Theory*. http://scholar.google.com/scholar?hl=en&btnG=Search&q=intitle:An+Inductive+Inference+Machine#0.

Specht, D. F. 1991. "A General Regression Neural Network." *Neural Networks*, IEEE Transactions on 2 (6): 568–76. doi:10.1109/72.97934.

Tompson, J., A. Jain, Y. LeCun, and C. Bregler. 2014. "Joint Training of a Convolutional Network and a Graphical Model for Human Pose Estimation." *Advances in Neural Information Processing Systems*, 1799–807. http://arxiv.org/abs/1406.2984.

Wang, Y., Y. Fan, P. Bhatt, and C. Davatzikos. 2010. "High-Dimensional Pattern Regression Using Machine Learning: From Medical Images to Continuous Clinical Variables." *NeuroImage* 50 (4). Elsevier Inc.: 1519–35. doi:10.1016/j.neuroimage.2009.12.092.

Yusuff, H., N. Mohamad, U. K. Ngah, and A. S. Yahaya. 2012. "Breast Cancer Analysis Using Logistic Regression 1,2." *International Journal of Recent Research and Applied Studies* 10 (1): 14–22. http://www.arpapress.com/volumes/vol10issue1/ijrras_10_1_02.pdf.

Zhang, M.-l., and Z.-h. Zhou. 2007. "M L-KNN: A Lazy Learning Approach to Multi-Label Learning" 40: 2038–48. doi:10.1016/j.patcog.2006.12.019.

14 The IoT Revolution

Adrish Bhattacharya and Denim Datta

CONTENTS

14.1 WHAT IS IoT?

> The most profound technologies are those that disappear. They weave themselves into the fabric of everyday life until they are indistinguishable from it.
>
> **Mark Weiser**

With the emergence of smart home solutions and the variety of smart health wearables in the market today, IoT (Ashton 2009) has brushed aside all skepticism to become the most disruptive technologies of the twenty-first century.

It is predicted that by 2020 about 7.6 billion people will be connected via 50 billion devices worldwide. Business analysts predict IoT will lead to an economic growth of 4.6 trillion dollars globally in the public sector by 2020. No wonder IoT has become the buzzword of the decade. It is perhaps the only technology which has successfully leveraged on the advancements in big data, artificial intelligence, and embedded systems (Wigmore 2014) to create new avenues for democratizing technology. Even preliminary discussions about IoT reveal its *behind the scene nature* with the simplest of IoT implementations focusing on smart systems.

Looking back we find that one of the first IoT devices was a Coke dispensing machine at the Carnegie Mellon University, which could not only keep records of its inventory, but could also determine if the drinks recently loaded were chilled. Besides being a pretty simple idea, it highlights the concept of an intelligent solution using the Internet. In fact it was Mark Weiser's 1991 paper on ubiquitous computing, "The Computer of the twenty-first Century," as well as academic venues such as UbiComp and PerCom who produced the contemporary vision of IoT (Weiser 1991; Mattern and Floerkemeier, 2010).

Although a variety of definitions exist, the term *Internet of things* or IoT was recognized officially in 2005, when the *ITU Internet Reports* defined it as:

> A new dimension has been added to the world of information and communication technologies (ICTs): from anytime, any place connectivity for anyone, we will now have connectivity for anything.... Connections will multiply and create an entirely new dynamic network of networks—an Internet of Things.

From these introductory definitions, we have come a long way since Kevin Ashton (then at Proctor and Gamble) coined the term "Internet of Things." Apart from IoT, similar concepts have been called *Internet of Everything (IoE), Internet for things,* or *Smart Things,* etc.

When a number of physical devices fuse into a network capable of synchronizing several sensors (or actuators) and software components together for data exchange, the complete system is dubbed as the "Internet of things."

They can be considered as an "inextricable mixture of hardware, software, data, and service." A smart thing (Santucci, 2015) is one that connects to the Internet and with other devices to exchange data to create an integrated environment (Vermesan and Friess 2013).

In simple terms, IoT provides an end-to-end solution, where the device acquires information continually (through sensors), processes them to produce meaningful data, which can be fed to other devices (as input to process further), or can be used directly by users. Due to the wide applicability of such an end-to-end solution, IoT will become an integral part of technology in the coming decades. Some argue that if all objects and people were equipped with identifiable tags such as RFID, computers would be able to manage and inventory them (Magrassi and Berg 2002; Wood 2015).

Since IoT is posed to be *the foundation for future infrastructure* and improve our quality of life in multiple areas (some of which are currently being envisioned, such as smart cities and IIoT), it is crucial we form a strong understanding of the challenges encountered and the advantages gained through an IoT-based process.

14.1.1 Advantages

The above definitions of IoT and the "things" in IoT provide us with a clear understanding of the core idea. After that brief discussion on the nature of IoT, we can form a structured idea about the nature of problems IoT can solve. Bearing in mind the characteristics of IoT, it is easy to understand that the way IoT interacts with other devices produces some unique advantages over other existing technology. Some of them have been mentioned below.

14.1.1.1 Smart Technology

The idea of smart technology has been around for some time. Computers are smart, our phones have becomes smarter, even our watches and TVs have become smart. Therefore, the claim of a newer, smarter technology can't account for its market disruptive behavior on just its claims of being smart. Speaking specifically of IoT, we are aware that machine learning (ML) and artificial intelligence (AI) form a key component of the technology. The main difference between traditional smart devices and AI powered devices is the latter evolves with the user. AI has has the potential of eliminating humans from the scene altogether—even though it may take another century, but who knows!

AI forms a fundamental part of IoT and when we consider a technology developed for evolving to future trends, we can truly say that finally we have something smart enough to be called truly *smart*.

14.1.1.2 User Engagement

Passive user engagement remains one of the key problems of analytics today. It not only leans on indirect methods of obtaining required data, but also suffers from incorrect conclusions due to the lack of exact data. These drawbacks are characteristic of any passive data collection system. The only method of overcoming these concerns is replacing passive data collection by "*active data collection.*" It removes blind spots in data collection by increasing the *scope* of collection. IoT adds new areas of machine-to-machine interaction and human-to-machine interaction, thereby adding new opportunities for data assimilation. In today's world, humans are not the only users of technology—machines have become the primary user with 50 billion machines expected to be connected to the Internet by 2020. With the ever-increasing number of devices in our lives and the advent of cheap and reliable sensors, operating machines has become more engaging and IoT has added a plethora of opportunities to improve upon them.

14.1.1.3 Enhanced Data Collection

Consider the following data:

1. Internet users send 204 million e-mails, tweet 278,000 times, and upload over 200,000 photos (just on Facebook) per minute.
2. By 2021, the estimated number of RFID tags sold is expected to reach 209 billion, from a mere 12 million in 2011.
3. Experts opine the *big data* industry is posed to reach US$54.3 billion (by 2017) in revenues, from US$10.2 billion (2013)—a fivefold increase (Figure 14.1).

The vastness of the information is mind blogging. It shows not only are we generating massive amounts of data every second, but ignoring this data can prove counterproductive. The kinds of conclusions that can be drawn from it have the potential of changing the very fabric of human–machine interaction. Also, we realize that the data is collected through active engagement with the users. It supports our previous claim of active user engagement for improving the scope of data analytics. The kind of data we are collecting each second has been available only due the efficient deployment of IoT solutions at specific locations. IoT solutions have made it simpler for machines to mine data from newer sources. In fact, IoT can be said to create sources of data collection. The IoT has thus created a more detailed and more exact picture of our world.

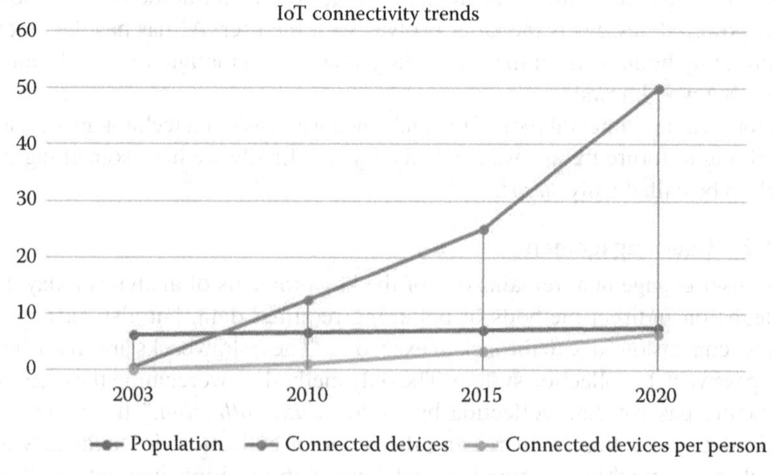

FIGURE 14.1 Growth of M2M connections since 2003.

14.1.2 Limitations

Like any other technology, IoT is no magic wand. Although it ventures into areas deemed unfathomable earlier, it brings along multifaceted solutions, and its very own set of limitations (Hendricks, 2015). If these are not understood completely, it can render the IoT revolution useless. Addressing these concerns, a study was conducted by Noura Aleisa and Karen Renaud at the University of Glasgow entitled "The Internet of things' potential for major privacy invasion is a concern." Also, Louis Basenese, an investment director at the *Wall Street Daily*, expressed that "Despite high-profile and alarming hacks, device manufacturers remain undeterred, focusing on profitability over security. Consumers need to have ultimate control over collected data, including the option to delete it if they choose… Without privacy assurances, wide-scale consumer adoption simply won't happen"(Basenese, 2012). According to the 2016 Accenture Digital Consumer Survey, which was taken by over 28,000 consumers across 28 countries, security "has moved from being a nagging problem to a top barrier as consumers are now choosing to abandon IoT devices and services over security concerns."

Therefore, we must open ourselves to these possible challenges and fully understand them. Among the many limitations in an IoT-based system, the most important are those involving security and privacy. Let us have a look at them one by one (Figure 14.2).

14.1.2.1 Security

Our basic understanding of IoT tells us it creates an ecosystem of continually connected devices communicating over numerous networks, which leaves the network exposed to various kinds of attacks: more devices lead to more opportunities for compromise. IoT systems' nature of *amorphous computing* is a major security concern for many. As it does not always allow the patch fixes to reach every subsystem in the entire net of devices (Franceschi-Bicchierai, 2015; Kingsley-Hughes, 2015). When security patches do not reach older, outdated systems, some estimate it renders 87% of active

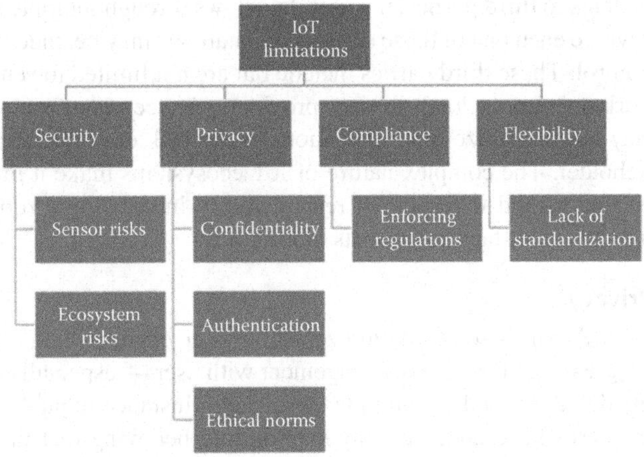

FIGURE 14.2 Major limitations of IoT.

devices vulnerable (Tung, 2015). Additionally, it allows for postsale manipulations by corporations (Walsh, 2016).

The lack of strong security protocols fuel increased risks of other types of compromise, namely, sensor-related risks and ecosystem risks.

Sensor-related risk: IoT devices are susceptible to various types of sensor-related risks such as:

1. *Counterfeit products*: Duplicate products ingrained with malicious programs. At times only certain components are modified to allow unauthorized access to the system or subsystems. Identity spoofing is a commonly used technique for gaining such access.
2. *Information extraction*: Malicious code designed to extract sensitive information from any sensor or connected device(s).

Let us consider a situation where the ABC Corporation wants to use sensors to track its equipment. Now, for ABC Corporation, it is convenient and economical to adapt existing sensors to the IoT system designed for its procedures. But the sensors themselves have intelligence; they can independently process information and take their own course of action. Since the sensors and systems are designed independently, the generic sensors may not employ the strict security protocols the corporation requires. If malicious actors were able to break into such a device, they may cause it to malfunction or fail. These actors might even make attempts to steal sensitive data, manipulate other sensors within the network, or adversely impact plant operations.

The above example showcases how the lack of proper planning regarding the deployment of IoT solutions—for increased efficiency in operation—can ultimately prove counterproductive. The other kind of security risk is the ecosystem risk a compromised system poses.

14.1.2.1.1 Ecosystem Risk

It is the weakest link that determines the strength of a chain. Therefore, as corporations extend IoT systems to third parties, the user data flows through multiple devices and/or databases where each one of those devices or databases may be under independent third party control. These third parties include but are not limited to vendors, supply chains, network operators, cloud services providers, device manufacturers, and end users who may not recognize the implications of the lack of secure strategies used by each stakeholder. The complex nature of IoT ecosystems make it imperative for businesses to understand their shared responsibility in creating a robust security infrastructure capable of tackling threats of all kinds.

14.1.2.2 Privacy

Although the IoT claims to revolutionize active user engagement, it has already proved to be a great tool for passive engagement with users—especially in situations where a single device is used by multiple users, as for instance in public places.

Privacy concerns have nudged many experts into believing that infrastructures based on big data, such as IoT, are inherently incompatible with the concept of privacy. The recent example of hoardings and billboards having cameras hidden in them—to track the demographics of all the commuters who showed a considerable interest in the specific advertisement, has been cited to support the claims of "invasion of public space." Considering the smart home systems, we can easily realize how the security and privacy of most households are susceptible to compromise by simple analysis of the smart home system traffic patterns.

The Internet of Things Council's comparison between Bentham's Panopticon and the increasing use of IoT-based digital monitoring systems is not unfounded. Editorials published in *WIRED* have expressed major concerns over this issue, one stating "*You aren't just going to lose your privacy, you're going to have to watch the very concept of privacy be rewritten under your nose*" (Webb, 2015). The American Civil Liberties Union (ACLU) has stated "*Chances are big data and the Internet of things will make it harder for us to control our own lives, as we grow increasingly transparent to powerful corporations and government institutions that are becoming more opaque to us*" (Howard, 2015).

Hence, we observe that despite the immense potential offered by IoT for empowering citizens and making governments transparent, there are mammoth-sized privacy threats such as the potential for *social control through political manipulation*. In fact, many fear the advent of IoT can lead to an Orwellian dystopia.

14.1.2.3 Flexibility

One of the major concerns regarding viability of an IoT-enabled system lies in its inability to integrate seamlessly into another system. Due to the lack of interoperability standards (Kovach, 2013; Raggett, 2016; Piedad, 2015), it is a common belief that the potential of IoT may never be realized. IoT systems suffer heavily from platform fragmentation and the lack of technical standards (Ardiri, 2014; Bauer et al., 2015; Wallace, 2016; Wieland, 2016).

The incompatible and often conflicting protocols employed make it difficult to extend an IoT system (Franceschi-Bicchierai, 2015; Kingsley-Hughes, 2015). The key aspect of this problem can be ascribed to platform fragmentation, which is a situation where the sheer variety in IoT systems makes the task of developing consistent applications working with inconsistent platforms difficult (Brown, 2016). With proprietary systems there is always the risk of the technology fading out. These issues pertaining to the flexibility of IoT-based systems is a major hurdle in achieving mass adoption.

14.1.2.4 Compliance

IoT aims to democratize technology but the vastness of government regulations make compliance for any technology a tedious job. Adding to it the range of security and privacy concerns, we end up with a potentially groundbreaking system tangled in legal knots. Confusing terminology in the field of IoT is another hassle, which makes compliance difficult. However, regarding IIoT, the Industrial Internet Consortium's Vocabulary Task Group has created a "common and reusable vocabulary of terms," which ensures "consistent terminology" in all publications by the Industrial Internet Consortium.

Owing to the increasingly tougher regulation in the fields of privacy, future technologies like IoT, big data, and AI must prove themselves against stringent moral and ethical questions, as well as legal norms.

14.2 COMPONENTS

The concept of IoT originated from the amalgamation of various technologies such as AI (and ML), real-time analytics, wireless communication, and embedded systems. The advances in all of these fields have contributed to the current boom in IoT.

14.2.1 AI and ML

The essence of IoT is in making objects "smart" and AI plays a pivotal role in achieving this, as AI and ML improve both accuracy and speed in big data analytics. Machine learning can be defined as "*a field that deals with the construction and study of systems that can learn from data, rather than follow only explicitly programmed instructions.*" It can help corporations and governments alike in taking millions of collected data points and reducing them to meaningful information. It aims to analyze data collected in order to detect patterns, which can be further used to be learned from and for making better decisions. The realization of the IoT vision can be successful only through leveraging its ability of gaining the hidden insights obtained from the enormous and growing sea of data available. Owing to the sheer volume of data available today, it's time to let the machines take the front seat in pointing out the areas where opportunities truly lie.

14.2.2 Connectivity

IoT creates small networks between subsystems for the exchange of information (Figure 14.3). Therefore, it becomes imperative to have knowledge about the various means and protocols of connectivity. There are various short, medium, and

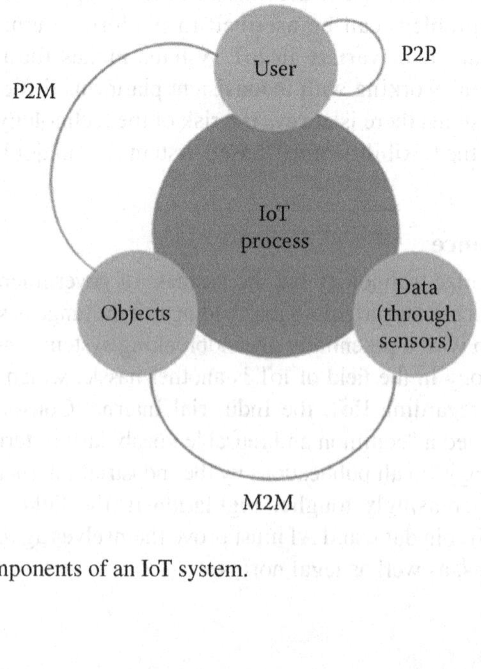

FIGURE 14.3 Components of an IoT system.

long-range connectivity solutions based on the needs of the users. A comprehensive list of commonly used wireless systems is provided below for quick reference.

- *Short-range systems*: Radio-frequency identification or RFID tags, near-field communication or NFC, Wi-Fi, Wi-Fi Direct and Li-Fi (Light-Fidelity), ZigBee, Bluetooth, QR codes, barcodes, and Z-Wave.
- *Medium-range systems*: LTE-Advanced, HaLow.
- *Long-range systems*: Long-range Wi-Fi connectivity, low-power wide-area networking (LPWAN), and very small aperture terminal (VSAT).

14.2.3 Sensors and Modules

Sensors form a key component of IoT systems, since in most designs they are the primary source of data collection. Without sensors it would be nearly impossible to collect the variety of data forming the basis of IoT systems. The following is a list of some of the most extensively used sensors in the IoT industry.

14.2.3.1 Optical Sensors

Measures different quantities simultaneously. The technology behind this sensor allows it to monitor electromagnetic energy, which includes electricity and light among others.

14.2.3.2 Chemical Sensor

Chemical sensors transform information regarding chemical properties such as the concentration of ions, chemical activity, etc. into useful signals.

Major uses of chemical sensors include CO detectors, glucose detectors, pregnancy test kits, and others.

14.2.3.3 Image Sensors

The image or imaging sensor determines and transmits information constituting any image. An image sensor gives an output of small packets of current.

14.2.3.4 Accelerometer Sensors

An accelerometer measures linear acceleration of any movement.

14.2.3.5 Gyroscope Sensors

The gyroscope, aka gyro, tracks information pertaining to rotation that an accelerometer can't determine. The output of a gyro is generally in the form of an angular rotational velocity.

14.2.3.6 Water Quality Sensor

These sensors are used to measure temperature, dissolved oxygen, conductivity, pH, and so on of the water sample. Major applications of water quality sensors include among others determining soil quality, monitoring water quality, water treatment, and modeling an ecosystem.

Other commonly used sensors include:

- Temperature sensors
- Proximity sensor
- Pressure sensor
- Gas sensor
- Smoke sensor
- IR sensors
- Motion detection sensors
- Humidity sensors

14.3 MICROCONTROLLERS FOR IoT

IoT prototyping boards play today an essential role in simplifying the prototyping process of IoT-based systems. These boards are *microcontrollers* and *microprocessors* with chipsets to handle wireless connections. These development boards with Cloud IoT platforms enable the quick deployment of IoT processes. The commonly used development boards include the following.

14.3.1 Arduino Uno

Arduino Uno is among the preferred development boards. It is an open-source development board based on ATmega328P.

It has both analog and digital pins. Another aspect of this board is it can be expanded using shields, that is, another board can be plugged into Arduino Uno to add new functionalities such as GSM, WIFI, etc.
Specifications:

Operating voltage: 5 V
8bit
16 MHz

14.3.2 Arduino MKR1000

Arduino MKR1000 is one of the latest boards based on the Atmel ATSAMW25.

Just like Arduino UNO, this board supports analog and digital pins. Additionally, it can be powered using an external Li-Po battery.
Specifications:

Operating voltage: 3.3 V 32bit
48 MHz WIFI

14.3.3 Raspberry Pi 2 Model B

Raspberry can be considered as a small computer owing to it's features. It is powered by Linux.

Specifications:

- Quad-core BCM2837, 900 MHz
- 1Gb RAM
- Four USB 2.0 ports
- 40 GPIO pins
- HDMI and RCA video output

14.3.4 Raspberry Pi 3

The newer Raspberry Pi 3 Model B has been built on its predecessors' features fused with a faster processor on board to enhance clock speed. It as the following specifications:

- Micro USB power source up to 2.5A
- 1.2 GHz 64-bit quad-core ARMv8 CPU
- 40 GPIO pins
- Four USB 2.0 ports
- 802.11n WLAN
- Bluetooth 4.1
- Bluetooth Low Energy

14.3.5 Intel Edison

Intel Edison is a very powerful IoT development board and is available in several variants.

Specifications:

- Intel Atom
- 500 MHz dual-core x86 1GB RAM
- Wi-Fi 802.11n
- Bluetooth v4.0

14.3.6 Pinguino 45K50

Built by Pinguino, it gives students and designers from the art community a powerful microprocessor. It has the following features:

- 8-bit 12 MIPS processor, 48 MHz
- 17 digital pins for both input and output
- 5 shared analog inputs
- 2 PWM outputs
- UART for serial data transfer

14.3.7 Teensy 2.0

The sole reason for developing Teensy 2.0 was to offer a great alternative to Arduino. It not only offers Arduino-like features but can also run the Arduino IDE allowing

Teensy 2.0 to access Arduino's extensive library. It flaunts 25 input and output pins and is powered by a 16 MHz AVR processor.

14.3.8 Nanode

The Nanode is a great alternative to the Arduino Uno, Mega, and Yun. The features of the Nanode include:

- ATMega328P, 16 MHz
- Red and green LEDs for program diagnostics
- Mini USB power connector
- ENC28J60 Ethernet controller, 25 MHz

14.3.9 Mediatek Linkit One

The Linkit One is based on the smallest SOC and comes with compatible Arduino pinout features. Linkit One can be used for rapid prototyping of connected IoT systems owing to its rich connectivity features. Its software development kit (SDK) comes with libraries for connecting the board to AWS and PubNub.

- MT2502A based chipset, 260 MHz
- GPS, GSM, GPRS, WiFi, and Bluetooth
- Supports Arduino shields

14.4 APPLICATIONS AND EVERYDAY ELECTRONICS

IoT has taken the center stage in making the 4th industrial revolution a reality. It can be exploited by almost every industry to deliver more efficient solutions, increase their scale of operation, and improve profitability (Westerlund et al., 2014).

Newer areas for deployment of IoT services are cropping up routinely, which undoubtedly means the *sky is the limit* for IoT applications.

The sheer variety of IoT applications creates a need to classify them. Although many different aspects can be considered for the classification of IoT applications, the most popular scheme is based on the end user, which is

- Consumer application—for the general consumer. For example: home automation solutions, wearable technology, etc.
- Enterprise (business) application—for providing business solutions to a myriad of enterprises. It is estimated that by 2019, enterprise IoT (EIoT) will account for 9.1 billion devices. For example: manufacture control, automatic shipment tracking, etc.
- Infrastructure applications application—for maintenance and improving infrastructure. For example: smart traffic control, automatic toll collection, etc.
- Health care application—for improving efficiency, accuracy, and precision of current medical instruments through responsive and integrated systems. For example: remote health monitoring system, wearable heart rate monitor, etc.

Listed below are some fields transformed by the ongoing IoT revolution. The IoT is finding newer applications every day with improvements in technology and innovative deployments. The given list is only meant as a reference to enable readers to grasp the scope of the practicality of IoT-based systems.

14.4.1 Home Automation

Smart home or home automation have been the most googled words last year in reference to IoT applications. An ever-growing number of consumers are excited about them, partly because almost every major electronics company, including Google and Amazon, have invested heavily in this area. Another strong reason for the general public to advocate for domotics (another name for home automation) is because of the plethora of opportunities it creates. The capabilities of even the most basic of smart home systems feel otherworldly. A recent study by Intel has revealed that an astonishing 71% of consumers expect to see "*at least one smart home device in every home as early as 2025.*" The buzz around Amazon Alexa and Google Home indicates we are at the cusp of a home automation revolution.

IoT-based home automation systems monitor and control the electrical and mechanical aspects of any place of residence including private and public areas. These aspects include lighting, air conditioning, and security solutions (Kang et al., 2017). The most appealing feature of these systems is their ease of use. However, the omnipresent security threats looming at the edges and the high initial establishment costs often outweigh the benefits.

Some commonly implemented features of home automation include:

- Providing assistance to the elderly by utilizing assistive technology to accommodate specific requirements such as voice control for users with limited sight and mobility, monitoring systems for medical emergencies, etc.
- Connecting the user with a home appliance regardless of geographical distance.
- Providing the user with the latest news while he/she is busy performing other chores (Figure 14.4).

14.4.2 Media and Advertisement

IoT has enabled a more ambitious targeted media era. The last few years have witnessed rapid development in the fields of conversation tracking, click rate, and so on. The data collected through these studies of consumer behavior has opened boundless opportunities for gaining meaningful insights into the salient traits of customers. The media industry is primarily concerned with maximizing the effect of advertisements and their content through known customer habits in line with their now predictable behaviors. IoT-based systems have allowed corporations to fine tune their advertisements according to individual needs, thereby improving customer experience. These detailed reports also enable business houses to make a better analysis of customer preferences and improve their own business strategies in the long run.

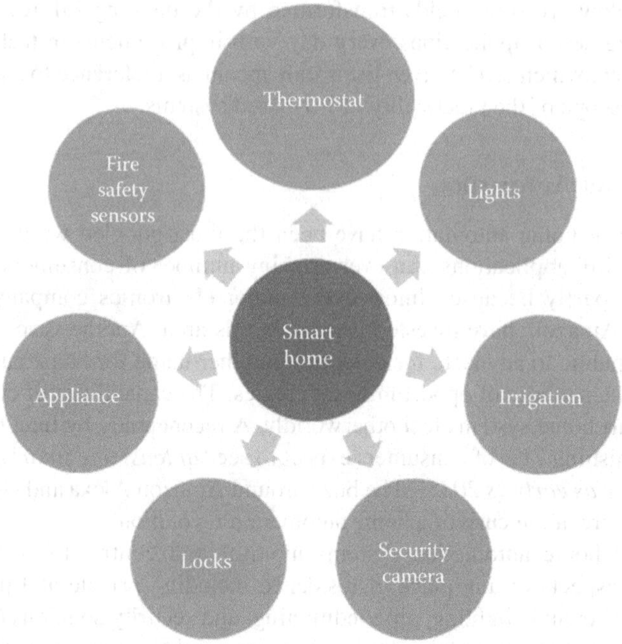

FIGURE 14.4 Common features of home automation.

Other than corporations, government agencies, TV soaps, and so on also stand to gain from this ever-growing range of insights into the masses. For instance, governments can now predict with far larger accuracy the behavior of their citizens during specific situation; TV soaps can alter the story of the show according to the perceptions of their viewers and so on.

14.4.3 Manufacturing and Industry

The manufacturing industry packs a series of processes within its scope of operation including maintaining logistical records of assets, supporting supply chain networks, dynamic time-response to manufacturing demands, and optimization of all of these tasks in real time. Control and management of these operations is within the purview of IoT systems used in the manufacturing industry. Digital control systems able to automate a myriad of process controls and successfully optimize manufacturing, industrial safety, and so forth through real-time analysis of critical parameters make IoT an indespensible aspect of future manufacturing industries.

The range of applications for IoT is so large that these IoT solutions and the systems delivering them have been dubbed the Industrial Internet of Things (IIoT). Basically, IIoT is the use of IoT technologies in the manufacturing sector. The IIoT holds immense potential for quality control, green practices, and profound improvements in supply chain dynamics. By some estimates, the potential for growth through IIoT is worth $12 trillion globally by 2030.

The IIoT combines a variety of recent trends including AI, big data, machine to machine (M2M) communication, and automation technologies, which have already existed in industrial settings for a long time (Perera et al., 2015). The replacement of humans by machines not only increases the efficiency of the process by limiting the scope of error, but also attempts to quantify the ineffective points in the system. The quantified information pertaining to the drawbacks of the systems in use allows companies to improve upon them easily and save valuable time and resources. In such a situation, IIoT is posed to become a key driver in growth and competitiveness of industries around the globe.

Although newer areas of application of IIoT are being discovered routinely, there is a common consensus about the main aspects of benefit from IIoT, namely:

1. Improving efficiency—IIoT has enabled enterprises to capture more data about industrial processes that provide critical information, which can help increase the efficiency of their business practices.
2. Innovative business models—due to the added capabilities of machines through improved connectivity and faster processing, new and innovative products have emerged. These new business models open up previously unexplored sources of revenue for corporations. As the industry has previously worked with the counterparts of IIoT, such as ML and AI, it can easily monetize its experience to create new streams of IoT services for their customers, requiring little external support.
3. Increasing safety—constant surveillance and making decisions on the fly based on the collected data is one of the key attributes of IoT. IIoT can thus help corporations to monitor manufacturing and related procedures, as well as produce alerts when sensing irregularities.

In today's economy where industry drives the governments' and countries' prosperities, it is essential to introduce smart machines to empower our markets to gather and analyze data in real time in order to make informed decisions.

14.4.4 Agriculture

Agriculture is a labor-intensive process relying largely on human intervention for successful production. The reason for this dependence is because the best farming techniques are decided based upon multiple environmental variables including temperature, humidity, humus content in soil, nutrient requirements, pests, etc. Now, with the use of various sensors linked to an IoT-based system, most of these parameters can be very accurately monitored for making informed decisions. Once decisions are made, various automation techniques can then be used to improve yields and reduce waste.

14.4.5 Energy Management

Integration of sensing and power-generating stations can potentially optimize power distribution and management through centralized control hubs. Experts predict IoT would enable the integration of all kinds of energy consuming devices including lights, fans, and A/Cs among others to enable remote control of these devices. Apart from this, the data pertaining to the energy usage patterns can prove extremely valuable for smart grids. This information can be exploited to improve efficiency and reliability of power grids while ensuring sustainability in the production of energy.

14.4.6 Environmental Monitoring

Environmental monitoring is one the toughest and exhaustive fields of work, and it has benefited greatly from the advent of IoT. Although the lack of standardized wireless protocols and interoperability standards have stunted development in this area, nevertheless, IoT systems have been deployed on a major scale. They are being used mainly for environmental protection by monitoring air and water quality, atmospheric and soil conditions, etc. (Li et al., 2011). IoT has also assisted in monitoring the movements of wildlife and their habitats. Active research is currently underway to leverage IoT for developing early warning systems against natural calamities such as tsunamis.

14.4.7 Medical and Health Care

The most fundamental contribution of IoT in the area of medical care has been to improve the accuracy and efficiency of current instruments by providing real-time analytics and more precise information. It has drastically enhanced the collection of critical patient information, which had been impossible to assemble earlier.

IoT devices have made remote health monitoring and emergency notification systems a reality for blood pressure and heart rate monitors among others. Some additional innovative uses of IoT in the health care industry are:

- Smart beds detecting occupancy and movement in order to adjust themselves for ensuring appropriate pressure and support to the patient.
- Specialized sensors installed within living areas for the surveillance and evaluation of the general well being of senior citizens and infants.
- Consumer devices to encourage healthy living, including wearable heart monitors (Istepanian et al., 2011) to help in managing health vitals and recurring medical needs (Swan, 2012).
- Battery-powered prosthetic arms, powered by myoelectricity (converting muscle group sensations into motor control).

14.4.8 Transportation

Simplifying transportation has been one of the most extensive and successful applications of IoT. The integration of communication systems, information processing units, and control mechanisms across various transportation systems has enabled a dynamic interaction between transportation components such as the vehicle, driver, and IoT-based system for smart traffic and parking control, safety assistance, automated toll collection, etc.

14.5 CONCLUSION

With the rapid pace of developments in IoT and the ever-increasing interest in this field, IoT as undoubtedly become the star child of the century. However, despite sincere dedication challenges such as standardization, security, and so on remain constant. Apart from these, perhaps the biggest challenge today is the overhype regarding IoT. We must understand the abilities and limitations of any technology, and focus instead on viable solutions, which are not a "source of confusion for the end user." As Mike Farley quotes, "Instead of convincing consumers that they need complex systems to serve needs they don't have, we should fix real problems people struggle with every day."

Identifying the problem is the first step towards any solution (Yarmoluk, 2012), especially with IoT. We must first dedicate ourselves to understanding this technology before wielding it in order to create an efficient and sustainable ecosystem reaping benefits for all of humankind.

REFERENCES

Ardiri, A. 8 July 2014. "Will fragmentation of standards only hinder the true potential of the IoT industry?" evothings.com.

Ashton, K. 22 June 2009. "That 'Internet of Things' Thing."

Basenese, L. 2012. "The Best Play on the Internet of Things Trend." *Wall Street Daily.*

Bauer, H., M. Patel, and J. Veira October 2015. "*Internet of Things: Opportunities and Challenges for Semiconductor Companies.*" McKinsey & Co.

Brown, E. 13 September 2016. "Who Needs the Internet of Things?" Linux.com.

Franceschi-Bicchierai, L. 2 August 2015. "Goodbye, Android." *Motherboard. Vice.*

Hendricks, D. 2015. "*The Trouble with the Internet of Things.*" London Datastore. Greater London Authority.

Howard, P. N. 2015. Pax Technica: "How the internet of things May Set Us Free, Or Lock Us."

Istepanian, R. S. H., S. Hu, N. Y. Philip, and A. Sungoor. 2011. "The potential of Internet of m-health Things 'm-IoT' for non-invasive glucose level sensing." In *Engineering in Medicine and Biology Society, EMBC, 2011 Annual International Conference of the IEEE*, pp. 5264–5266. IEEE.

Kang, W. M., S. Y. Moon, and J. H. Park. 2017. "An enhanced security framework for home appliances in smart home." *Human-centric Computing and Information Sciences* 7, no. 1: 6.

Kingsley-Hughes, A. 2015. "The toxic hellstew survival guide." ZDnet.

Kovach, S. 30 July 2013. "Android Fragmentation Report." Business Insider.

Li, S., H. Wang, T. Xu, and G. Zhou. 2011. "Application Study on Internet of Things in Environment Protection Field." *Lecture Notes in Electrical Engineering* Volume 133: 99–106. ISBN 978-3-642-25991-3.

Magrassi, P., and T. Berg. 12 August 2002. "A World of Smart Objects." Gartner research report R-17-2243.

Mattern, F., and C. Floerkemeier. 2010. "From the Internet of Computers to the Internet of Things." *From Active Data Management to Event-based Systems and More*: 242–259.

Perera, C., C. H. Liu, and S. Jayawardena. 2015. "The emerging internet of things marketplace from an industrial perspective: A survey." *IEEE Transactions on Emerging Topics in Computing* 3, no. 4: 585–598.

Piedad, F. N. 2015. "Will Android fragmentation spoil its IoT appeal?" *TechBeacon.*

Raggett, D. 27 April 2016. "Countering Fragmentation with the Web of Things: Interoperability across IoT Platforms." W3C.

Santucci, G. 2015. "The internet of things: Between the revolution of the internet and the metamorphosis of objects." *Vision and Challenges for Realising the Internet of Things*: 11–24.

Swan, M. 2012. "Sensor mania! the internet of things, wearable computing, objective metrics, and the quantified self 2.0." *Journal of Sensor and Actuator Networks* 1, no. 3: 217–253.

Tung, L. 13 October 2015. "Android security a 'market for lemons' that leaves 87 percent vulnerable." zdnet.com.

Vermesan, O., and P. Friess, eds. 2013 "Internet of Things: Converging Technologies for Smart Environments and Integrated Ecosystems." River Publishers.

Wallace, M. 19 February 2016. "Fragmentation is the enemy of the Internet of Things." Qualcomm.com.

Walsh, K. 5 April 2016. "Nest Reminds Customers That Ownership Isn't What It Used to Be." Electronic Frontier Foundation.

Webb, G. 5 February 2015. "Say Goodbye to Privacy." WIRED.

Weiser, M. 1991. "The computer for the 21st century." *Scientific American* 265, no. 3: 94–104.

Westerlund, M., S. Leminen, and M. Rajahonka 2014. "Designing business models for the internet of things." *Technology Innovation Management Review* 4, no. 7: 5.

Wieland, K. 25 February 2016. "IoT experts fret over fragmentation." *Mobile World.*

Wigmore, I. June 2014. "Internet of Things (IoT)." TechTarget.

Wood, A. 31 March 2015. "The internet of things is revolutionizing our lives, but standards are a must." The Guardian.

Yarmoluk, D. 2012. "5 Barriers to IIoT Adoption & How to Overcome Them." ATEK Access Technologies.

15 Healthcare IoT (H-IoT) *Applications and Ethical Concerns*

Srijita Banerjee, Adrish Bhattacharya, and Shampa Sen

CONTENTS

15.1 INTRODUCTION

Internet connects various government-private-public and academic-business networks of local to global scope using optical networking technologies, via a broad array of electronic and wireless communication gadgets. It is mainly considered to be the storehouse of various information interlinked hypertext documents, applications of the World Wide Web (WWW), electronic mail, telephony, file sharing and many more. The evolution of the web and Internet has been vast, and can be categorized into five stages:

- *Advance Research Project Agency Network* (APRANET) was primarily designed for researchers and academicians.
- *The Gold Rush for domain names* had the objective of sharing product or service-related information.

- *The boom and bust of the dot com bubble*: An access was obtained for purchasing goods and services through Internet where companies like Amazon and eBay came into the picture.
- *The social and experience web*: The stage which customized Internet into a platform for social interaction (Facebook, Twitter, Instagram, WhatsApp).
- *IoT*: This concept deals with a hyperconnected society whereby an object is usually connected to various interconnected set-ups.

IoT, which usually incorporates traditional fields like embedded systems, control systems and automation, and wireless sensors to facilitate device-to-device communications through Internet was first coined in 1999 by Kevin Ashton (Kulkarni and Sathe 2014). The IoT can be chronicled as connecting everyday objects like smartphones, Internet, television sensors, and actuators to the Internet. Here, the devices are intelligently linked together enabling new forms of communication between objects and people, and between objects themselves. IoT can be applied to various areas, a prominent example being the health care sector. IoT has the potential to enhance varieties of medical applications such as remote monitoring, chronic diseases, organ donation, hygiene monitoring, real-time location services, and many more (Riazul Islam et al. 2015). With the advent of IoT, new data streams can be gathered, recorded, and analyzed faster and more accurately by enabling devices to collect and share information directly with the cloud and with each other, which impacts several application domains. These domains are classified based on the type of network availability, heterogeneity, scale, coverage, repeatability, and user involvement (Gluhak et al. 2011).

15.2 ROLE IN HEALTH CARE

Applying IoT to the health care sector has a number of advantages, such as reduced cost, device downtime, enhanced quality of services, and enriched experience for the user. This IoT-driven health care area is expected to help in the early diagnosis of various diseases and various medical emergencies (Riazul Islam et al. 2015). At the same time, medical servers are seen to have a cardinal importance in conceiving health records and delivering on demand health services to authorized stakeholders (Riazul Islam et al. 2015). Such systems not only allow faster delivery of health care solutions but also ensure accurate treatment for the patient.

The IoT-based health care network (H-IoT) supports access to the IoT backbone, which in turn facilitates the transmission and reception of medical data, thus enabling the use of health care-tailored communications. The H-IoT topology mainly deals with the arrangement of the various elements as described in (Figure 15.1).

H-IoT is developed in such a way it can contribute to profiling users as "health impaired" or "at risk" when needed (Rigby 2007).

15.2.1 Real-Time Location Services

Real-time location systems (RTLS) are tracking devices are used for identifying the location of an asset or person in real time, or near real time. Small ID badges or tags

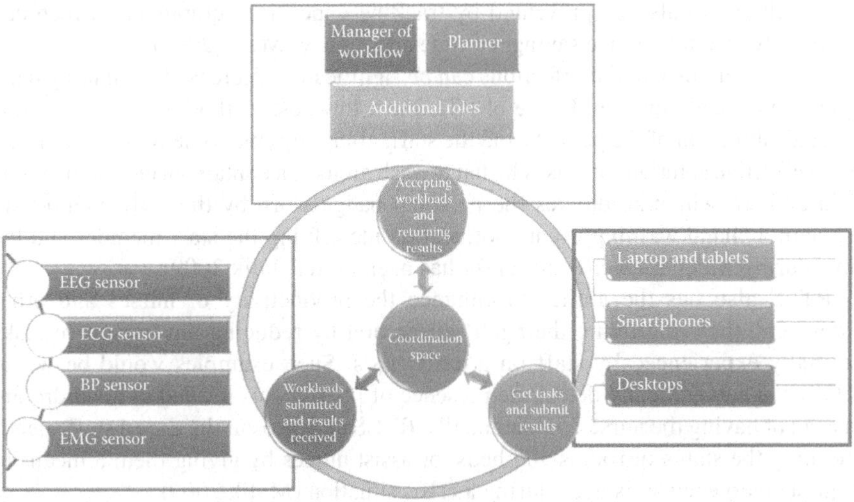

FIGURE 15.1 Working of the H-IoT.

are attached to the object or the person of interest, which in turn remain connected through a wireless signal receiver to determine the location of the tagged entity. (Malik 2009) (Figure 15.2).

RTLS can be used to find and alert health care staff quickly in large facilities when any staff member or patient in the hospital is found in danger or during medical emergencies. For instance, patients suffering from Alzheimer's and dementia usually have a tendency to wander away, and this can be monitored using RTLS. It pinpoints the location of the patients and alerts the staff members. Newer technologies have also come up in which the doors are automatically locked when patiens come close. Some outdoor tracking devices related to this purpose also come up make use of a GPS (Boulos et al. 2011; Chang et al. 2005).

Devices have been developed in which RTLS analyses the overall well being of an individual by recording his/her mobility inside the house. This method includes calculating the daily distance walked by the individual based on the distance between each sensor he/she passes during a lapse of time (Graham 2012). Patient flow nowadays is also being tracked for throughput management, which is simplifying problems such as extended waiting time, overcrowding, boarding in outpatient clinics and emergency departments/rooms (ED/ER), postanesthesia care units (PACUs), and bumped and late surgeries (Drazen and Rhoads 2011; Stahl et al. 2011). Accidental

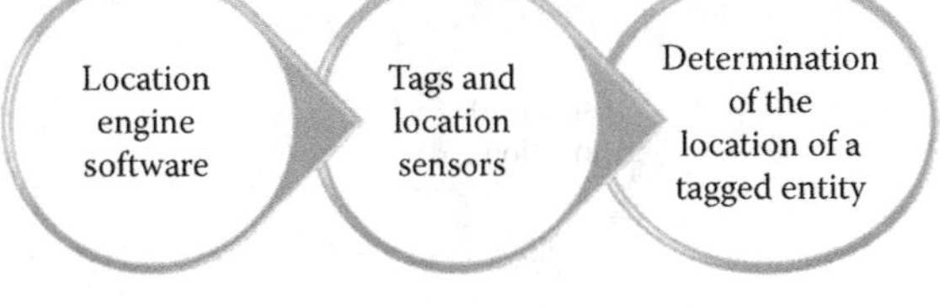

FIGURE 15.2 Real-time location device.

loss or theft can also be prevented by tracking expensive equipments, which can potentially lead to massive savings of time and money (Malik 2009).

Chances of nosocomial infections can be significantly decreased by making hand hygiene protocols smarter. Usage of RTLS in a low-cost method to record the time of hand-sanitation of the patients and the staff, after they enter or leave the room, can be an effective solution for this. The hand-wash areas can contain an electromagnetic field emitter, which stimulates the personal badge worn by the staff member, to transmit a "hand-washing event" notification identifying the staff member and the time during which a specific dispenser has been used (Malik 2009).

RTLS also has the ability to enhance the productivity of nurses and other caregivers, thus improving their job satisfaction by reducing any repetitive tasks normally experienced by staff on a daily basis. Such examples would be where RTLS can automatically detect the presence of instruments needed in a wardroom, instead of having the nurse do so manually. RTLS can also cut the time a staff spends checking the status of rooms and beds, or assist nurses by giving them a means to request emergency assistance during a crisis situation (Malik 2009).

Such systems can and also improve a patient's family/visitors' satisfaction by increasing their awareness on the patient's location and condition. Other than hospitals, RTLS was found to be beneficial in other applications as well. Fingerprint techniques have been developed (Lin et al. 2014), which deal with the matching of a fingerprint to some location dependent characteristic signal. This technique was mainly developed to provide safety services for the workers who work in the tunnels of high arch dam sites. It involved the determination of the position of a person with the help of a fingerprint map, which was developed using a K-nearest neighbor method. The fingerprint of the workers remains in the database while the position of the worker is determined (Lin et al. 2014) (Figure 15.3).

On conducting various surveys in US hospitals in 2008, 15% of administrators indicated their hospitals already had RTLS in place, and another 43% expressed their intent to purchase a system within the next two years (HIMSS [Health Care

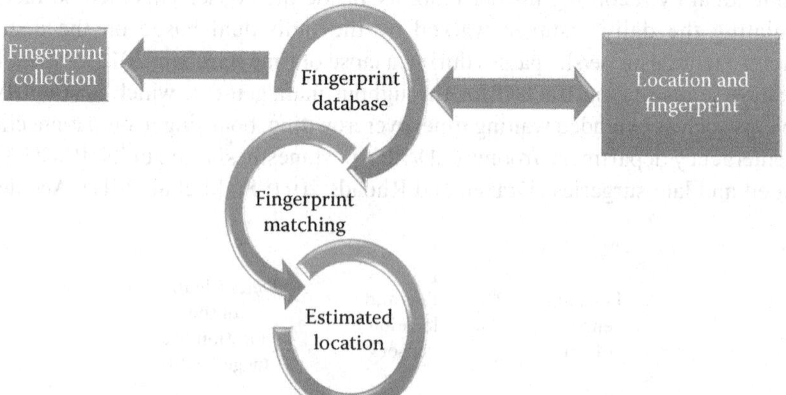

FIGURE 15.3 Fingerprint matching system.

Information and Management Systems Society], 19th Annual—2008 HIMSS Leadership Survey 2008).

Among the numerous advantages of using RTLS in hospitals, certain drawbacks still prevail, such as the financial concerns and the lack of understanding of how RTLS can be integrated within the existing information technological investments (electronic medical record systems, for example) (Bouet and Pujolle 2010; Chung et al. 2010). Moreover, in a study conducted by Fisher and Monahan (2012), it was stated that:

- There was substandard functionality of most RTLS in hospitals.
- Several serious obstacles are faced by the hospitals during the effective deployment of the systems due to material and organizational constraints.

Various hospitals tend to consist of old buildings often arranged in complexes and unintuitive configurations. Considering these factors, the implementation of RTLS must go beyond the simple deployment of technology to be effective and useful (Dash 2009; Lorenzi et al. 2008). Before selecting a tracking technology, plans need to be designed detailing what goals the RTLS will meet, who will manage and operate the system, and how buy-in will be obtained from personnel(s) (Murphy 2006). RTLS should be chosen in such a way that specific goals are matched in the best way considering their own facilities, given their unique material and organizational constraints (Fisher and Monahan 2012). If we prioritize from the *hospital and health care context*, RTLS can perhaps be influenced to better serve the health care industry.

15.2.2 Hygiene Monitoring

On a statistical basis, there are 1.7 million cases of public health care-acquired illnesses each year in the United States. These cases result in 100,000 deaths each year, which ranks as the fourth-leading cause of U.S. deaths and cost hospitals approximately $30 billion annually. Studies show that poor hand hygiene practices are the most likely cause of the spread of various bacterial diseases (Centers for Disease Control and Prevention 2015). Hand hygiene, when maintained properly, is the single biggest defense against the spread of numerous diseases, especially in public health care institutions such as hospitals and nursing homes (World Health Organization (WHO) 2017). For all these reasons it is important for hospitals, business authorities, and public health agencies to consistently monitor hand hygiene among workers to control the spread of various bacterial diseases.

The methods involved in IoT hygiene monitoring range from simple hygiene dispensers and sprays to more sophisticated systems using advanced technologies such as *wireless communication* and *machine learning* in electronic monitoring of hand hygiene.

One of the simplest IoT-based systems included a battery-powered recording device incorporating a force-sensitive resistor and a microcontroller. The microcontroller in this unit has the function of recording the reduction in the amount of wall-mounted soaps and alcohol dispensers in hospitals. This device calculated the number of times the soap dispensers were used. However, this system was found to be disadvantageous,

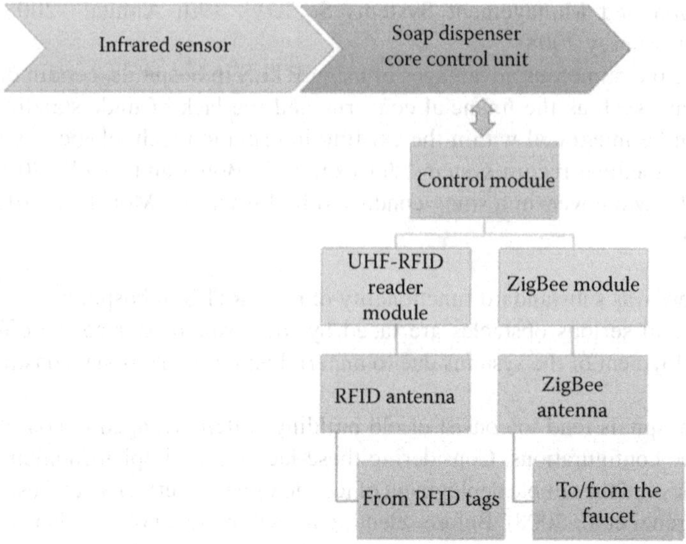

FIGURE 15.4 Components of hygiene dispenser.

as it was not able to assess hand-cleaning episodes delivered from other nonfixed dispensers (i.e., personal gel dispensers), thus giving only a partial picture of the total hygiene (Kinsella et al. 2007).

Another system presented in 2013 had a function for monitoring the hand-hygiene compliance of caregivers, or individuals in hospitals, by using a passive RFID wearable wristband and wall-mounted dispensers. The sanitizer dispensers are equipped with RFID readers and ZigBee wireless transceivers. On activation of the dispenser the action gets registered by a centralized software system, which is then used for monitoring the hand-hygiene compliance (Meydanci et al. 2013) (Figure 15.4).

"*Smart hospital room*" is another technique using the RFID approach to monitor hand hygiene in hospital rooms. RFID tags are used in this system to record the movements of caregivers as they enter the patient rooms. It works by activation or deactivation based on the caregivers' location, and thus behaves as a hand hygiene reminder. Sometimes, it also logs the time used for the hand hygiene activity (Luen and Ellen 2009).

In the study presented by Rhodes (2014), the hand hygiene compliance was developed in such a way that duration and techniques of hand hygiene activities were recorded on a particular hand hygiene station using wrist-worn accelerometers (using machine learning techniques).

Nowadays, in addition to monitoring, systems have been developed giving importance to enforcement of the hand hygiene compliance. Biovigil Hygiene Technologies has developed a system involving changes in color of a badge (used by the staff members) based on whether they have washed their hands or not. The badge electronically measures the alcohol and turns bright green if alcohol is detected (and red if not) (Rhodes 2014) (Figure 15.5).

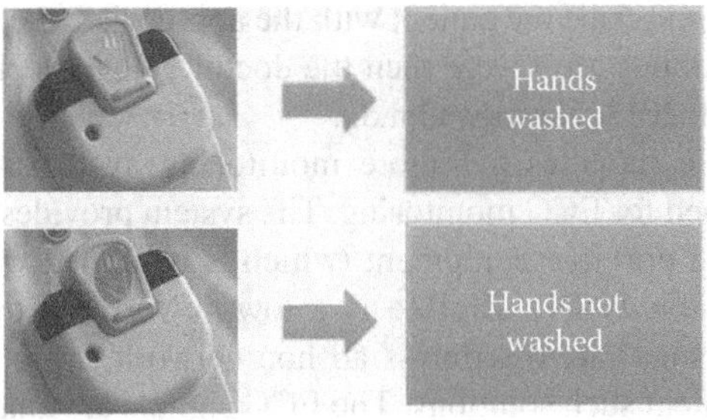

FIGURE 15.5 Hygiene Monitoring system.

A different product from Pure Hold Ltd has an automated sanitizer gel dispenser installed within a standard pull door handle with a specific design where the gel is emitted through a valve upon a pull grip. After operating a patient or opening any operating room, the user clean their hands with gel thus preventing staff, other patients, and visitors from transmitting and acquiring infections through touching door handles (Emery 2015).

A system was constructed by Bal and Abrishambaf (2017) consisting of a node in a wireless network sensor (WNS). The node represents the hand wash station and the data concerning the use of hand washing or hand-sanitizing gets transferred to the data center, which serves as a cloud to monitor the usage statistics of the hand washing stations and generate reports in real time. Health care facilities usually have an access to these reports. The basic form of the soap dispenser is automated and wall mounted. The system uses an infrared sensor to detect the presence of a hand and dispenses a controlled amount of hand soap or gel hand sanitizer. The soap dispenser module contains an RFID reader and an antenna encapsulated in the housing. These readers scan personalized passive RFID name tags, which are carried by the user associated with the hospital (Bal and Abrishambaf 2017).

15.2.3 Remote Monitoring

Remote monitoring in IoT mainly deals with the patients who fail to visit doctors regularly. In this case, the collected data is taken from the patients' home and sent to the cloud for doctors to review. For example, we can consider the presence of an ECG monitoring application downloaded on a mobile device collecting the biosignal data using a microcontroller, and then uploading it to the cloud for keeping a record of the unstructured data. Benefits are a reduction in waiting time for triage at the hospital, minimizing visits, reduction in the cost of the personnel and administrative operations, etc. Another word for this can be *telemedicine*, which can be explained as usage of telecommunication technologies as means to provide medical diagnosis, treatment, education, and health care for patients (Istepanian 1999).

A device able to check the health parameters of a patient at any place or time (Pont 2002) is actually a reality with the smart application of IoT solutions to health care. The only thing the person needs to do is place his finger on the device, which will sense the necessary parameters and notify the doctor through the wireless network;

the doctor then can prescribe the patient with the appropriate medicines. Sometimes, if the medical conditions are severe then the doctor gets to decide how to treat the patient. (Shelar et al. 2013; Gayakwad n.d.).

Another example includes a remote monitoring system developed in 2008 specifically developed for ECG monitoring. The system provides remote monitoring for patients wearing portable equipment (which is equipped with ZigBee module connectivity based on WSN). ZigBee is a low-power, low data rate, and close proximity (i.e., personal area) wireless ad hoc network, which can be efficiently employed to implement such solutions. The ECG signals are analyzed and recorded in an online database. On detection of any serious aberrations in heartbeat, an alarm is sent to the authorized medical staff through a telecommunication network (Khanja et al. 2008).

A.M. Cheriyan developed an H-IoT design embedded within a system capable of tracking biosignals from a person in real time. This system facilitated a dependable decision-making process in real time by incorporating the use of electroencephalogram (EEG) and oxygen saturation (SpO_2) signals, which provided alerts for possible changes in brain activity (Cheriyan et al. 2009).

Another group of scientists developed an original ECG measurement system based on a web service-oriented architecture capable of monitoring the heart conditions of patients. The device was developed in such a way to provide personalized diagnosis by using the personal data and clinical history of the patients being monitored (De Capua et al. 2010).

Although these models are efficient, poor bandwidth have limited the number of channels and signals obtained, which were degraded due to noise.

Yuan and Herbert (2011) presented a new setup, CARA, a health care architecture essentially focusing on web-based health monitoring systems. It is constituted of four parts—*the wireless monitoring device (WMD), the home monitoring system, the remote clinical monitoring, and the health care reasoning system*. This whole system included the real-time analysis of vital signs of patients resulting in automated responses. It was developed in such a way as to avoid inherent problems of data errors in wearable sensors (Yuan and Herbert 2011).

15.2.4 Organ Donation and Transplantation

Organ transplantation, or organ donation, is a very complex process requiring automation in different services. IoT-related technologies like RFID, RTLS, Real Time communication technologies, etc. are nowadays being heavily implemented to improve patient safety and enhance clinical operations. In the case of organ transplantation, the demand for organs far outweighs supply (Talmale and Shrawankar 2017) (Figure 15.6).

Previously, communications related to human organ transplantation processes relied on phone or fax. This system suffered from several drawbacks, for instance when difficulties arose in keeping track of patients in need of vital organs, where it led to a gap between supply of organs and their demand. Other than that, the increase in time to allocate organs due communication gaps between different stockholders of the organ transplantation processes were prevalent (Asberg et al. 2011).

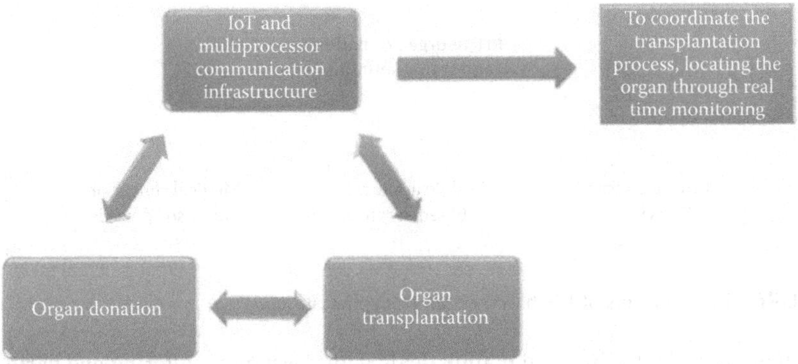

FIGURE 15.6 Role of IoT in organ transplantation.

Therefore, it's clear that the human organ transplantation process comprises many services, which can be implemented using IoT. Some of these services are mentioned below:

- *Blood, organ and tissue cell banking*: An online database system can be maintained for blood, organ, tissue, and cell banks, which is accessible from any part of the world. It should have the capability to manage a national transplant waiting list, and match the donor details to those of the recipient (Bastoni et al. 2010). It should hold all kinds of data related to every transplantation event. To ensure the effective utilization of the limited supply of organs, this should be a common platform for all stakeholders (Hahm et al. 2015). Constant monitoring of the organ match process should be used to ensure proper working of the system. Nowadays, even sensors have been infused in such systems able to monitor the different parameters simultaneously (Talmale and Shrawankar 2017).
- *Real-Time Information to Physician*: This is also an important aspect of organ donation, where the physician is constantly kept updated about the status of the recipient and the organ. This is usually maintained by real-time monitoring of ambulatory patients (Talmale and Shrawankar 2017).
- *Drug and Medical devices monitoring*: It entails the monitoring of medical devices, biological and prescription drugs, and the patients' food. It involves identifying the risks of various drugs and medication for patients and notifying them, which enhances patient surveillance and produces a reduction in errors caused by medical instruments. It also deals with the increase in efficiency of product management with product validation and tracking. In this regard, H-IoT can help in managing medical sponges used during organ transplants, surgical processes, etc. and alerts can be given if no sponges are issued or matches are not found (Talmale and Shrawankar 2017).

An IoT transplantation system consists of a number of subsystems like pretransplant, inpatient, clinicians, organ provider organization, donor care, recipient care etc. The

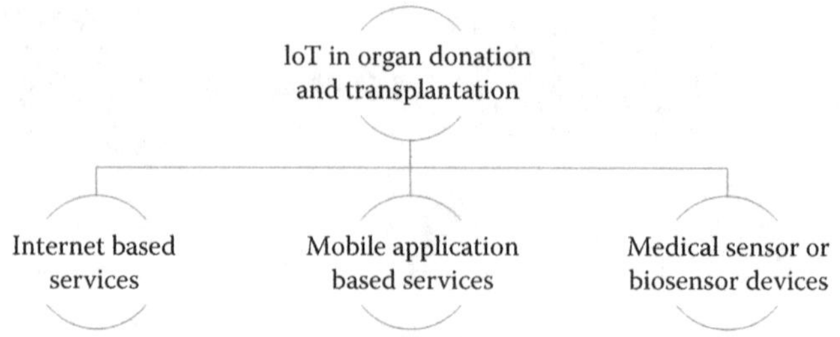

FIGURE 15.7 Aspects of IoT-based organ transplantation.

inpatient transplant subsystem mainly deals with constant monitoring and caring of inpatients to prevent the organ rejection or failure.

It also includes tasks performed to manage dosages of medicines required in organ transplantation. Moving on, the posttransplantation subsystem consists of care and monitoring units including systems for managing the long-term medication after transplantation (Rosenkranz et al. 2015). The posttransplantation subsystem also accommodates tasks where real-time patient data is evaluated. These systems notify the doctor whenever changes in the lab settings are required and help to remotely monitor medications such as antiviral drugs, etc (Adjih et al. 2015) (Figure 15.7).

15.3 ETHICAL OBSTACLES

Despite the numerous groundbreaking applications of IoT in the health care sector, it raises a host of ethical concerns pertaining to the sensitive nature of any health-related data generated and analyzed. Furthermore, it becomes impossible for the users to retain control over the data produced and analyzed owing to the scale, scope, and complexity of such systems (Edgar 2005; Pellegrino 1993).

15.3.1 Privacy

The majority of H-IoT applications empower users to control collection and storage of personal data thereby limiting the opportunities for unwanted profiteering from sensitive information (Wel and Royakkers 2004).

This sensitivity of the health care data has been prominently recognized in international data protection and privacy laws (Terry 2014). Due to continued pressure from human rights' groups, protection of the user's privacy has become a critical concern for the design of H-IoT (Shilton 2012). For all these reasons, techniques such as anonymization and aggregation are being included, which are thought to reduce the risk of reidentification, and thus guarantee the privacy of a user (Joly et al. 2012). Policies allow access to identifiable data, only when the purpose(s) is (are) acceptable (Beaudin et al. 2006; Chakraborty et al. 2011; Giannotti and Saygin 2010; Massacci et al. 2009). Other than privacy, another essential feature of H-IoT infrastructure is security. Security in health care systems is mainly concerned with preventing hacks or breaches, thus also helping to protect privacy (Peppet 2014). Although reasonable exceptions are allowed by the system manufacturer to the user or to any third party, with regard to storage and distribution of information, the key factor required for such provisions is the trust between users and the manufacturers (Liyanage et al. 2014). After taking all these factors into account, several privacy principles have been developed.

- *Proactive not reactive, preventative not remedial*: This is aimed at anticipation and prevention of security hacks and privacy infractions. These principles discourage spyware infiltrations from materializing by placing robust security infrastructure.
- *Privacy as the default setting*: Building IT systems and business practices such that privacy protection becomes a "by default" operation and not an additional feature by embedding privacy into the design, such systems integrate privacy safeguards into the design of IT systems, and make them a core component of business practices.
- *Full functionality*: Accommodating all the legitimate interests and the objectives without compromising efficiency.
- *End-to-end security—full lifestyle protection*: This principle talks about embedding the privacy protection into the system prior to the first element of information being collected in order to avoid leaks of even the most seemingly trivial information.
- *Visibility and* transparency*—keeping it open*: Assuring all the stakeholders through independent verification that the relevant business practices or technologies are "operating according to the stated promises and objectives."
- *Respect for user privacy—keeping it user-centric*: This ensures that the operators are keeping the interests of the individuals at the center by offering measures such as strong security measures (Cavoukian 2011).

15.3.2 Sensitivity

When it comes to the generation of data and their transfer, a lot of ethical concerns arise. The users ought to be included in the H-IoT architecture design wherever possible (Nissenbaum 2004). H-IoT designing should be done in such a manner benefiting the interests of specific groups of people, while simultaneously allowing individual control of features and privacy policies. For avoiding marginalization or exclusion of certain groups using H-IoT, enhancement in the accessibility scheme of the device and protocols can be put in place. These devices are generally so developed that an equity of access is maintained for the sake of commercial interests, and the interests of affluent users are ignored to prevent marginalization (Mittelstadt and Floridi 2016).

H-IoT can raise a host of ethical problems arising from the inherent risks of Internet-enabled devices, the sensitivity of health-related data and their impact on the delivery of health care. Being scientifically and technologically reliable are two of the most important criteria, or rather challenges, of H-IoT. Along with this, users also expect it to be ethically responsible and trustworthy. Other than this, the device needs to be made in such a way that it strikes a balance between preserving the autonomy of the individuals on their life (Mittelstadt 2017). Ultimately, access to information should be under the control of the patient or their appointed guardian to prevent misuse by any third party.

15.3.3 Ensuring Data Safety

Data safety can be brought out in several different ways.

- *Maintaining proper trust and confidentiality between the users and providers*: To contribute to the trustworthiness of the users and providers, proper design considering the key interest of the individuals, groups, and society is the priority for developing the product. Privacy, confidentiality, safety, and efficacy are closely linked to trust (Kosta et al. 2010). A user can always expect distress and anguish if their providers are untrustworthy, even if the system enhances privacy, confidentiality, safety, and effectiveness (Bagüés et al. 2010). Confidentiality is one of the critical elements for establishing a trustworthy relationship between the H-IoT users and providers. To handle the data responsibly and to ensure it's safety, users commonly place their trust in the devices and service providers (Little and Briggs 2009). Establishing trust is so important that a lack of trust has been linked to the reluctance to adopt H-IoT (Hildebrandt and Koops 2010).
- *Transparency and accountability of the data processing protocols*: H-IoT has got the capability to collect large amounts of data and process it in a complex and opaque manner. If the data is being processed and exchanged without properly letting the data subject (I.e., the user) know, a critical explanation is required about the handling. The social impact of decisions made based on the data is increasing, as well as the need for transparency when the rights of the data subjects are being kept constant (Hildebrandt and Koops 2010). When IoT is brought into picture, usually the guidelines proposed by TechUK, is followed. One school of thought argues that obtaining data using IoT without the involvement of users is unethical. Sometimes, in order to earn proper rights to derive value from IoT data, data processors are required to demonstrate the transparency, integrity, and security in processing to the general public. This in turn empowers the users to believe the impact of H-IoT on their medical care and quality of life. This impact can be assessed by checking the degree to which the providers adhere to the preceding eight principles in designing and deploying H-IoT devices and data protocols. Transparency is sometimes required to explain how the data produced might influence the medical care of the user (Wachter et al. 2017). These explanations can be directed to both the user and the care team, who may be better informed with technical knowledge to understand and modify the user's treatment (Giannotti and Saygin 2010).

15.3.4 Monitoring Consent

Manufacturers should always consider the potential value of the data generated even if there is uncertainty about the future value of the data in academic research and commercial analytics. Sometimes participants are de-identified (Ioannidis 2013), or data generated are repurposed for medical applications (Taddeo 2016), owing to lack of consent between H-IoT manufacturers and users. For this reason, device manufacturers are advised to design user agreements to represent fairly the uncertain value of the data generated and the amount of aggregation and linkages to any third party for research and commercial purposes. Consent is usually considered when there is the participation for a single study and which does not include unrelated investigation resulting in sharing or aggregation or even repurposing of data within the wider research community (Choudhury et al. 2014). Ethical problems occur when H-IoT leads to psychological disappearance, for instance, when used in personal spaces or residential care surroundings (Ebersold and Glass 2016). Some H-IoTs' also involve the use of sensors inside the private sphere. For these reasons, consent is required in case the user forgets that monitoring is occurring inside their homes. Nowadays, occasional renewal of consents has been developed, which ensures monitoring has not been forgotten. This kind of system is mainly required for the cognitive impaired users who are unable to grant consent. Concerns are extended to the guests, which usually suggests the possibility of careless monitoring (Kenner 2008).

15.4 FUTURE PROSPECTS

By actuating the potential of existing services of IoT, various technological solutions have been designed to enhance health care provisions in multiple ways. Till now abundant research and development efforts have been carried out in IoT-driven health care appositeness and applicability (Riazul Islam et al. 2015). From all these we can safely conclude that IoT will change our society and bring coherent personalized health care and monitoring over fast, reliable networks. This in all likelihood indicates that the world is approaching towards the end of the present divide between digital, virtual, and physical worlds. Standard web service technology is one of the most widely adapted technologies for the Internet right now (Kulkarni and Sathe 2014). Similar kinds of efforts are going on for the development of networks with wireless identifiable embedded health care systems. Various plans have been made for the development of WSNs and ubiquitous networks, where the sensors will be connected to and controlled by embedded systems, and where services encapsulate the functionality and provide unified access to the system. These components produce, consume, and process information in different health care environments such as hospitals, households, and nursing homes. This is also the case in the work and everyday lives of people (Sundmaeker et al. 2010), thus making round the clock, affordable health care a reality for the masses.

REFERENCES

Adjih, C., Baccelli, E., Fleury, E., Harter, G., Mitton, N., Noel, T., Pissard-Gibollet, R. et al. 2015. "FIT IoT-LAB: A Large Scale Open Experimental IoT Testbed." *The 2nd IEEE World Forum on Internet of Things (WF-IoT).*

Asberg, M., Nolte, T., and Kato, S. 2011. "A Loadable Task Execution Recorder for Hierarchical Scheduling in Linux." *17th IEEE International Conference on Embedded and Real-Time Computing Systems and Applications* 380–387.

Bagüés, S. A., Zeidler, A., Klein, C., Valdivielso, F., and Matias R. 2010. "Enabling personal privacy for pervasive computing environments." *Journal of Universal Computer Science* 16, 341–371.

Bal, M., and Abrishambaf, R. 2017. *"A System for Monitoring Hand Hygiene Compliance based-on Internet-of-Things."* DOI: 10.1109/ICIT.2017.7915560.

Bastoni, A., Brandenburg, B. B., and Anderson, J. H. 2010. "An empirical comparison of global, partitioned, and clustered multiprocessor edf schedulers." *31st IEEE Real-Time Systems Symposium, RTSS '10.* USA: IEEE. 14–24.

Beaudin, J., Intille, S., and Morris, M. 2006. "To track or not to track: User reactions to concepts in longitudinal health monitoring." *Journal of Medical Internet Research* 8, e29.

Bouet, M., and Pujolle, G. 2010. "RFID in ehealth systems: Applications challenges, and perspectives." *Annals of Telecommunications* 65, 497–500.

Boulos, M. N. K., Anastasiou, A., Bekiaris, E., and Panou, M. 2011. "Geo-enabled technologies for independent living: Examples from four European projects." *Technol Disabil* 23, 7–17.

Cavoukian, A. 2011. *The Privacy by Design: 7 Foundational Principles.*

Centers for Disease Control and Prevention. 2015. *Healthcare associated Infections (HAIs): Data and Statistics.* https://www.cdc.gov/hai/surveillance/index.html.

Chakraborty, S., Choi, H., and Srivastava, M. B. 2011. "Demystifying privacy in sensory data: A QoI based approach." *9th IEEE International Conference on Pervasive Computing and Communications Workshops, PERCOM Workshops.* Seattle. 38–43.

Chang, I. C., Cheng-Yaw, L., Yu-Chuan, L., Chia-Cheng, C., Chien-Tsai, L., and Chieh-Feng, C. et al. 2005. "Pervasive observation medicine: The application of RFID to improve patient safety in observation unit of hospital emergency department." *Connecting Medical Informatics and Bio-informatics Proceedings of MIE* 2005, 311–320.

Cheriyan, A. M. et al. 2009. "Pervasive embedded real time monitoring of EEG & SpO2." *3rd International Conference on PervasiveHealth 2009.* Pervasive Computing Technologies for Healthcare. 1–4.

Choudhury, S., Fishman, J. R., McGowan, M. L., and Juengst, E. T. 2014. "Big data, open science and the brain: Lessons learned from genomics." *Frontiers in Human Neuroscience* 8, 239.

Chung, Y.-C., Shih, P.-C., Li, K.-C., Yang, C.-T., Hsu, C.-H., and Hsu, F.-R. et al. 2010. "Medicare-grid: New trends on the development of E-health system based on grid technology." *E-health*, Springer, Boston. 335, 148–150.

Dash, A. 2009. "Lost + found: Making the right choice in equipment location systems." *Health Facilities Management* 22, 19–20.

De Capua, C., Meduri, A., and Morello, R. 2010. "A smart ECG measurement system based on web-service-oriented architecture for telemedicine applications." *Instrumentation and Measurement, IEEE Transactions* 59, 2530–2538.

Drazen, E., and Rhoads, J. 2011. *"Using tracking tools to improve patient flow in hospitals."* California Health Care Foundation.

Ebersold, K., and Glass, R. 2016. "The internet of things: A cause for ethical concern." *Issues in Information Systems.*

Edgar, A. 2005. "The expert patient: Illness as practice." *Medicine, Health Care, and Philosophy* 8, 165–171.

Emery, G. 2015. *Enforcing Rather than Encouraging Hand Hygiene in Healthcare.*

Fisher, J. A., and Monahan, T. 2012. "Evaluation of real-time location systems in their hospital context." *International Journal of Medical Informatics* 81, 705–712.

Gayakwad, R. A. n.d. In *Op-Amps and Linear Integrated Circuits*, 4th Edition, 342, 417, 455. New York: Prentice-Hall.

Giannotti, F., and Saygin, Y. 2010. "Privacy and security in ubiquitous knowledge discovery." *Lecture Notes in Computer Science* 6202, 75–89. *Berlin.*

Gluhak, A., Krco, S., Nati, M., Pfisterer, D., Mitton, N., and Razafindralambo, T. 2011. "A survey on facilities for experimental Internet of Things Research." *IEEE Communications Magazine* 58–67.

Graham, E. 2012. "My Amego relationship centred telecare: Using location-based technology to build relationships and manage risk in care homes." *Online Proceedings of 'Location-Based Technologies in Health & Care Services'.*

Hahm, O., Baccelli, E., Petersen, H., and Tsiftes, N. 2015. "Operating system for low-end device in internet of things: A servey." *IEEE Internet of Things Journal* 3, 720–734.

Hildebrandt, M., and Koops, B. J. 2010. "The challenges of ambient law and legal protection in the profiling era." *The Modern Law Review* 73, 428–460.

HIMSS. 2008. "HIMSS (Healthcare Information and Management Systems Society), 19th Annual – 2008 HIMSS Leadership Survey." *Healthcare CIO.*

Ioannidis, J. P. A. 2013. "Informed consent, big data, and the oxymoron of research that is not research." *American Journal of Bioethics* 13, 40–42.

Istepanian, R. S. H. 1999. "Telemedicine in the United Kingdom: Current status and future prospects." *IEEE Transactions on Information Technology in Biomedicine* 3, 158–159.

Joly, Y., Dove, E. S., Knoppers, B. M., Bobrow, M., and Chalmers, D. 2012. "Data sharing in the post-genomic world: The experience of the International Cancer Genome Consortium (ICGC) Data Access Compliance Office." *PLoS Computational Biology* 8, e1002549.

Kenner, A. M. 2008. "Securing the elderly body: Dementia, surveillance, and the politics of 'aging in place'." *Surveillance & Society* 5, 252–269.

Khanja, P., Wattanasirichaigoon, S., Natwichai, J., Ramingwong, L., and Noimanee, S. 2008. "A web base system for ecg data transferred using zigbee/IEEE technology." *The 3rd International Symposium on Biomedical Engineering.*

Kinsella, G., Thomas, A. N., and Taylor, R. J. 2007. "Electronic surveillance of wall mounted soap and alcohol gel dispensers in an intensive care unit." *Journal of Hospital Infection* 66, 34–39.

Kosta, E., Pitkänen, O., Niemelä, M., and Kaasinen, E. 2010. "Mobile-centric ambient intelligence in Health- and Homecare-anticipating ethical and legal challenges." *Science and Engineering Ethics* 16, 303–323.

Kulkarni, A., and Sathe, S. 2014. "Healthcare applications of the Internet of Things: A Review." *International Journal of Computer Science and Information Technologies (IJCSIT)* 5, 6229–6232.

Lin, P., Li, Q., Fan, Q., Gao, X., and Hu, S. 2014. "*A Real-Time Location-Based Services System Using WiFi Fingerprinting Algorithm for Safety Risk Assessment of Workers in Tunnels.*" Hindawi.

Little, L., and Briggs, P. 2009. "Pervasive healthcare: The elderly perspectives." In *Proceedings of the 2nd International Conference on PErvasive Technologies Related to Assistive Environments.* Corfu.

Liyanage, H., de Lusignan, S., Liaw, S. T., Kuziemsky, C. E., Mold, F., Krause, P., Fleming, D., and Jones, S. 2014. "Big data usage patterns in the health care domain: A use case driven approach applied to the assessment of vaccination benefits and risks." *Yearbook of Medical Informatics* 9, 27–35.

Lorenzi, N. M., Novak, L. L., Weiss, J. B., Gadd, C. S., and Unertl, K. M. 2008. "Crossing the implementation chasm: A proposal for bold action." *Journal of the American Medical Informatics Association* 15, 290–300.

Luen, D., and Ellen, Y. 2009. "Technological interventions for hand hygiene adherence: Research and intervention for smart patient room." *In joining Languages, Cultures and Visions: Proceedings of the 13th International CAAD Futures Conference.* Canada: Les Presses de lUniversité de Montréal. 303–313.

Malik, A. 2009. "*RTLS For Dummies*." Hoboken: Wiley Publishing.

Massacci, F., Nguyen, V. H., and Saidane, A. 2009. "No purpose, no data: Goal-oriented access control forambient assisted living." *1st ACM Workshop on Security and Privacy in Medical and Home-Care Systems, SPIMACS'09, Co-Located with the 16th ACM Computer and Communications Security Conference*. Chicago. 53–57.

Meydanci, M. A., Adali, C., Ertas, M., Dizbay, M., and Akan, A. 2013. "RFID based hand hygiene compliance monitoring station." *2013 IEE International Conference on Control System, Computing and Engineering*. Mindeb. 573–576.

Mittelstadt, B. 2017. "Designing the health-related internet of things: Ethical principles and guidelines." *Information* 8, 22–25.

Mittelstadt, B., and Floridi, L. 2016. "The ethics of big data: Current and foreseeable issues in biomedical contexts." *Science and Engineering Ethics* 22, 303–341.

Murphy, D. 2006. "Is RFID right for your organization?" *Materials Management in Health Care* 15, 28–30.

Nissenbaum, H. 2004. "Privacy as Contextual Integrity." *Social Science Research Network: Rochester*, NY, USA, 22–25.

Pellegrino, T. 1993. *The Virtues in Medical Practice*. New York: Oxford University Press.

Peppet, S. R. 2014. "Regulating the internet of things: First steps toward managing discrimination, privacy, security and consent." *Texas Law Review* 93, 85–178.

Pont, M. J. 2002. In *Embedded C, Edition 2002*, 57–87, 217. New York: Addison-Wesley.

Rhodes, M. 2014. "*A Gadget Designed to Finally Make Doctors Wash Their Hands Enough*."

Riazul Islam, S. M., Kwak, D., Humaun Kabir, M. D., Hossain, M., and Kwak, K.-S. 2015. "The internet of things for health care: A comprehansive survey." *IEEE Access* 3, 678–708.

Rigby, M. 2007. "Applying emergent ubiquitous technologies in health: The need to respond to new challenges of opportunity, expectation, and responsibility." *International Journal of Medical Informatics* 76, 349–352.

Rosenkranz, P., Wählisch, M., Baccelli, E., and Ortmann, L. 2015. "A distributed test system architecture for open-source IoT software." *ACM MobiSys Workshop on IoT Challenges in Mobile and Industrial Systems (IoT-Sys)*. DOI: 10.1145/2753476.2753481.

Shelar, M., Singh, J., and Tiwari, M. 2013. "Wireless patient health monitoring system." *International Journal of Computer Application* 62, 975–8887.

Shilton, K. 2012. "Participatory personal data: An emerging research challenge for the information sciences." *Journal of the American Society for Information Science and Technology* 63, 1905–1915.

Stahl, J. E., Drew, M. A., Leone, D., Crowley, R. S. 2011. "Measuring process change in primary care using real-time location systems: Feasibility and the results of a natural experiment." *Technol Health Care* 19, 415–421.

Sundmaeker, H., Guillemin, P., Friess, P., Woelffle, S. 2010. "Outlook on Future IoT applications," *Vision and Challenges for Realising the Internet of Things: European Commission*. Brussels: *Information Society and Media* 189.

Taddeo, M. 2016. "Data philanthropy and the design of the infraethics for information societies." *Philosophical Transactions of the Royal Society*. DOI: 10.1098/rsta.2016.0113

Talmale, G., and Shrawankar, U. 2017. "Dynamic clustered hierarchical real time scheduling for IoT based human organ transplantation." *International Journal of Control Theory and Applications* 10, 239–249.

Terry, N. 2014. "Health privacy is difficult but not impossible in a post-hipaa data-driven world." *Chest* 146, 835–840.

Wachter, S., Mittelstadt, B., and Floridi, L. 2017. "Why a right to explanation of automated decision-making does not exist in the general data protection regulation." *International Data Privacy Law* 7, 76–99.

Wel, L., and Royakkers, L. 2004. "Ethical issues in web data mining." *Ethics and Information Technology* 6, 129–140.
World Health Organization (WHO) 2017. "World Health Organization (WHO) guidelines on hand hygiene in health care."
Yuan, B., and Herbert, J. 2011. "Web-based Real-time Remote Monitoring for Pervasive Healthcare." *Smart Environments to Enhance Health Care* 625–629.

16 Brain–Computer Interface

Abhishek Mukherjee, Madhurima Gupta, and Shampa Sen

CONTENTS

16.1 INTRODUCTION

In all primates, the human brain is considered to be most advanced and complex. It has a fascinating structure with a variety of functions. It allows for remembering past events, processing all the present sensory inputs, and integrating the information, thinking, analyzing situations, creating memories, displaying emotions, and making estimations into the future. For a while now, scientists from various fields seek to understand the fine intricate workings of the brain and channel it in a way as to enhance the overall productivity and efficiency of the task at hand. BCI or brain–computer interface is a buzzing technology in the world, which aims to do this. It can be understood as a mechanism of commanding various devices just by the use of signals generated by the brain, and also using those to comprehend the intention of the brain to perform a certain activity. The microelectrical signals generated by the neurons are assembled as the concept of EEG (electroencelagraphy), which can be translated as a pattern of data generated on by a brain stimulus (Nicolas-alonso and Gomez-gil 2012). A variety of algorithms can be used to understand and comprehend these patterns, and they can be used to build generic models to interpret signals of similar nature obtained in future readings. This can be further used to trigger computer functionalities as per requirement. A plethora of applications have the potential to find their roots with the advent of BCI and its subsidiary technologies. Starting from motor-related activities of the body to IoT applications and automation, everything can have a widespread use in the domain of BCI. This chapter discusses the concepts of inter-connection of the knowledge domain corresponding to the biological brain and the computer & computational related technical aspects. This will involve the conjugation of various aspects of brain anatomy, understanding of various perceived signals, algorithmic approaches, and finally some of its popular applications.

The primary focus is to understand the key elements of this technology and its workings so as to pave the way for researchers in this domain.

16.2 BASIC BRAIN ANATOMY

Before discussing the brain we need to understand the nervous system. The nervous system is essentially the electrical wiring of the body. It is made of trillions of nerve cells (neurons and glial cells). Neurons conduct impulses and cause sensations throughout the body, while glial cells take part in protective functions. In general terms, the nervous system takes in signals from your senses and processes the information to provide an appropriate response. On the whole this system is divided into (1) the PNS, or peripheral nervous system, which has nerves spanning and connecting all the body parts to the spinal cord; and (2) the CNS, or the central nervous system, consisting of the spinal cord and the brain.

The brain is the command center of the nervous system in all vertebrates and most invertebrates. It can be viewed as the biological computer; because just like an electronic computer it is involved in taking inputs from the environment in the form of sensory signals, storing the information (memory), processing and integrating it (learning), and producing an output (motor and hormonal signals) to exert control over the rest of the body. It occupies a space of up to 2% of the total body volume. The size of the brain differs slightly among the sexes and also from one individual to the other. However, this difference does not reflect on the cognitive ability of a person. The brain is constituted of mainly of three components: nerve cells, vascular system, and interstitial system, while it is divided into three functionally distinct but connected regions: cerebrum, brainstem, and cerebellum. The cerebrum is the largest part of the brain occupying 80% of the total brain volume. This region is involved in motor control, cognitive thinking, and behavior and processing of sensory inputs; the brainstem connects the cerebrum to the cerebellum and consists of three regions:

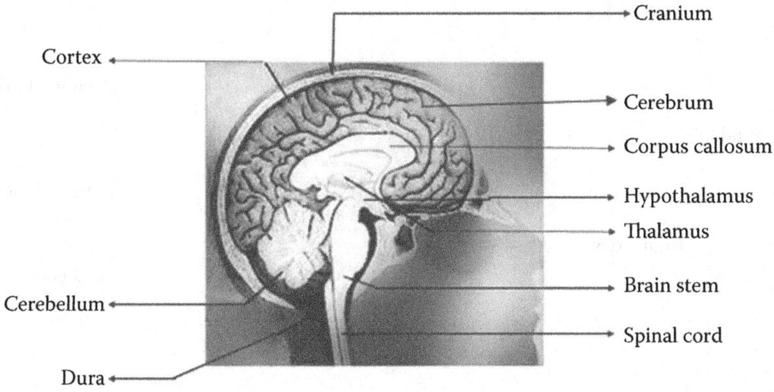

FIGURE 16.1 Schematic representation of the human brain.

mid brain, pons, and medulla (not shown in Figure 16.1), and continues after that into the spinal cord. The brainstem is the control center for processes such as sleep cycles, respiration rate, heart rate, temperature, blood pressure and volume, etc. while the small bulb-like structure of the cerebellum is involved in maintaining posture and balance of the body, as well as finely coordinating muscle movements. The basic structure can be understood from Figure 16.1; it is situated in the head region, and protected by the skull, or brain box, and the meninges.

In the course of this chapter our main focus shall be on the functioning of the cerebral cortex.

16.2.1 Cortex

As mentioned before, the largest part of the brain is the cerebrum. It is divided into two symmetrical hemispheres—right and left hemispheres—connected together by the calossum, as observed in Figure 16.1. The cortex is composed of grey and white matter, and these are nothing but bundles of neurons having special and distinctive orientations. The white matter is just the axons with myelin sheaths, while the grey matter is composed of everything else pertaining to the neurons. The cerebral cortex is the region of cortex associated with the cerebrum. It is composed of a greater number of layers compared to the cortex associated to the cerebellum. The cerebral cortex is a highly convoluted structure, composed of ridges known as gyri and fissures called sulci. Functional demarcation results in dividing each hemisphere into parietal, temporal, occipital, and frontal lobes. They all have distinct functions and yet are integrated which each other. The schematic structure and position of the lobes can be observed in Figure 16.2. As we can see, the frontal lobe is the largest of the four; it is our emotional control center and home to our personality. Thus, as expected, it is involved in a multitude of activities such as cognition (memory, problem solving, information processing, etc.), behavioral responses, and motor functioning and control. The occipital lobe is involved in visual processing while the temporal lobe takes care of sound and language. This lobe also harbors the amygdala and hippocampus, regions involved in emotions and memory, respectively. Parietal lobes function as an

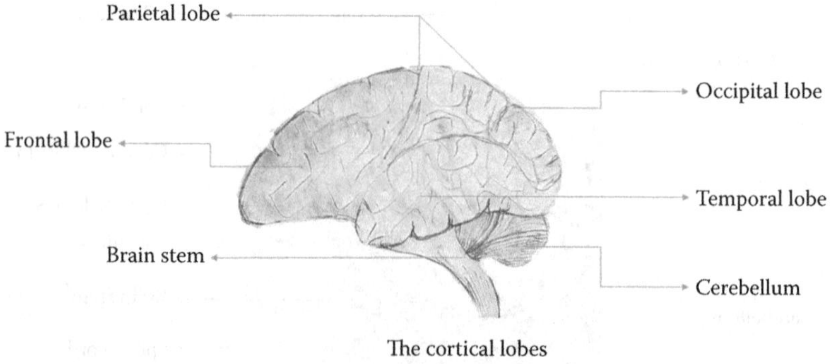

FIGURE 16.2 Cortical lobes.

integration center for sensory inputs from different senses, and are also involved in spatial orientation, awareness, and sensation of self, as well as navigation. Between the cerebrum and brainstem lie the thalamus and hypothalamus. The hypothalamus acts as a relay center between the nervous system and the endocrine system. The thalamus communicates motor and sensory data to the cortex; also, it comprehends wakefulness and sleeping patterns.

16.2.2 Functioning of the Brain

Our brain is never asleep. It works 24/7. Some activities are voluntary like walking, writing, talking, etc. whereas others like breathing, heartbeat rate, balancing, etc. are involuntary. Both of these are monitored and controlled by different regions of the brain. However, in the course of this chapter we shall focus on the activities and functioning of the cortex.

Typically, the functioning domain of the brain can be divided into active (conscious state) and the passive (subconscious) stages. The conscious state is when we are awake and knowingly making decisions, thinking and involved in motor movements and assessing our environment, whereas the subconscious mind is a background phenomenon, in which we are always involved although we are not actively aware of it. The workings of the subconscious mind is not yet understood, however, it is accepted that all of our conscious decisions and our individual personalities and responses are greatly affected by it.

Further in this chapter we shall come across methods and technologies to study brain activity, and in turn the applications for their results. These studies are based on cortical activity.

The majority of our understanding of the functioning of the cortex and the task assignment to various regions comes from knock out studies and observing subjects with mental deformities and diseases (Dikmen, Machamer, and Temkin 2017). Take for instance the frontal cortex, which given its size and location is the most vulnerable to injury. An injury to the frontal lobe can cause drastic behavioral and personality changes in the patient, as well as it can permanently completely impair the fine-tuning of motor skills and may cause a loss of function in the limbs (Kraus et al. 2007).

The brain typically functions by means of electrical impulse transmission. The chemical signals (hormones) act in response to the internal environment; while electrical signals (transmitted by nerves) act based on both internal and external stimuli. For BCI applications, our focus is more on the electric stimuli, although understanding the fine control mechanisms hormones have over the motor responses and thought processes is also important. For this purpose, data from techniques such as PET, can be combined with EEG data to draw conclusions and identify unique patterns concerning the particular stimulus. As will be explained in the coming units, the EEG is applicable only for picking up voltage fluctuations from synchronized cortical pyramidal neurons of similar spatial orientation.

16.2.3 Signal Reception

So far, we have understood that the neuron is the functional unit of sensation and brain activity; in this context we shall introduce the pyramidal neurons, which are a special type of neurons found in all areas of the cortex, hippocampus, and the amygdala. Their orientation is perpendicular to the cortical surface; the cell body is oriented towards the grey matter while dendrites extend towards the white matter. The reason why their electrical fields are this strong and stable lies in their peculiar anatomy with multiple branches and unique orientation.

In the cortex, the synaptic activity of these neurons is dynamic in nature and keeps on changing in strength during the course of development and learning process. The dendrites conduct the input signal while the axons are involved in conducting output signals. Both dendrites and axons are highly branched in this case. The specialty of these cells, and one of the possible reasons for their abundance in the cortex, lies in the existence of dendritic domains with distinct excitabilities, modulation abilities, synaptic inputs, and plasticities. These distinctions allows the synapses in the dendritic tree to be dynamic while maintaining structural and functional connectivity. The synapse is the functional element related to changes in electric flux and the generation of signals. Synapses are present all over the body; however, in the cortex an extra complexity is added, where the synaptic strength is different in strength for different synapses (think about long-term and short-term memories).

The action potentials are generated when these cells fire, that is, when there is synapse formation. The synapse formation is initiated by an external stimuli, for example, a picture flickers in front of your eyes and then the neurons in the visual cortex are stimulated and start firing. However, how long the region will fire depends on factors such as the duration for which the image is shown, the attention span of the respondent, and the purpose for which the picture in question is being studied. It must be understood that in the cortex, when a particular type of stimuli predominates and is more focused than the others, the neurons pertaining to that stimuli fire more strongly and actively; however, that doesn't mean that the other regions are not active or are not firing simultaneously as well. The EEG, which shall be discussed later, also gets focused with a typical pattern when a particular stimulus is being forced upon the respondent. The encephalogram will detect more pulses over the region dedicated to that stimulus, yet small signals will be captured from other parts as well during the experiment.

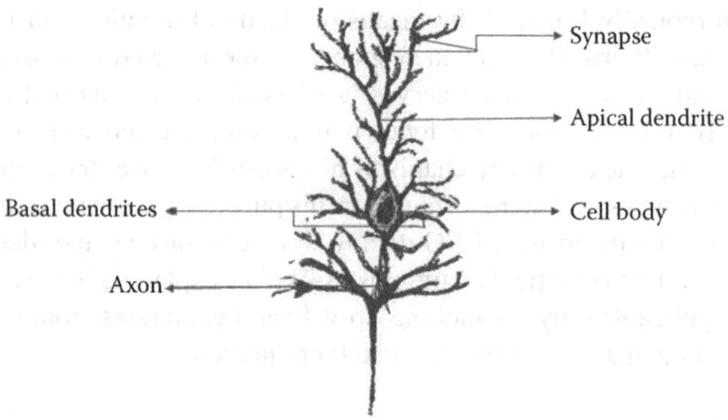

FIGURE 16.3 Pyramidal neuron.

For a detailed account on the signal transduction in pyramidal cells, the reader is referred to Spruston (2008) (Figure 16.3).

16.3 EEG

16.3.1 Functioning

An EEG is a noninvasive method commonly used for recording the electrical activity of the brain. The EEG data is collected by making a subject wear an electrode cap. There are several electrodes placed at strategic locations around the scalp. Each electrode is individually connected to a separate wire, which in turn is connected to the monitor. The EEG records the spontaneous electrical field of the brain over a period of time. It does so by measuring the voltage fluctuation resulting from ionic current within the neurons. However, at this point it is critical to mention that the voltage fluctuation thus measured is not of a single neuron, rather it is a cumulative effect of activity of millions of neurons located in that area. Let's visualize it in this way, when an impulse is generated in one neuron, the disturbance causes changes in the resting potential and ion distribution. Tremors near one group of neurons causes tremors in the neighboring neurons, which in turn disturbs the next group of neighboring cells and so on. This creates a wave, similar to creating ripples on the surface of a body of water. And when the wave of ions reaches the scalp, the electrodes capture the disturbances and are displayed on the screen. The graph generated can either be voltage versus time or voltage versus frequency. In case of the voltage versus time graph the readings are represented in the form of pulses.

As mentioned above, a single wave is the result of the synchronous activity of neurons of similar spatial orientations. The firing patterns of the cortical neurons are exceptionally complex. Hence, to make sense and draw conclusions from the EEG data it is necessary to characterize the frequency ranges and the cognitive activity they characterize in turn. Based on individual factors, stimulus properties, and internal states, scientists classify the band frequencies into five ranges: delta (1–4 Hz), theta (4–8 Hz), alpha (8–12 Hz), beta (13–25 Hz), and gamma (>25 Hz).

16.3.2 Data Acquisition from a Subject

The experiment is designed keeping in mind the specific cortex region we need to examine, and the cognitive activity to be studied. For example, if we wish to differentiate between patients with epilepsy and normal brain graphs, we plan different awareness states; sleep, awake, while reading, meditation, etc. The final results for the two different types of subjects are then summarized in order to draw common patterns and compare and contrast them with each other.

The different frequency bands, elaborated in Section 16.3.6, are detected under different stimuli cases and cognitive awareness states. Hence, we see that based on our target stimuli and activity, the specific band frequencies can be concentrated on effectively to draw patterns. Some of the studies and their specific band frequencies are mentioned here:

Delta bands, which are present only during sleep state. More prominent in the right hemisphere, these are observed for studies concerning sleep and sleep disorders, alcoholism, and its effects on sleeping patterns. These frequencies play an important role in characterizing memory consolidation and rewiring activities of the brain, typical of the asleep state.

Theta bands are observed when the person is awake and involved in task solving. These are recorded from all over the cortex, hence pointing towards the fact that to solve a particular problem all the cortical regions are active simultaneously, groping for information which may help to solve the task at hand. This can be used to understand the task-learning pattern of the brain, and hence design the circuits to be used for implantation.

Alpha bands are generated in the posterior cortical regions and are involved in correlation activities between sensory inputs, motor function, and memory; in other words, it is the brain activity when one is in a meditating state, awake with eyes closed, and no other voluntary physical activity. The alpha activities are effectively monitored for meditation studies and understanding the attention levels of respondents during different tasks; in case of latter high alpha power is a feature of distraction that is to say, that in case of complex and attention involving tasks, high alpha power corresponds to active blocking of distractions by the subject (Bonnefond and Jensen 2013).

Beta bands are observed in both the posterior and frontal cortex. These are generated when the subject is involved in active, busy, anxious thinking, that is when the person is involved in planning and decision-making activities. These frequencies can also be associated to learning and perfecting the fine nuances of a task, motor movement coordination, and fine-tuning; for example, learning a particularly difficult dance step or sketching a detailed picture, etc.

Gamma bands, the exact regions of generation or function are not known; however, these high frequency bands are suspected to be involved in events of long-term and short-term memory formation, and segregation of the information gathered from the external environment, for grouping under these two memory pathways. One of the studies involving these are microsaccade studies involving coupling of the EEG to eye movements, and thus drawing inferences for attention and visual information processing based on tracking subtle eye movements.

As pointed earlier, to collect EEG data, the subjects wear electrode caps. While generating the data and tuning the electrodes, care must be taken to generate sensitive and noise free data. Collecting clean data is of prime importance, as there is no algorithm that can "correct" the EEG readings once generated, that is, if your initial data is of bad quality and unclear you cannot improve upon it and then reanalyze it better. Each electrode should be shown to be active on the monitor and must be fine-tuned. The positioning of the electrodes around the scalp must be precise, and hence the size of the cap and the number of electrodes incorporated must be carefully chosen (Samy 2016).

The EEG electrodes are cleaned with 70% alcohol before and after use. The subject's scalp must be grease free and hair should be washed well before the experiment. There should not be any hair accessories. Generally, we have the option for both dry and wet electrode-based systems, however wet electrodes are preferred as the sharpness and sensitivity of the data is greater in this case. This type of electrodes forms a capacitor-like arrangement with the skin; here, the electrode comes in contact with the skin via a saline-based conductive solution (gel, cream, or paste). This capacitor analogy system attenuates any background low-frequency signals (Herron et al. 2017).

As discussed above, the respondent is subjected to various stimuli based on the subject matter of the study. Each test includes several reruns to gain confidence on the patterns detected.

However, the electrical data we gain needs to be assessed for miscellaneous nonrelated signals (contaminants) arising due to factors such as muscle movements (clenching of jaws, facial muscles, neck movement, etc.), eye movements, blinking, movement of the electrodes, or headset movements. These contaminants can mislead the interpretations, for example, if during a visual stimuli the respondent blinks too much the EEG may fail to record any activity pertaining to that stimuli in the visual cortex, giving rise to a blind spot. For each of the factors mentioned above, the researcher should try to minimize these variations as much as possible. As for the eye movements and blinking, the number of trials is kept high (at least hundred or above) and these movements are separately tracked, and their effect is attenuated using statistical tools from the raw EEG data (Singh, Singh, and Sandel 2014).

16.3.3 The 10–20 System

As previously mentioned, in order to extract the EEG data from the subject data acquisition should be done properly and the steps, guidelines, and rules should be strictly adhered to. It is understood that for the signals to be transmitted the cap studded with electrodes must be worn by the subject. Hence, it is crucial to have a fixed system, which will ensure there is a standardized way of getting the electrodes placed inside so that there is a uniform way of collecting data irrespective of the size, structure, or shape of the subject's head (Soutar 2013).

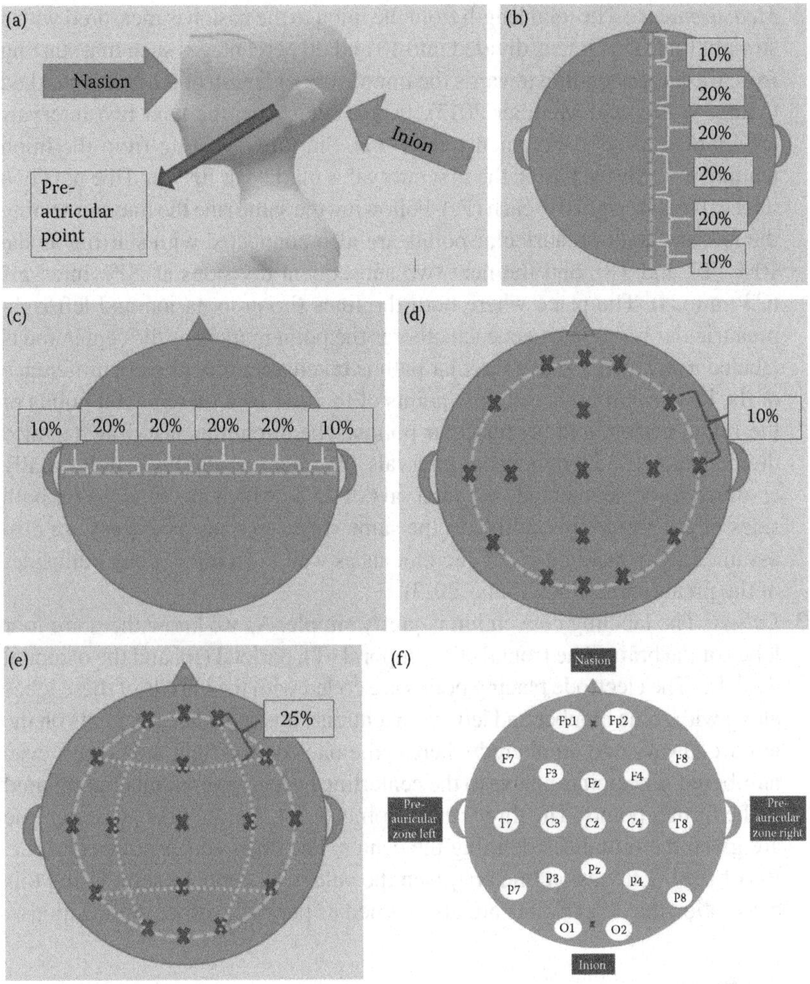

FIGURE 16.4 (a) Nasion, inion, and preauricular points. (b) Nasion to inion 10–20 division. (c) Left-right pre-auricular 10–20 division. (d) 10–20 circumferential division. (e) 10–20 parasagittal and transverse. (f) Electrode locations.

For this reason, there is a system popularly known as the *International 10–20 System,* which provides a standard arrangement of placing the electrodes on the cap. The procedure is explained in the following steps.

1. *Location*: First the head is marked by four primary locations known as nasion, inion, left preauricular point, and right preauricular point. These are the points, which are the primary locations for the 10–20 system. The nasion is the point just between the eyebrows or a little below that. The inion is the little bony protrusion at the center back of the head where the neck joins the skull. The left and right preauricular zones are the points just in front of the left and right ears, or near the regions of sideburns (Figure 16.4).

2. *Measurements*: The total length from the inion to the nasion is measured with a straight line. This line is divided into 10 and 20 percentages such that starting from the nasion tending towards the inion, 10% of length of the line is marked (Yang, Wang, and Member 2017). In a similar way, the next two intervals are marked at 20% increments each (Fz). Similarly starting from the inion tending towards the nasion the first interval is marked at 10%, and the next two intervals marked at 20% each (Pz). Following the same rule the line connecting the left and right preauricular points are also connected with starting as the 10% (T7 and T8), and the next two subsequent divisions at 20% intervals (C3 and C4). The place where both the lines (inion to nasion and left/right preauricular lines) intersect each other is the point marked as the center and is labeled as Cz. After that a circular path is taken assuming the circumference of the head running through the points of the first 10% incremental points of the inion, nasion, and preauricular points. The circumference drawn is also divided into 10% increments or intervals. Two parasagittal lines are virtually assumed to be segmented into intervals of 25%, which are running on both sides of the nasion–inion line. In the same way two transverse lines are also assumed segmented into 25% increments as well as running along both sides of the preauricular line (Soutar 2013).
3. *Labels*: The labeling convention is pretty simple. As we know there are four lobes of the brain. The frontal (F), temporal (T), parietal (P), and the occipital (O) lobe. The electrode placing points are coded with the initials of these lobes along with some number and letter conventions. It is such that the points on the left are always odd numbered whereas the ones on the right are always even numbered. The points nearer to the centerline have a lower number compared to the farther points. The three center points (C3, Cz, C4) of the auricular line are given the initial (C) defining the center. The three center points (Fz, Cz, Pz) of the inion–nasion line are given the values (z) defining zero. Using this convention the other points are also named as per their respective locations.

16.3.4 Benefits

With advancement in technology a number of technologies are now available to study brain activity, such as functional magnetic resonance imaging (fMRI), positron emission tomography (PET), magnetoencephlaography (MEG), to name a few. However EEG remains the primary choice in both clinical and research-based applications as hardware costs are significantly lower than for the other techniques, as well as its light weight, mobile, robust, and flexible nature (in contrast to techniques such as MRI, which need to be conducted in a shielded room and the respondent has to be strapped to the observation table or chair). Hence, EEG can be used in conveniently for real-time experimentation. It is a noninvasive technique with very high temporal resolution in the order of milliseconds. EEG is safe and nonhazardous and does not cause any sort of discomfort to the respondents; the experimentation is carried in a relatively open and comfortable environment, thus the subject does not feel claustrophobic. Also, the machine is silent and hence gives better results for auditory stimuli-based studies. Another very important real-time advantage is it can be used for respondents who are incapable of making motor responses. Also, some of the event related potential (ERP) components can be studied when the subject is not actively attending to stimuli.

16.3.5 Drawback

Despite its many advantages, one of the major challenges faced with EEG is its poor spatial resolution. It cannot precisely tell which part of the brain is primarily active in response to a particular stimuli; such inferences can be drawn only on the basis of an hypothesis. Another major drawback is that EEG can be used only to study the cortex; for deeper structures and subcortical layers, assistance from other techniques such as fMRI is required. Additionally, EEG cannot detect the specific location of neurotransmitters or drugs, which is possible with PET.

Though the experimental setup is comparatively simple, it is time consuming, as sample size has to be kept very large and the number of trials for each stage of the experiment has to be kept sufficiently large in order to extract reliable information.

16.3.6 Waveforms and Common EEG Patterns

EEG is all about detecting patterns and signals, which are interpreted into waveforms from the EEG signals. For a clearer understanding of these concepts, it is important to have some fundamental knowledge of the waveforms and common EEG patterns, as well as what they imply.

The EEG signals are categorized as EEG rhythms, or frequencies, in technical words. These frequencies are nothing but oscillations of the microelectrical signals obtained from the postsynaptic potentials. There are a couple of frequency types distributed into bands, or in simple words, which can be understood as a range of frequencies.

- *Delta frequency band*: The frequencies in the delta band range between 1 and 4 Hz. This band of signals is primarily encountered during sleep. These frequencies are low and their amplitude is high. These are mainly used by sleep specialists and physicians to assess the depth and quality of sleep, and are also useful in detecting symptoms of sleep apnea. The right part of the brain shows a higher occurrence of such frequencies in comparison to its left counterpart.
- *Theta frequency band*: These frequencies range from 4 to 8 Hz. These frequencies are detected when the brain is performing deep memory-related activities, such as remembering or trying to recall some information, learning something, and storing information. Primarily it is associated with the frontal and prefrontal cores of the brain. Although these kinds of frequencies are found all over the cortices, its intensity has been observed mainly when the brain is processing difficult tasks, as well as mental and memory intensive operations.
- *Alpha frequency band*: These range from 8 Hz to 12 Hz. They are associated with relaxed wakefulness, when the brain is awake and during focused attentiveness. Also, they are sometimes noticeable when the mind and body are relaxed and awake, while the alpha wave diminishes when there is high mental or physical activity. They typically signify motor, sensory, and other related brain activities. They mainly originate from the posterior

cortical regions such as the parietal, occipital, and temporal zones. These frequencies seem to fire up when the brain is involved in activities requiring the senses to acquire attention to a particular thing.
- *Beta frequency band*: Ranging from 12 to 25 Hz, these frequencies depict subtle body motion such as micro limb movements, which require undivided focus, as well as excitement or being busy in some activity. These are ideally found in the posterior and frontal zones. They also show activity when other people initiate limb movements.
- *Gamma frequency band*: Frequencies ranging above 20 Hz are known as gamma bands. It still is a highly debated theory among the researchers regarding the location of the origin of the gamma signals. Some believe these are close to the theta frequencies, whereas others feel they originate as a residue of other neural functions.

The brainwaves generated are often a combination of all these kinds of waves where there can be hybrid frequencies, such as one of the frequencies (alpha, beta, gamma, and theta) superimposed on the other. These represent different cortical activities and can be assimilated as combinations of brain functions as well. However, over the years researchers have figured out some common EEG patterns, which are related to meaningful information.

The waves are acquired from the electrodes as a differential values of montages running through the adjacent electrodes. There are various montages of travelling patterns which are known as bipolar montages. A differential amplifier shows the differences between two identical looking waves and only portrays the difference between the two signals. There are sequences observed how their montages travel from brain cortex like anterior to posterior, temporal wise, etc. For example a sequence obtained from a difference of Fp2–F8, F8–T8, T8–P8, and P8–O2, which depicts a chain moving posteriorly picking up brain activities from the right frontal, temporal, parietal, and occipital regions (right temporal chain).

Let us now discuss some common EEG patterns useful in detecting common brain activities and used in medical sciences.

One of the most commonly studied EEG patterns is an eye blink. The two main parts of the eye are the retina and the cornea. The retina is situated at the back of the eye and is negatively charged, whereas the cornea is at the front containing a positive polarity. When an eye blink occurs, according to Bell's phenomenon, the eyeball rolls upwards to the forehead. Now, since regions like Fpl and Fp2 are nearest to the eye, the positively charged cornea induces some charge on Fpl and Fp2, which causes a significant deflection in the signals of the frontal polar electrodes.

Alpha rhythm is another popular signal, as discussed earlier, and is generally obtained during wakefulness with eyes closed.

A common pattern, which is largely used in medical sciences, is the focal epileptiform discharge. This occurs when there is a simultaneous unusual activity by the neurons in the cortex of the brain. These can originate in any region be it temporal, frontal, occipital, or parietal and range anywhere from anterior to posterior.

Another form of this is the generalized epileptiform discharges. These forms of frequencies are found in patients with epilepsy syndromes. As per some studies, when they occur all the regions oscillate in a 3 Hz frequency generally in adults in a sudden time frame. Studying the results of EEG recordings in epileptic patients reveal that records vary from the time to time when attacks take place. If the recording is done within 24 hours of the occurrence there are 50 percent chances of finding abnormalities in the readings, If the recording done routinely or even non routinely in a particular epilepsy attack period there is a high chance of observing epileptiform discharges, whereas if the recording is done after a 24-hour time gap the chances of such discharges is reduced significantly.

16.4 CLASSIFICATION OF EEG SIGNALS

So far we have seen how the electroencephalogram signals are processed as to extract meaningful brainwave information from the subjects. We have also seen the supporting systems and a few standard extraction techniques to accomplish this. Common patterns can be systematically rendered so as to associate them with a specific brain–body activity. Now, we will discuss how the EEG signals are interpreted and the algorithms with which they are classified through the process of extraction (Poorna, Sai Baba, and Ramya 2016).

16.4.1 Support Vector Machine

The support vector machine algorithm is a classification algorithm initially proposed by Vapnik. This classification algorithm is just like any other classification algorithm which labels samples and categorizes them into classes. In the initial version of this algorithm it performed only linear classification. The linear classification process is such that, suppose there are two "attributes" of a sample using which the sample has to be "categorized" into either of the classes, there is a straight line that is termed as a hyperplane which is generated in such a way that it acts as a segregation border for the classification or so called categorization. There is a concept of maximum margin, which means that the distance of the nearest occurrence (technically called a "datapoint") of both the classes from the hyperplane should be maximum in order to get the most accurate classification and to be considered into one of the categories (Mehmood and Lee 2015) (Figure 16.5).

In Figure 16.5 we can see that there are two hyperplanes, the shaded one is "A" and the non-shaded one "B." Both hyperplanes A and B segregate the classes accurately but the maximum margin is left by the hyperplane "A," as it is evident that it has the maximum gap from the nearest elements of both classes compared to "B." Hence, it proves to be a better classifier. This algorithm focuses first on the accuracy by taking care of 2 aspects: (i) whether is able to classify all the data points accurately and (ii) margin maximization. In some cases there may be outliers of each class, which fall into other classes and are very difficult for linear SVM classification (Rahimi 2016). For that reason, Vapnik, Guyor, and Boser proposed a method to create a nonlinear classifier for SVM using the kernel trick for high dimensional planes. SVM is a training-based algorithm, which uses a training set for producing accurate results. In BCI, this is useful when there is a deviation of a signal received from the electrodes and distinguished by a differential amplifier (Ilyas et al. 2016).

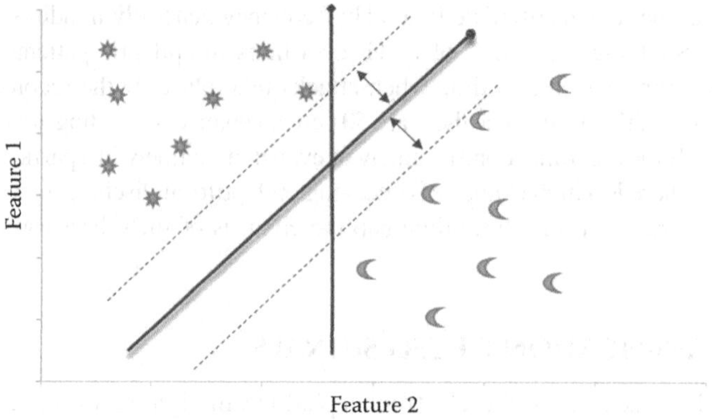

FIGURE 16.5 SVM hyperplane.

16.4.2 KNN (K-Nearest Neighbors)

K-nearest neighbors is considered one of the simplest machine learning algorithms and is a nonparametric algorithm widely used for classification and regression. The algorithm falls under the supervised learning category and makes use of training data for classification of data. The working of the algorithm is such that a new sample can be classified into either classes based on the majority of the k-nearest attribute values. The k value is a number, which tells how many nearest data points of various attributes or features have to be considered so as to allot a class to the new sample. For the nearest neighbors, the Euclidian distance formula is used to evaluate the distances from the neighbors (Fayzrakhmanov and Bakunov 2016).

$$\text{Euclidian Distance}(x,y) = \sqrt{(x_1 - y_1)^2 + (x_2 - y_2)^2 + \cdots (x_n - y_n)^2}$$

The above formula for Euclidian distance is used to calculate the distance between two points x and y. The nearest neighbors are weighted or ranked in ascending order, and the first k values of that order are selected for analysis. Further among those selected, the class which shows the majority is considered as the class label for the feature vector (Figure 16.6).

Example: [Class {Sun: Moon}]: For k value = 3 the shape triangle will be classified as a shape "sun," as among the nearest three neighbors the majority of up-votes are obtained by the shape "sun."

It is conceptually described as a lazy learner algorithm, which does not have any generalized model, whereas it uses the training values of the various instances in the training set to generate its own model. K-NN does not have a very high success rate for high-dimensional data, and hence it is considered efficient for low-dimensional scenarios (Manjusha and Harikumar 2016). The k value is appropriately chosen for cross validation and multiclass environments (Subasi and Erçelebi 2005). K-NN thereby produces a confusion matrix, which depicts the accuracy for a given data, while the accuracy of K-NN varies from subject to subject in BCI (Awan 2016; Ilyas et al. 2016).

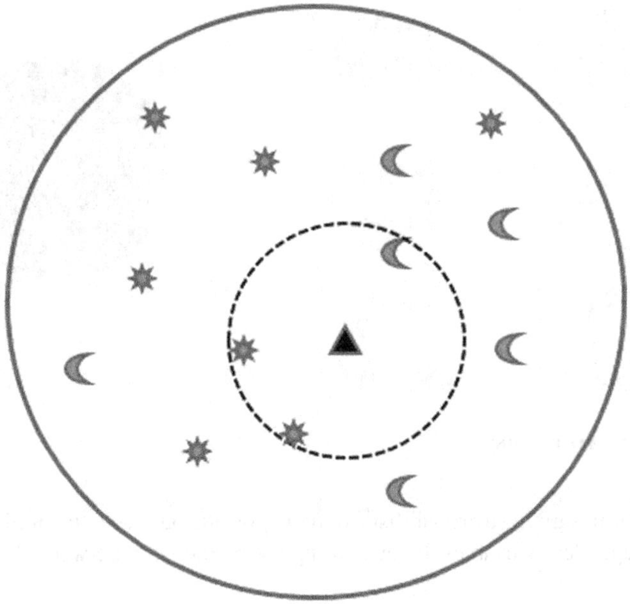

FIGURE 16.6 K nearest neighbor.

16.4.3 MLP-ANN (Multilayer Perceptron-Artificial Neural Networks)

The multilayer perceptron is an algorithm that makes use of the artificial neural network (ANN) method of classification. ANN is a multidimensional and nonlinear model for classification, which is used in multiple disciplines for computational and classification purposes. The single layer perceptron may be used for linear classification, but often linear classification cannot accurately classify all the feature vectors of complex data sets. The EEG signals can be one such example. In those cases, the multilayer perceptron is required. In ANN, the working principle in single layer is such that each of the neurons transmit the signals to the output neuron, which computes the results based on the inputs of its preceding neurons. In the multilayer perceptron method there are multiple hidden layers present between the input and the output neurons. Inputs received by the neurons are assigned weights in each iteration of data flow, which are summed up while obtaining the bias as well. Now, the final output is measured against the desired output and the errors are recorded. Thereafter, for subsequent iteration of the algorithm the weights of the neurons are adjusted so as to reach the highly accurate and desired results. This feature of MLP-ANN is known as back-propagation (Figure 16.7).

It is used widely for training of neural networks and can be efficiently used for classification of motor signal images. The advantage lies in its flexibility and apparent multidimensionality (Ilyas et al. 2016).

16.4.4 Logistic Regression

Logistic regression helps in building models. These models make for a classification tool. The models are linear logistic regression models, which can be considered as

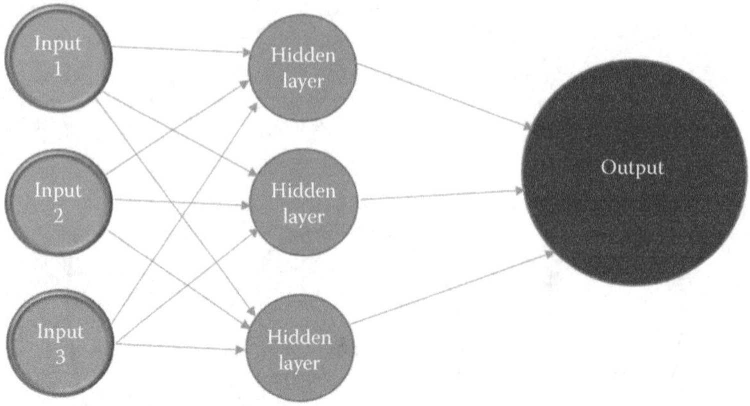

FIGURE 16.7 MLP-ANN.

learner algorithms giving a probabilistic output upon attribute selection for independent, as well as dependent variables, by evaluating the relationship between them.

16.5 DISCUSSION

Before we start the comparison and analysis of these algorithms we need to understand how these algorithms capture the data and initiate a procedure for classification. In the previous sections, we have seen how the brain waves generated are captured by the electrodes during the data acquisition process. The waves generated are passed on from the electrodes to the computer. Any raw data received from the subject is highly contaminated in the sense that the data acquired may pick-up signals from other external sources (other than the microvoltages generated from the combined postsynapses of the different zones of the lobes of the brain). These unwanted signals are known as artifacts. Artifacts are signals from electrodes other than the electrode we are focusing on to receive data for a specific functionality (Šťastný, Sovka, and Stančák 2003). They can also be other signals such as from electronic equipment in the environment, ambient noise, etc. After the raw data is received it is cleaned and preprocessed with methods such as manual cleansing of artifacts such as blinks of the eye, movements of the eye, muscle twitches, and bad channels as well. Also, methods like independent component analysis can be used to clean the data for each subject. Now, when the data is cleaned the uncontaminated EEG signals are prominent. Each channel connected to the array of electrodes supplies continuous analogous data where the EEG receptors take readings of that data at regular intervals. The efficiency of the receptors is measured by the number of samples collected per unit time, also known as *resolution*. The process of classification involves categorization of deviation of the frequencies from their usual readings, as received from the array of electrodes into two possible cases (i) Considerable deviation has occurred and (ii) Deviation has not occurred. The deviation may occur when a stimulus is given to the subject. These algorithms tend to build a learning model, which trains how a signal is classified as a “deviation” or a “nondeviation” with respect to a given stimulus. While the model is being built, the various combinations of the patterns from the various channels are considered in consolidation for a particular stimulus. This then becomes a reference pattern associated to a stimulus. Once the algorithms are done building their classification models they can predict the actions for new samples to be classified so as to conclude if they mean any deviation, which in turn may refer to some activity based upon the previous pattern of signals. There is a considerable amount of research in this field to determine how data is classified and the various steps involved in it in granular steps (Samanta 2017).

These algorithms perform differently for different scenarios when classification of EEG signals is involved. As discussed by M.Z. Ilyas et al. (Ilyas et al. 2016), information obtained from the “Berlin BCI group” where they had collected BCI data for motor imagery related to left and right limb movements upon controlled visual stimulus. It was obtained from various subjects. A stimulus provided resulted in some motor imagery tasks which were recorded. Conversion of those signals from time domain to a frequency domain was performed using Fourier transformations. It was further converted into an attribute relation file format using WEKA, which is capable of performing classification algorithms.

The results obtained from each of them indicate that no particular algorithm can be considered as the “perfect one” for a particular dataset of EEG specific to a subject. Each of them show varying results for different subjects. Logistic regression and support vector machine performed well if looked at from a broad perspective posing as highly efficient classifiers, on the contrary k-Nearest Neighbor and MLP- Artificial Neural network performed better for some instances where the feature vector was highly correlated but also deteriorated for other instances. The performance of k-NN seems to be poor statistically, as it apparently fails to work efficiently if the data contains high dimensionality (Zhang et al. 2015).

In Section 16.6, we will see how EEG and BCI can be used for various application-oriented utilities and practical usage in a variety of fields.

16.6 APPLICATIONS

Humans love their machines and a technician is as good as his tools. The software developers have claimed the virtual world with their algorithms; the genetic engineers' job is eased with gene sequencing tools and CRISPER (clustered regularly interspaced short palindromic repeats). Doctors and clinicians have too many electronic helpers at their disposal to be named here.

These days, science has advanced enough to enable us to print functional organs (3D bioprinting); develop computers able to detect and record the brain activities on paper (EEG and related technologies); create robots that can do our manual work (robotic arms and process belts); and the list goes on (Hasan et al. 2016).

However, all these remarkable discoveries and inventions are employed separately, hence we are now faced with the opportunity to build a platform to integrate it all in one. AI (artificial intelligence) is one such project. Yet, it merely seeks to mimic the workings of the mind and does not connect with it. Hence, human intelligence still reigns supreme outside of narrowly defined tasks. To go beyond these limitations, it has been sought to integrate the human brain and its synapses with the Internet in the ever-expanding domain of IoT. In separate endeavors, Bryan Johnson, founder and CEO of Kernel, Elon Musk, and the US military's research arm, DARPA, among others have begun to look for ways to link mind and machine.

This quest has a number of applications, as discussed below.

16.6.1 Aiding Invalids

Diseases such as lower body paralysis, muscular dystrophy, Parkinson's, limb amputation, etc. can render the patient incapable of making appropriate motor responses (Zafar et al. 2017). A person with severe motor disabilities can feel helpless and go into a depression as they become imprisoned in their own bodies. As the disease progresses, the person loses the ability to interact with their environment, or lose the ability to communicate—which is terrifying. Take for instance ALS (amyotrophic lateral sclerosis), a condition in which the neuromuscular connections progressively degrade to the extent that the only way to stay alive is to be connected to a life support systems. However, the fact is that the patient, even though physically impaired beyond resurrection, is not brain dead and can still hope for some normalcy, if a technology is introduced, which can work on brain synapses rather than neuromuscular connections. This was demonstrated by Kübler et al. where they used EEG based BCIs to help four ALS patients (Kübler et al. 2005).

The basic idea is that the brain activity can be transformed into a digital form, which can then act as an input command for the computer. The BCI tools can be used in a multitude of areas, such as education, gaming, industry, astrophysics, medical fields, etc. (Hadi, Sholahuddin, and Rahmawati 2016).

It has been confirmed by neuroscientists that, even when the limb is no longer functional due to some accidental damage, the area of the brain associated with control of that organ still fires signals. In extreme cases, we have cases of phantom limbs (Nikolajsen and Jensen 2001); these findings show that there is an innate neural circuit associated with a particular motor response. Hence, to control the robot to perform the corresponding physical task, it can be made to recognize the specific firing pattern with the help of EEG readings and data from healthy individuals. This way it forms an algorithm for the specific command (Obeidat, Campbell, and Kong 2017).

The machine can be trained to perform a specific task by tuning in to the thought process (in other words conscious decision to do the work) associated with the task (Arjestan and Vali 2016). This means that BCIs can not only be used as a means of communication but also to develop implantable chips with clean and safe surgical procedures. Also remote controls can be designed which can be operated with least amount of muscular input and be employed to control an external exoskeleton or a robotic limb. As had been reviewed by Carmen in her 2012 review paper under the section for IOTs for physical disabilities (Carmen 2012).

16.6.2 IoT-Based Applications

IoT involves connecting various sensors, receivers, transmitters connected to computing devices so as to process, analyze, visualize, and provide necessary responses to input information.

Brainwaves controlling physical and electronic devices was still science fiction in the recent past. But with the evolution of various BCI techniques and their enhancements using brainwaves to control various devices, it may soon be a reality finding uses in medical and home-automation fields.

IoT involves three basic elements—the sensors, the processors, and the actuators. These three devices work in unison to fulfill the activities of IoT-based environments. The IoT has several layers in its interface. The first is the perception layer, which involves the sensors or the actors. Next, is the network layer or the layer involving the connectivity. Then comes the middle layer which acts as a facilitator for various types of different devices to communicate without any hindrances. The top two layers consist of the application layer and the business layer, which consist of the user interaction and deep analysis of data involving machine learning for data analytics and inferring meaningful information (Mathe and Spyrou 2016).

One of the popular devices in the market is the NeuroSky Mindware Mobile device. These are used as we use normal headsets and it mainly perceives brainwaves generated at the center of the forehead. It also has an auxiliary sensor near the ear. This product has its own e-sense algorithm, which can be used to translate the brainwaves into values such as calm or attentive depending upon the frequency readings of alpha, beta, gamma, theta, or delta frequencies (Jiang et al. 2016). It can infer important functions such as eye closing/opening, etc. These can be used for a variety of purposes (Arora, Choudhury, and Kumar 2016; Arora et al. 2016).

One of the best uses of BCI in IoT is home automation and control. Although it is still at a developmental stage, there are a variety of functionalities which can be performed. For example, the Mindwave device can be linked to a computer, which can be in turn linked to the controller unit. A popular controller unit is the Arduino Uno, which can be coupled with a Bluetooth module. Upon receiving the signals the Bluetooth transceiver passes on the signals to the Arduino controller to enable real automation. The Arduino controller is linked with various electronic devices or appliances (Nanditha and Persya 2017).

It can be used to turn off/on home lighting, or help prepare a mood lighting setting for each individual user. It can be used in setting the air-conditioning temperature of the home. It is found that beta frequencies become activated when the subject is brought in contact with high levels of sound or light stimulus. These can be used to set toggle lighting or the music player in the house. Since the alpha waves depict concentration, such sensors can support applications, which performs analytics using various algorithms that may be used to track effective meditation patterns as well as sleep disorders. This brings us to medical utilities. In the medical field, it can be used to alert paramedics in case a patient undergoes some kind of epileptic attack, or to monitor postepileptic symptoms of a patient. Alongside aiding invalids with sensory motor utilities, it has been observed that such frequencies can used a good deal in virtual reality. Virtual reality being the next big thing, this has huge applications in multimedia, interactive media, andgaming control.

There may be a plethora of applications developed further, as and when the BCI and its supporting technologies are enhanced and become even more advanced.

16.7 CONCLUSION

The brain and its interface with the computer is a field with immense potential. We aimed to provide the various nuances of the concept of Brain-Computer-Interface using a holistic approach. In the course of this chapter we explored briefly the brain anatomy and how its functional aspects are exploited and integrated with the machines. The EEG is our most powerful and vital tool for achieving this integration. As we have seen, electrical synapse is generated by the neurons and then transmitted to one another, thus creating a field of electrical discharge succeeding the postsynaptic activity. These are then detected and recorded by the EEG. The international 10–20 system is the most commonly followed system for collection of this data. EEG will always be the first choice for the applications of BCI and IoT because of its ease of application and robustness. However, it is still not perfect, nor an all-around approach, as there are many gaps. Nevertheless, scientists are now coming up with supporting technologies, algorithms, and sister techniques (FMRI, PET etc.), which can attenuate these gaps to a great extent and open more arenas for us to explore. We hope there may be a day when we can finally understand the incredible functioning of the human brain and tap into its many potentials that still hidden from us.

REFERENCES

Arjestan, M. A., and M. Vali. 2016. "*Brain Computer Interface Design and Implementation to Identify Overt and Covert Speech.*" no. November: 23–25.

Arora, B., T. Choudhury, and P. Kumar. 2016. "*An Intelligent Way to Play Music By Brain Activity Using Brain Computer Interface.*" no. October: 223–28.

Arora, B., T. Choudhury, S. Sabitha, and P. Kumar. 2016. "*An Exhaustive and Comparative Analysis of Brain Computer Interface Algorithms.*" no. October: 202–7.

Awan, U. I. 2016. "*Effective Classification of EEG Signals Using K-Nearest Neighbor Algorithm.*" 120–24. https://doi.org/10.1109/FIT.2016.28.

Bonnefond, M. and O. Jensen. 2013. "The Role of Gamma and Alpha Oscillations for Blocking out Distraction," 20–22.

Carmen, M. 2012. "Journal of Network and Computer Applications An Overview of the Internet of Things for People with Disabilities." *Journal of Network and Computer Applications* 35 (2). Elsevie: 584–96. https://doi.org/10.1016/j.jnca.2011.10.015.

Dikmen, S., J. Machamer, and N. Temkin. 2017. "Mild Traumatic Brain Injury: Longitudinal Study of Cognition, Functional Status, and Post-Traumatic Symptoms." *Journal of Neurotrauma* 34 (8): 1524–30. https://doi.org/10.1089/neu.2016.4618.

Fayzrakhmanov, R. A., and R. R. Bakunov. 2016. "*The Reduction of Learning Sample in Information- Measuring and Control Systems Based on Brain- Computer Interface Technology.*" 0–3.

Hadi, S., A. Sholahuddin, and L. Rahmawati. 2016. "*The Design and Preliminary Implementation of Low-Cost Brain-Computer Interface for Enable Moving of Rolling Robot.*" no. Icic: 8–12.

Hasan, K., S. M. Hasnat Ullah, S. S. Gupta, and M. Ahmad. 2016. "*Drowsiness Detection for the Perfection of Brain Computer Interface Using Viola-Jones Algorithm.*"

Herron, J. A., M. C. Thompson, T. Brown, H. J. Chizeck, J. G. Ojemann, and A. L. Ko. 2017. "Cortical Brain Computer Interface for Closed-Loop Deep Brain Stimulation." 4320 (c): 1–7. https://doi.org/10.1109/TNSRE.2017.2705661.

Ilyas, M. Z., P. Saad, M. I. Ahmad, and A. R. I. Ghani. 2016. "Classification of EEG Signals for Brain-Computer Interface Applications: Performance Comparison." *2016 International Conference on Robotics, Automation and Sciences (ICORAS)*, 1–4. https://doi.org/10.1109/ICORAS.2016.7872610.

Jiang, M., T. N. Gia, A. Anzanpour, A. M. Rahmani, T. Westerlund, S. Salantera, P. Liljeberg, and H. Tenhunen. 2016. "IoT-Based Remote Facial Expression Monitoring System with sEMG Signal." *SAS 2016 - Sensors Applications Symposium, Proceedings*, 211–16. https://doi.org/10.1109/SAS.2016.7479847.

Kraus, M. F., T. Susmaras, B. P. Caughlin, C. J. Walker, J. A. Sweeney, and D. M. Little. 2007. "White Matter Integrity and Cognition in Chronic Traumatic Brain Injury: A Diffusion Tensor Imaging Study." *Brain* 130 (10): 2508–19. https://doi.org/10.1093/brain/awm216.

Kübler, A., F. Nijboer, J. Mellinger, T. M. Vaughan, H. Pawelzik, G. Schalk, D. J. McFarland, N. Birbaumer, and J. R. Wolpaw. 2005. "Patients with ALS Can Use Sensorimotor Rhythms to Operate a Brain-Computer Interface." *Neurology* 64 (10): 1775–77. https://doi.org/10.1212/01.WNL.0000158616.43002.6D.

Manjusha, M., and R. Harikumar. 2016. "Performance Analysis of KNN Classifier and K-Means Clustering for Robust Classification of Epilepsy from EEG Signals." *Proceedings of the 2016 IEEE International Conference on Wireless Communications, Signal Processing and Networking, WiSPNET 2016*, 2412–16. https://doi.org/10.1109/WiSPNET.2016.7566575.

Mathe, E., and E. Spyrou. 2016. "Connecting a Consumer Brain-Computer Interface to an Internet-of-Things Ecosystem." *Proceedings of the 9th ACM International Conference on PErvasive Technologies Related to Assistive Environments - PETRA '16*, 1–2. https://doi.org/10.1145/2910674.2935844.

Mehmood, R. M., and H. J. Lee. 2015. "Emotion Classification of EEG Brain Signal Using SVM and KNN." *International Journal of Bio-Science and Bio-Technology* 7: 23–32. https://doi.org/10.1109/ICMEW.2015.7169786.

Nanditha, M., and S. C. Persya A. 2017. "EEG-Based Brain Controlled Robo and Home Appliances." *International Journal of Engineering Trends and Technology (IJETT)* 47 (3): 161–69.

Nicolas-alonso, L. F., and J. Gomez-gil. 2012. "*Brain Computer Interfaces, a Review.*" 1211–79. https://doi.org/10.3390/s120201211.

Nikolajsen, L., and T. S. Jensen. 2001. "Phantom Limb Pain." *British Journal of Anaesthesia* 87 (1): 107–16.

Obeidat, Q., T. Campbell, and J. Kong. 2017. "Spelling with a Small Mobile Brain-Computer Interface in a Moving Wheelchair." 4320 (c): 1–11. https://doi.org/10.1109/TNSRE.2017.2700025.

Poorna, S. S., P. M. V. D. Sai Baba, and L. Ramya. 2016. "*Classification of EEG Based Control Using ANN and KNN- A Comparison.*" 0–5.

Rahimi, M. 2016. "*Ensemble Methods Combination for Motor Imagery Tasks in Brain Computer Interface.*" no. November: 23–25.

Samanta, B. 2017. "*Empirical Mode Decomposition of EEG Signals for Brain Computer Interface.*" 4–9.

Samy, A. A. 2016. "*Extracting and Discriminating Selective Brain Signals in Non-Invasive Manner and Using Them for Controlling a Device : A Cost-Efficient Approach to Brain Computer Interface (BCI).*"

Singh, M., M. Singh, and A. Sandel. 2014. "Data Acquisition Technique for EEG Based Emotion Classification." *IJITKM* 7 (2): 133–42.

Soutar, R. 2013. "An Introductory Perspective on the Emerging Application of qEEG in Neurofeedback." *Clinical Neurotherapy: Application of Techniques for Treatment* 19–54. https://doi.org/10.1016/B978-0-12-396988-0.00002-7.

Spruston, N. 2008. "Pyramidal Neurons: Dendritic Structure and Synaptic Integration." *Nature Reviews Neuroscience* 9 (3): 206–21. https://doi.org/10.1038/nrn2286.

Šťastný, J., P. Sovka, and A. Stančák. 2003. "EEG Signal Classification: Introduction to the Problem." *Radioengineering* 12 (3): 51–55. https://doi.org/10.1109/IEMBS.2001.1020628.

Subasi, A., and E. Erçelebi. 2005. "Classification of EEG Signals Using Neural Network and Logistic Regression." *Computer Methods and Programs in Biomedicine* 78 (2): 87–99. https://doi.org/10.1016/j.cmpb.2004.10.009.

Yang, Y., H. Wang, and S. Member. 2017. "*A Brain-Computer Interface Based on a Few-Channel EEG-fNIRS Bimodal System.*" 5.

Zafar, M. B., K. A. Shah, H. A. Malik, and A. Principles. 2017. "*Prospects of Sustainable ADHD Treatment Through Brain-Computer Interface Systems.*" 1–6.

Zhang, R., P. Xu, R. Chen, T. Ma, X. Lv, F. Li, P. Li, T. Liu, and D. Yao. 2015. "An Adaptive Motion-Onset VEP-Based Brain-Computer Interface." 604 (c). https://doi.org/10.1109/TAMD.2015.2426176.

17 IoT-Based Wearable Medical Devices

Avitaj Mitra, Abanish Roy, Harshit Nanda, Riddhi Srivastava, and Gayathri M.

CONTENTS

17.1 INTRODUCTION

In the current day and age, with increasing usage of technology, the majority of people in urban areas work white-collar jobs. These jobs, although well paid, involve working in front of a screen for extended periods of time. The lifestyle of these people is becoming increasingly sedentary. This, combined with a propensity for consuming large quantities of junk food, leaves them susceptible to various diseases like diabetes, hypertension, cholesterol, heart disorders, and many more.

The Internet of things (IoT) is a concept, which in simple terms is the interconnectivity of sensors, actuators, and microcontrollers, etc. along with smart devices and Internet connectivity to exchange and modify large sets of data. A subset of IoT is wearable medical devices (WMDs). As the name suggests, these are small contraptions like a band or a small microchip, which can be worn on any part of the body and can help to monitor certain parameters and send warning signals if necessary. The basic components of a WMD are the chip/band, a sensor, a detector, a Bluetooth system for data transmission, and an alarm detector system. Since the simplest method to treat diseases is to detect the symptoms as early as possible, WMD's can prove to be hugely beneficial in this aspect.

Apart from the usual problems faced in the initial concept design formulation, obtaining funds for components, etc. possibly the biggest challenge preventing the large-scale commercialization of medical devices is the time taken for FDA approval. A probable reason for this is that these devices target the core physiological functions of the body. Any kind of operational failure can, in the worst possible scenario, lead to the death of the patient.

There is no doubt about the fact that WMDs, though currently at a nascent stage in their development, is the technology for the future. With the advent of smartphone technology and with increasing and widespread access to Internet services for the common people, health care professionals will come to depend upon wearable devices as a means to improve health care and reduce patient mortality rates.

17.1.1 The History of Medical Devices

WMDs are a classification of devices generally attached to the surface of the skin of an individual. They usually assist in monitoring certain parameters such as heartbeat, blood pressure, and oxygenation levels. An example of this is a heartbeat monitor or an SpO2 sensor. The other major function of a WMD is to improve the efficiency of any kind of cognitive function, which has degenerated in the human body. An example of this kind of medical device is a hearing aid/implant. The earliest examples of medical devices can be considered to be stethoscopes, sphygmomanometers, earlier forms of CT and MRI scanners, etc. However, the shift in medical device manufacturing received a big boost with the advent of the IC, semiconductors, and BioMEMS or biological microelectromechanical systems technology in the 1970s and 1980s. This led to devices small in size, increasing their portability and overall utility. The major impact has been on cardiovascular devices with a number of heart valves and stents coming up in the market. All this has increased not only the lifespan of patients but also provided comfort and improved their daily lives.

17.1.2 The Advent of IoT

The IoT is a global system facilitating the exchange of information in a secure manner. The primary architecture is based on data communication, facilitated by the use of RFID (Radio Frequency Identification) tags (Weber and Weber 2010). In the current industrial scenario, the most popular method is that of an EPC (Electronic Product Code). In simple terms, this means that a unique code is given to any object with an RFID tag, which helps to track the object remotely, without causing any discomfort to the individual carrying the object. The entire data is not saved within the RFID tag, but the information is made available to servers on the Internet by cross-linking with the help of an ONS (Object Naming System).

The ONS is a centralized system linking both metadata and services and is operated by a private company, VeriSign. The ONS architecture is designed to be multifunctional and can be used for computation, recognition and identification of objects, and retrieval of datasets from the Internet.

The ONS is based on the more commonly used and well-known domain name system (DNS). In order to retrieve information about an object, the EPC of the object should be converted into a form the DNS can understand. This is done by using the usual format of a domain name ("dot," delimited, "domain name from left to right"). The EPC is usually encoded within the domain name and is used within the existing architecture of the DNS. ONS can be considered to be a subset of the DNS.

Thus, using IoT, the data collected by the WMDs can be transmitted to a storage device and can be accessed by a medical practitioner in real time. This has enabled the development of new age WMDs, which are aimed at improving health care delivery, with minimal effect on patient comfort and mobility.

17.1.3 The Age of WMDs

The recent advances in wireless sensor technology and information technology have resulted in an unprecedented number of technological advancements in the field of WMDs based on IoT (IoT). The use of WMDs for both monitoring and diagnostic purposes is rapidly becoming popular. With an increasing number of people owning smartphones enabled with Bluetooth, global positioning system (GPS), and Internet service it has become much easier to link WMDs, which measure physiological parameters for a software. This has opened up new avenues in medical devices currently being explored.

Round-the-clock monitoring of certain physiological parameters are of paramount importance in the detection and control of cardiovascular diseases. With the development of advanced wireless sensors, we no longer need to confine a patient in a hospital bed to monitor his/her vitals; it can be done remotely as the patients carry on with their daily lives. The power of remotely monitoring the vitals of a patient can go a long way in exponentially increasing the reach of proper health care facilities to semirural and rural locations. With the advent of IoT, these physiological data can be stored and transferred to the desktop of a physician for monitoring in real time.

17.2 MECHANISM OF WMDs

The heart of any WMD is data. In this case, data refers to patient history of the vital signs such as core body temperature, heart rate, and blood pressure. The functioning of any WMD can be broken down into the following: collection of raw data, transmitting the raw data using Ethernet, Bluetooth, or wireless technology, uploading the data into a secure server linked to a research facility and/or a hospital, postprocessing of the data using mathematical algorithms and statistical analysis, and finally displaying and storing the processed data using encrypted files, which can then be sent to individual patients.

The detection, diagnosis, and treatment of diseases using WMDs depends upon various factors such as the type of bio signal extracted, previous patient history, any genetic aspects which may interfere with electronic signals from the wearable device, the age of the person, work history (for a working professional, round-the-clock remote monitoring may not be a feasible option), and any kind of lifestyle changes, which need to be incorporated, etc. In short, it can be said there is no perfect wearable device, or no ideal form of treatment existing for an individual, and the diagnosis and treatment need to be carried out on a case-by-case basis.

17.3 DETECTION, DIAGNOSIS, AND TREATMENT

17.3.1 Hypertension

With the rise in the percentage of the population suffering from obesity, the number of people suffering from hypertension is on the rise. High blood pressure is often asymptomatic, and thus hypertension goes on unnoticed in many patients before it is too late. Recent studies have shown that blood pressure variability, or BPV, is an important factor for determining the cardiovascular health of an individual. Traditionally, blood pressure was measured in a clinic by a trained technician, nurse, or a doctor. In that scenario, constant monitoring of blood pressure was not possible. To tackle this problem, medical researchers and engineers have come up with a few types of WMD (Yilmaz et al. 2010).

The first type of medical device designed for home use by nontrained personnel was based on the oscillometric method. It consists of an armband, much like the one used in clinics attached to a monitoring device with a display. This device could measure the blood pressure with respect to the brachial artery. A similar type of device was later designed able to measure the blood pressure using the radial artery (Teng et al. 2008).

This new device used a wrist cuff instead of an armband. This enabled the development of smart wristwatches for blood pressure monitoring and display. The smartwatch could be connected to a smartphone to store and relay the data as per needed. The wrist cuff or smartwatch system made it easier to use the device outdoors with ease. It reduced the bulkiness of the previous devices (Yilmaz et al. 2010).

Although these devices enabled people to monitor their blood pressure without the aid of a doctor or a trained clinician, there were certain drawbacks. One of the major drawbacks was the sensitivity and data reliability. Second, these methods do not allow for a continuous monitoring of blood pressure; at best these devices could allow the monitoring of blood pressure at a regular interval of around 15–30 minutes. These methods can also lead to wrist or arm pain, sleep disturbance, and can even cause skin allergy (Bobade and Walli 2015).

Vasotrac (Medwave Inc., Arden Hills, MN) is a popular wristwatch-based ambulatory noninvasive blood pressure sensor. This medical device consists of a circular sensor placed on top of the radial artery (marked using a disposable adhesive tape) and a display unit. For reliable measurement, an external pressure needs to be applied. The data is collected from the sensor, processed by a control unit, and then displayed on the screen. Vasotrac is a good and reliable device for periodic measurement of blood pressure, but not for continuous blood pressure monitoring (Findlay et al. 2006).

17.3.2 Epilepsy

Epilepsy is a functional pathophysiology present in the cerebrum part of the brain, which arises in almost all mammalian species. It's a medical condition classified with a broader range of different types of seizures and activation from one individual to another. Different types of brain seizures in epilepsy lead to the development of brain tumors.

In epilepsy, people face different types of brain seizures. This marks the development of brain tumors. A person suffering with epilepsy has higher chances of suffering from a depression. This is due to higher exposure to chronic stress (Duyn et al. 2007). During sleep nocturnal seizures occur, which worsen the condition of the patient. These seizures mostly occur at specific sleep stages or during sleep instability (Bernhardt et al. 2015). The recent advance in the pathogenesis of epilepsy is the involvement of genetic testing. Next generation sequencing has proven to be effective for revealing epilepsy causing gene mutations (Witt et al. 2013).

Epilepsy is initially treated with medications where approximately 70% of patients get complete seizure control (French 2007). However, this is not possible for patients who have poor seizure control, and for them several other methods are used such as medical management, surgery, vagus nerve stimulation (VNS), ketogenic diets, and complementary therapies.

To detect brain abnormalities, several techniques are used such as electroencephalogram (EEG), computerized tomography (CT) scan (Duyn et al. 2007), magnetic resonance imaging (MRI) (Jewells and Shin 2014), functional MRI (Fang et al. 2017), positron emission tomography (PET) (You et al. 2012), and single photon emission computerized tomography (SPECT) (Błaszczyk and Czuczwar 2016).

There are various WMDs being developed for the timely detection of epileptic seizures. One of the latest devices is SeizeIT, which is being developed in KU Leuven (Vandecasteele et al. 2017). It is based on designing a WMD that detects the electrocardiography (ECG) and photoplethysmography (PPG) in real time. The wearable ECG device proved to be as efficient as the wired hospital ECG systems, and thus allows for greater patient mobility and comfort. The sensitivity of the wearable ECG device was 70%, whereas that of the hospital system was 57%, and it had a slightly greater false alarm rate (2.11 per hour as compared to 1.92 per hour rate of the hospital system) making it an ideal choice for people suffering from epilepsy.

The Brain Sentinel is an interventional device for the detection of tonic clonic seizures. It is based on measurement of electromyography (EMG) signals from the concerned subject and then analyzing and transmitting the data to look for identifiable patterns. The subjects were monitored for seizure-based activity for a period of 4.5 years and the test results are currently being analyzed for any favorable outcomes. However, one disadvantage is that the study excludes pregnant females from participation.

The Embrace Watch is an IoT-based device providing regular monitoring of vital signs for epilepsy-afflicted patients. It is considered to be a therapeutic rather than a diagnostic device, which transmits the dataset to a smartphone using Bluetooth technology. The data is then uploaded to Empatica servers where further analysis and processing is performed.

The neurospace responsive neurostimulator (RNS) is a recently FDA-approved device for epilepsy treatment. The principle behind this device is monitoring the brain activity and thus delivering an electrical stimulus when there is a detection of abnormal brain activity in different regions of brain (Thomas and Jobst 2015).

The RNS system is approved in the United States as a medical therapy for treatment of partial onset seizures of epilepsy. The different parts of this system include a stimulator, implanted leads, and wireless programming wands connected to a complex computer hardware and software system. The stimulator works on a design where wires connected to electrodes are implanted in the hippocampus area of the brain. The machine targets the resistant partial seizure onset zone.

The neurotransmitter monitors the ECoG (electrocorticography) activity and utilizes various methods for detection of abnormal brain activity. Measurement of changes in electrical activity and frequencies is done with this machine. The four proposed theories for mechanism of action includes:

- Depolarization blockage, which refers to changes in voltage-gated channels leading to process of inhibition in excitability (Beurrier et al. 2001).
- Synaptic inhibition refers to depolarization effects of distal axon.
- Synaptic depression, which causes the decrease in release of neurotransmitters (McIntyre et al. 2004).
- Electrical stimulation by modulating the action of pathologic networks (Durand 1986).

The RNS system does not include the allowance for external recordings and includes only limited sampling for selective ECoG. A comprehensive file of the data obtained through this system is very expensive and is difficult to track the efficacy of seizures. Further research and experiments are currently carried out for provision of better understanding and effects of this system on the human body.

17.3.3 Cancer

Cancer is among the deadliest diseases in the world, for which there is no comprehensive cure till now. Early detection of cancer can enable its treatment is most cases. However, a large percentage of the cancer cases are detected in the late stages, where nothing much can be done except to ease the pain and accept the inevitable outcome. Countless research groups across the globe are working on developing methods to enable early detection of cancers to give patients a fighting chance, and to hopefully defeat this deadly disease.

The advent of IoT and WMDs has brought about great improvements in the cancer care delivery system. Real-time data collection and analysis has undergone massive improvements in the last few years (Sledge et al. 2013). CancerLinQ, an initiative of the American Society of Clinical Oncology focuses on collecting and analyzing patient data from electronic health data to drive the process of developing better cancer care technologies.

Patient generated health data (PGHD) is an important tool in the cancer care delivery system. It includes the data collected from the patient's regular environment in form of WMD data, other local sensor data, and outcomes reported by the patients. PGHD enables us to have a holistic view on the real-time health of the patient (Chung et al. 2016). IoT technologies can be used to collect and transmit these PGHD to oncologists and other researchers in real time. Because of IoT and WMDs, it is possible to collect such health data outside the clinic in real time.

There are a few sensor-based devices having been developed to detect various types of cancer in their early stages. A majority of these devices have focused on skin cancer and breast cancer detection. One such device was developed by S. K. Attili et al., which is a lightweight, organic light-emitting diode using photo-dynamic therapy (PDT) to detect skin cancer (Attili et al. 2009). Some of the other devices for skin cancer detection include a wearable designed by D. D. Godoy et al. which has a series of smart sensors able to detect the ambient temperature, humidity, light, and UV and notifies user through bluetooth about possible UV-harm using a smartphone-based application (Ray et al. 2017). My UV Patch and Violet Plus are two more WMDs to be worn around the wrist, which detects the UV radiation of the sun on the skin—one of the primary causes of skin cancer. The former is designed as an adhesive patch, whereas the latter is designed as a clip attached module.

In case of breast cancer, A. Rahman et al. designed a compact and ultrawide band antenna placed on a flexible substrate, which was able to measure breast cancer using microwave imaging (Rahman et al. 2016). Teng et al. 2017, developed a diffusion-based optical probe working on the principle of using continuous waves to supply the hemodynamic response during neoadjuvant chemotherapy infusions (Teng et al. 2017). H. Bahrami et al. developed a lightweight, flexible, and cheap microwave array of antennas able to detect breast cancer (Bahramiabarghouei et al. 2015).

Beyond these few devices, researchers and engineers around the world are developing better wearable devices integrated with IoT to detect cancer at an early stage. Although most of the devices discussed focused on skin and breast cancer, some research is being done to develop devices to detect prostate cancer (Ray et al. 2017).

17.3.4 Migraines

Migraine is a disorder of the neural as well as the vascular systems of the body (Silberstein and Silberstein 2004; Lipton et al. 2001). It is typically considered as a form of headache, though certain migraines can occur even without any kind of definitive pain.

In the general population, migraine is more prevalent among women, and nearly three times as many women suffer from migraine compared to adult men (Durand 1986). However, the onset of migraines is earlier among boys, which implies that until puberty, boys are likely to suffer from migraines more than girls (Abel 2009). Currently, three genes have been found for migraine and it is anticipated that further genes will be discovered, providing a genetic basis for explaining the disorder.

Migraines are likely to be a major distraction from work and familial responsibilities, since they have a debilitating effect on the energy and activity levels of an individual. Roughly one-third of migraineurs say they have missed school and/or work due to migraines, while 52% to 73% of the people feel their lives are being adversely affected by migraine attacks.

There has been extensive research carried out on the pathophysiology of migraines, however a definite cause or pathway has not yet been identified. The older theory, as propounded by Wolff, was that migraines were caused by constriction and dilation of cranial blood vessels.

However, current research points to the trigeminovascular system, or more specifically, the interactions between the ophthalmic area of the trigeminal nerve with the dura mater and the cranial blood vessels. A more recent theory postulates that the sensitization of neurons, specifically peripheral sensitization, may be responsible for the throbbing pain usually associated with migraines.

Treatment for migraine mostly involves either pharmacological drugs or sustained care under a neurologist and/or optometrist. Ensuring regular sleep and having meals on time may prevent migraines. Specific techniques, which have proven to be effective, include relaxation training, biofeedback training, and cognitive behavioral therapy.

Cefaly is an FDA-approved noninvasive method of treatment for individuals affected by frequent migraines. It is an external trigeminal nerve stimulation device, wherein a self-adhesive electrode is attached to the forehead and the Cefaly device is connected with it. Microimpulses are sent to the two branches of the trigeminal nerve to either relieve the pain in the head region or to prevent future migraine attacks.

17.4 COMPUTATIONAL TESTING METHODS

17.4.1 Epilepsy

Computational analysis provides a time- and cost-effective approach for the assessment of cognitive functions in patients suffering from epilepsy. The computerized testing is advantageous for the evaluation process in attenuation and they are also easily administrable.

17.4.1.1 Computerized Cognitive Testing in Epilepsy (CCTE)

This technique comprises mainly of performing eight tasks including digit span forward and backward, focused attention, incidental recollection issues, verbal learning, and figular memory (Kurzbuch et al. 2013). The performance in various subsets of CCTE shows a match with the results obtained in paper-pencil tests and thus, validates epilepsy (Orsini et al. 2016).

17.4.1.2 Cognitive Drug Research Computerized Assessment System (CDR)

The different assessment techniques of core aspects of cognitive function sustain the ability to conduct various activities in day-to-day life. The major mechanism underlying this technique is change over time in cognitive functions. This technique is simple to administer and is applicable in the studies on psychomotor effects, which are produced by remacenide and carbamazepine in the diagnostic treatment of newly developed epilepsy (Jan 2007).

17.4.1.3 Cognitive Neurophysiological Test (CNT)

The major advantage of this technique is it incorporates the speed and precision of task performance, as well as type of neural resources and the efforts to provide levels of performance and alertness are increased by this test. The additional knowledge is therefore increased to the specific effects of drugs and their in-depth pharmacokinetics and pharmacodynamics in the human body (Tufenkjian and Lüders 2012; Packer et al. 2015). Validation of this test is in the intellectual effects of antiepileptic drugs (Stufflebeam 2011; Hasegawa 2016).

17.4.1.4 FePsy

It comprises the eleven computerized tests for neurophysiological functions and is a relative database system for storing all the obtained results. It includes various subsets of tests for the measurement of different side-effects produced by the drugs. The simultaneous EEG recording is used to assess the effects of frequency and short and epileptiform EEG discharges in the absence of seizure (Menicucci et al. 2015; Stufflebeam 2011).

17.4.1.5 Neurocog FX

The key feature of this computerized software is the assessment of patients suffering from epilepsy and neurological diseases with respective experimental setups. This specific test was designed to be used in detection of CNS diseases and monitoring neuro-oncology and neurosurgery (Helmstaedter et al. 1996). This software is available only in Germany and the reaction of the patient is monitored in real time.

17.4.2 Cancer

In recent years, there has been several developments in the field of detecting cancer in its early stages using computer programs and applications. There have been some major breakthroughs in this regard.

In 2015, a team of scientists and engineers from the Technical University of Denmark developed a set of algorithms, dubbed as TumorTracer, able to detect the exact location of cancer with a nearly 85% success rate. It enabled the proper identification of the tissue of origin of the cancer, which is crucial for designing a proper treatment plan. This was a game changer for the people who suffered from cancer, whose primary origin was difficult to locate using conventional methods (Marquard et al. 2015).

In March 2017, a team of researchers from the University of California, Los Angeles designed a computer program dubbed CancerLocator, which could detect the tumor's DNA and from where it was coming from using blood samples of the patients. This program works by looking for specific molecular patterns in cancer DNA. It exploits the theory that cell free DNA from cancer cells are often found in the blood stream. By comparing their sequence with a known database of tumor cell sequences it was possible to both detect the cancer and locate its origin (Kang et al. 2017).

Beyond these two, there are two important algorithms used by several research groups to diagnose cancer. One of them was random forest. Cuong Nguyen et al., in 2013, designed a system based on this algorithm able to differentiate benign breast cancer tumors from malignant ones using a random forest algorithm. It was able to do so with an accuracy rate greater than 99% (Nguyen et al. 2013). This method was obtained to solve diagnosis problems by classifying the Wisconsin Breast Cancer Diagnosis Dataset (WBCDD) and to solve prognosis problems by classifying it.

Another algorithm popularly used in cancer detection is support vector machine (Sweilam et al. 2010).

17.5 CONCLUSION

To summarize, it is clear that WMDs have certainly proven to be hugely beneficial for the health care community. Not only have they significantly improved the lives of people suffering from various disorders, they have also made the job of medical professionals easier by introducing the concept of remote monitoring. The potential drawbacks, however minor, cannot be entirely ignored. Security and confidentiality of patient data is absolutely crucial and forms the framework of wearable devices. The unreasonably high cost of certain devices can sometimes act as a deterrent, however with increasingly newer technologies being developed, it can be hoped that wearable devices will become affordable for a majority of the population in the near future.

REFERENCES

Abel, H. 2009. "Migraine Headaches: Diagnosis and Management." *Optometry* 80 (3). Mosby, Inc:138–48. https://doi.org/10.1016/j.optm.2008.06.008.

Attili, S. K., A. Lesar, A. McNeill, M. Camacho-Lopez, H. Moseley, S. Ibbotson, I. D. W. Samuel, and J. Ferguson. 2009. "An Open Pilot Study of Ambulatory Photodynamic Therapy Using a Wearable Low-Irradiance Organic Light-Emitting Diode Light Source in the Treatment of Nonmelanoma Skin Cancer." *British Journal of Dermatology* 161 (1):170–73. https://doi.org/10.1111/j.1365-2133.2009.09096.x.

Bahramiabarghouei, H., E. Porter, A. Santorelli, B. Gosselin, M. Popović, and L. A. Rusch. 2015. "Flexible 16 Antenna Array for Microwave Breast Cancer Detection." *IEEE Transactions on Biomedical Engineering* 62 (10):2516–25. https://doi.org/10.1109/TBME.2015.2434956.

Bernhardt, B. C., S.-J. Hong, A. Bernasconi, and N. Bernasconi. 2015. "Magnetic Resonance Imaging Pattern Learning in Temporal Lobe Epilepsy: Classification and Prognostics." *Annals of Neurology* 77 (3):436–46. https://doi.org/10.1002/ana.24341.

Beurrier, C., B. Bioulac, J. Audin, and C. Hammond. 2001. "High-Frequency Stimulation Produces a Transient Blockade of Voltage-Gated Currents in Subthalamic Neurons." *J Neurophysiol* 85 (4):1351–56. http://www.ncbi.nlm.nih.gov/pubmed/11287459%5Cnhttp://jn.physiology.org/content/jn/85/4/1351.full.pdf.

Błaszczyk, B., and S. J. Czuczwar. 2016. "Epilepsy Coexisting with Depression." *Pharmacological Reports* 68 (5):1084–92. https://doi.org/10.1016/j.pharep.2016.06.011.

Bobade, Y., and R. M. Walli. 2015. "A Review of Wearable Health Monitoring Systems." *International Journal of Advanced Research in Electrical, Electronics and Instrumentation Engineering* 4 (10):8386–90. https://doi.org/10.15662/IJAREEIE.2015.0410076.

Chung, A. E., R. E. Jensen, and E. M. Basch. 2016. "Leveraging Emerging Technologies and the 'Internet of Things' to Improve the Quality of Cancer Care." *Journal of Oncology Practice* 12 (10):863–66. https://doi.org/10.1200/JOP.2016.015784.

Durand, D. 1986. "Electrical Stimulation Can Inhibit Synchronized Neuronal Activity." *Brain Research* 382 (1):139–44. https://doi.org/10.1016/0006-8993(86)90121-6.

Duyn, J. H., P. van Gelderen, T.-Q. Li, J. A. de Zwart, A. P. Koretsky, and M. Fukunaga. 2007. "High-Field MRI of Brain Cortical Substructure Based on Signal Phase." *Proceedings of the National Academy of Sciences* 104 (28):11796–801. https://doi.org/10.1073/pnas.0610821104.

Fang, Z., Y. Yang, X. Chen, W. Zhang, Y. Xie, Y. Chen, Z. Liu, and W. Yuan. 2017. "Advances in Autoimmune Epilepsy Associated with Antibodies, Their Potential Pathogenic Molecular Mechanisms, and Current Recommended Immunotherapies." *Frontiers in Immunology* 8 (APR). https://doi.org/10.3389/fimmu.2017.00395.

Findlay, J. Y., B. Gali, M. T. Keegan, C. M. Burkle, and D. J. Plevak. 2006. "Vasotrac® Arterial Blood Pressure and Direct Arterial Blood Pressure Monitoring during Liver Transplantation." *Anesthesia and Analgesia* 102 (3):690–93. https://doi.org/10.1213/01.ane.0000196512.96019.e4.

French, J. A. 2007. "Refractory Epilepsy: Clinical Overview." *Epilepsia* 48 (SUPPL. 1):3–7. https://doi.org/10.1111/j.1528-1167.2007.00992.x.

Hasegawa, D. 2016. "Diagnostic Techniques to Detect the Epileptogenic Zone: Pathophysiological and Presurgical Analysis of Epilepsy in Dogs and Cats." *Veterinary Journal* 215. Elsevier Ltd:64–75. https://doi.org/10.1016/j.tvjl.2016.03.005.

Helmstaedter, C., B. Kemper, and C. E. Elger. 1996. "Neuropsychological Aspects of Frontal Lobe Epilepsy." *Neuropsychologia* 34 (5):399–406. https://doi.org/10.1016/0028-3932(95)00121-2.

Jan, M. M. 2007. "The Value of Seizure Semiology in Lateralizing and Localizing Partially Originating Seizures." *Neurosciences (Riyadh, Saudi Arabia)* 12 (3):185–90. http://www.ncbi.nlm.nih.gov/pubmed/21857567.

Jewells, V., H. W. Shin. 2014. "Review of Epilepsy—Etiology, Diagnostic Evaluation and Treatment." *International Journal of Neurorehabilitation* 1 (3):1–8. https://doi.org/10.4172/2376-0281.1000130.

Kang, S., Q. Li, Q. Chen, Y. Zhou, S. Park, G. Lee, B. Grimes et al. 2017. "CancerLocator: Non-Invasive Cancer Diagnosis and Tissue-of-Origin Prediction Using Methylation Profiles of Cell-Free DNA." *Genome Biology* 18 (1):53. https://doi.org/10.1186/s13059-017-1191-5.

Kurzbuch, K., E. Pauli, L. Gaál, F. Kerling, B. S. Kasper, H. Stefan, H. Hamer, and W. Graf. 2013. "Computerized Cognitive Testing in Epilepsy (CCTE): A New Method for Cognitive Screening." *Seizure* 22 (6):424–32. https://doi.org/10.1016/j.seizure.2012.08.011.

Lipton, R. B., F. Stewart, S. Diamond, M. L. Diamond, and M. Reed. 2001. "Prevalence and Burden of Migraine in the United States: Data from the American Migraine Study II." *Headache* 41 (7):646–57. https://doi.org/10.1046/j.1526-4610.2001.041007646.x.

Marquard, A. M., N. J. Birkbak, C. E. Thomas, F. Favero, M. Krzystanek, C. Lefebvre, C. Ferté et al. 2015. "TumorTracer: A Method to Identify the Tissue of Origin from the Somatic Mutations of a Tumor Specimen." *BMC Medical Genomics* 8 (1):58. https://doi.org/10.1186/s12920-015-0130-0.

McIntyre, C. C., M. Savasta, L. K.-L. Goff, and J. L. Vitek. 2004. "Uncovering the Mechanism(s) of Action of Deep Brain Stimulation: Activation, Inhibition, or Both." *Clinical Neurophysiology* 115 (6):1239–48. https://doi.org/10.1016/j.clinph.2003.12.024.

Stufflebeam, S. M. 2011. "Clinical Magnetoencephalography for Neurosurgery." *Neurosurgery Clinics of North America* 22 (2):153–67. https://doi.org/http://dx.doi.org/10.1016/j.nec.2010.11.006.

Menicucci, D., A. Piarulli, P. Allegrini, R. Bedini, M. Bergamasco, M. Laurino, L. Sebastiani, and A. Gemignani. 2015. "Looking for a Precursor of Spontaneous Sleep Slow Oscillations in Human Sleep: The Role of the Sigma Activity." *International Journal of Psychophysiology* 97 (2). Elsevier B.V.:99–107. https://doi.org/10.1016/j.ijpsycho.2015.05.006.

Nguyen, C., Y. Wang, and H. N. Nguyen. 2013. "Random Forest Classifier Combined with Feature Selection for Breast Cancer Diagnosis and Prognostic." *Journal of Biomedical Science and Engineering* 6 (5):551–60. https://doi.org/10.4236/jbise.2013.65070.

Orsini, A., F. Zara, and P. Striano. 2016. "Recent Advances in Epilepsy Genetics." *Neuroscience Letters*. Elsevier Ireland Ltd. https://doi.org/10.1016/j.neulet.2017.05.014.

Packer, R., M. Berendt, S. Bhatti, M. Charalambous, S. Cizinauskas, L. D. Risio, R. Farquhar et al. 2015. "Inter-Observer Agreement of Canine and Feline Paroxysmal Event Semiology and Classification by Veterinary Neurology Specialists and Non-Specialists." *BMC Veterinary Research* 11 (1):39. https://doi.org/10.1186/s12917-015-0356-2.

Rahman, A., M. T. Islam, M. J. Singh, S. Kibria, and M. Akhtaruzzaman. 2016. "Electromagnetic Performances Analysis of an Ultra-Wideband and Flexible Material Antenna in Microwave Breast Imaging: To Implement A Wearable Medical Bra." *Scientific Reports* 6 (November). Nature Publishing Group:1–11. https://doi.org/10.1038/srep38906.

Ray, P. P., D. Dash, and D. De. 2017. "A Systematic Review of Wearable Systems for Cancer Detection: Current State and Challenges." *Journal of Medical Systems* 41 (11). https://doi.org/10.1007/s10916-017-0828-y.

Silberstein, S., and S. D. Silberstein. 2004. "Migraine Pathophysiology and Its Clinical Implications." *Blackwell Science* 24:2–7. https://doi.org/10.1111/j.1468-2982.2004.00892.x.

Sledge, G. W., C. A. Hudis, S. M. Swain, P. M. Yu, J. T. Mann, R. S. Hauser, and A. S. Lichter. 2013. "ASCO's Approach to a Learning Health Care System in Oncology." *Journal of Oncology Practice / American Society of Clinical Oncology* 9 (3):145–48. https://doi.org/10.1200/JOP.2013.000957.

Sweilam, N. H., A. A. Tharwat, and N. K. Abdel Moniem. 2010. "Support Vector Machine for Diagnosis Cancer Disease: A Comparative Study." *Egyptian Informatics Journal* 11 (2): 81–92. https://doi.org/10.1016/j.eij.2010.10.005.

Teng, F., T. Cormier, A. Sauer-Budge, R. Chaudhury, V. Pera, R. Istfan, D. Chargin, S. Brookfield, N. Y. Ko, and D M. Roblyer. 2017. "Wearable Near-Infrared Optical Probe for Continuous Monitoring During Breast Cancer Neoadjuvant Chemotherapy Infusions." *Journal of Biomedical Optics* 22 (1):14001. https://doi.org/10.1117/1.JBO.22.1.014001.

Teng, X.-F., Y.-T. Zhang, C. C. Y. Poon, and P. Bonato. 2008. "Wearable Medical Systems for P-Health." *IEEE Reviews in Biomedical Engineering* 1:62–74. https://doi.org/10.1109/RBME.2008.2008248.

Thomas, G. P., and B C. Jobst. 2015. "Critical Review of the Responsive Neurostimulator System for Epilepsy." *Medical Devices: Evidence and Research* 8:405–11. https://doi.org/10.2147/MDER.S62853.

Tufenkjian, Kr., and H. O. Lüders. 2012. "Seizure Semiology: Its Value and Limitations in Localizing the Epileptogenic Zone." *Journal of Clinical Neurology (Korea)* 8 (4): 243–50. https://doi.org/10.3988/jcn.2012.8.4.243.

Vandecasteele, K., T. D. Cooman, Y. Gu, E. Cleeren, K. Claes, W. V. Paesschen, S. V. Huffel, and B. Hunyadi. 2017. "Automated Epileptic Seizure Detection Based on Wearable ECG and PPG in a Hospital Environment." *Sensors (Switzerland)* 17 (10):1–12. https://doi.org/10.3390/s17102338.

Weber, R. H., and R. Weber. 2010. "Internet of Things." *Development*, 1–8. https://doi.org/10.1007/978-3-642-11710-7.

Witt, J. A., W. Alpherts, and C. Helmstaedter. 2013. "Computerized Neuropsychological Testing in Epilepsy: Overview of Available Tools." *Seizure* 22 (6). BEA Trading Ltd:416–23. https://doi.org/10.1016/j.seizure.2013.04.004.

Yilmaz, T., R. Foster, and Y. Hao. 2010. "Detecting Vital Signs with Wearablewireless Sensors." *Sensors* 10 (12):10837–62. https://doi.org/10.3390/s101210837.

You, G., Z. Sha, and T. Jiang. 2012. "The Pathogenesis of Tumor-Related Epilepsy and Its Implications for Clinical Treatment." *Seizure* 21 (3). BEA Trading Ltd:153–59. https://doi.org/10.1016/j.seizure.2011.12.016.

18 People with Disabilities
The Helping Hand of IoT

Ashmita Das, Sayak Mitra, and Shampa Sen

CONTENTS

18.1 SURVIVING WITH A DISABILITY

The World Health Organization (WHO) describes disability as an inability to lead a normal lifestyle according to their age, gender, social, and cultural state (Sen and Yurtsever 2007). It is an umbrella term covering impairments, boundary of participation, and restriction of activities. Currently, almost 15% of the world's population is estimated to be living with one form of disability or another.

We generally classify disabilities into four categories (World Health Organization (WHO 2003). They are:

1. Physical disability
2. Visual disability
3. Hearing disability
4. Cognitive disability

In today's world, the invention of various aids such as special glasses, walking sticks, and many more provide people support to lead an independent lifestyle under any circumstances (Stucki 2005). Hearing aids help deaf people to become connected. An affordable and effective alternative for deaf people is the American sign language (ASL) (Keating and Mirus 2004), which has evolved several means of communication among deaf people. The innovation of the wheelchair was as miraculous as the invention of the wheel itself, as it saved physically disabled people from being confined to a single place for a long time. But where a wheelchair causes problem, the crutches offer an alternative. Similarly, many other technologies have evolved to help people with disabilities and make their lives a bit easier.

The rapid development of technology has, however, brought more help to disabled people. One such boon is Internet of things (IoT). In essence, IoT promises that everyone and everything could be connected to each other. This makes it easier for people with disabilities to perform tasks they are usually unable to perform, simply by issuing a command to a smart device. Thanks to IoT, disabled people can change the thermostat settings of their home when they feel cold, summon a glass of water when they feel thirsty, open the door by means of a voice command, and even receive proper medications at the precise moment. Perhaps the most revolutionary application of IoT has been the automated smart home, which ensures, above all, the security of people with disabilities. IoT also allows concerned caregivers to monitor the well-being of disabled people and be notified immediately in case of emergencies.

This chapter focuses on the many ways IoT has helped people with disabilities gain confidence and become more sociable and communicable, while ensuring their proper health and well-being. It ends on an optimistic note, urging authorities to make these technologies more easily accessible to everybody, and not for profit, thus ensuring nobody is left out of society simply because they are unable to live like the majority.

18.2 EXISTING TECHNOLOGIES TO HELP DISABLED PEOPLE

Assistive technology includes all technologies designed to assist disabled people, starting from the wheelchair to hearing aids (Figure 18.1).

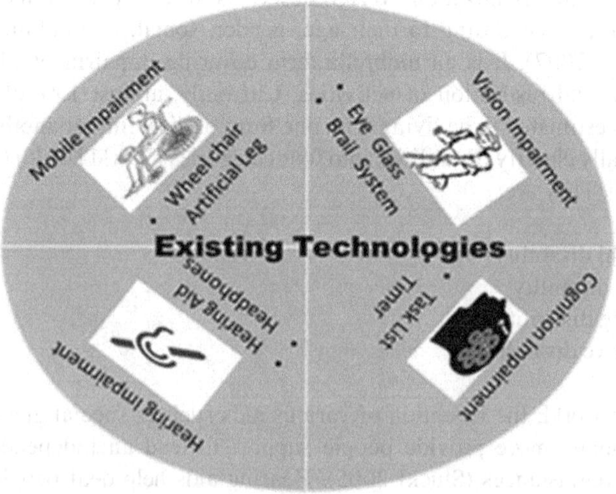

FIGURE 18.1 Existing technologies to help disabled people.

For the deaf, assistive devices such as hearing aids play a significant role. They are divided into two categories, external and internal devices. External assistive devices are worn by the patients, while internal ones are implanted inside the ear.

Among external devices, hearing aids (a battery operated electronic device) (Cox and Alexander 1983), assistive listening devices (ALDs) (Zanin and Rance 2016) amplify the incoming sounds to make them audible for the patients suffering from superficial or middle hearing loss problems. Specific telephone equipment helps the patients to communicate over the telephone. Other than that, several automated circuits are externally or internally implanted in different parts for producing profound sound (Garin et al. 2004). RetroX, Vibrant Sound Bridge (VSB) (Vincent et al. 2004), and cochlear implant devices (Minoda et al. 2004) are also helpful for severe hearing loss problems. An online sign language translator was invented converting English news or related articles to sign language by using a Natural Language Processor (Saleem and Alagha 2016).

Several technologies have been developed to help blind people. Screen readers plays a significant role to help the low vision or visually impaired patients (Neto and Fonseca 2014). Braille watches and braille printers are some of those latest inventions offering patients an opportunity to lead a normal life (Kurze 1996).

Wheelchairs, crutches, and artificial legs help the physically challenged people and make them more confident and independent. Above all, speech recognition, special keyboards such as "SmartNav," "Intellikeys," and "Sesame enable apps" are considered to be most effective and innovative solution.

Visually impaired people benefit from devices allowing them to use computers. They also present an alternative way to access the contents of the screen, process data, and command the computer. It is basically either a device or equipment which is able to read Braille language, or a software service (e.g., screen reading software) (Arrigo 2005).

18.3 DISABLED TO IoT-ABLED

The IoT provides the most effective and reliable solution for all problems related to disability. Nowadays, all things in the physical world achieve interconnection with each other through data exchange via the Internet. IoT highlights a complex paradigm, which also builds strong communication between things, things and humans, and even between humans (Coetzee and Olivrin 2012). It also has potential to solve problems related to remote health monitoring, fitness programs, etc. Various medical devices, sensors, and imaging devices are considered to be a core part of the IoT. IoT-based medical devices are expected to provide several advantages such as reduced costs, and increased quality of life and user experience (Ngu et al. 2017).

IoT health care networks are an essential part of IoT in the health care domain. They support the access to the IoT backbone and also facilitates the transmission of medical data (Figure 18.2).

IoT provides real-time access to device information and also allows for the remote management of devices. A combined system of IoT and cloud computing features a specialized scalable and globally accessible web application, which also provides an interface for effective and enhanced communication amongst the devices and also between the device and its user.

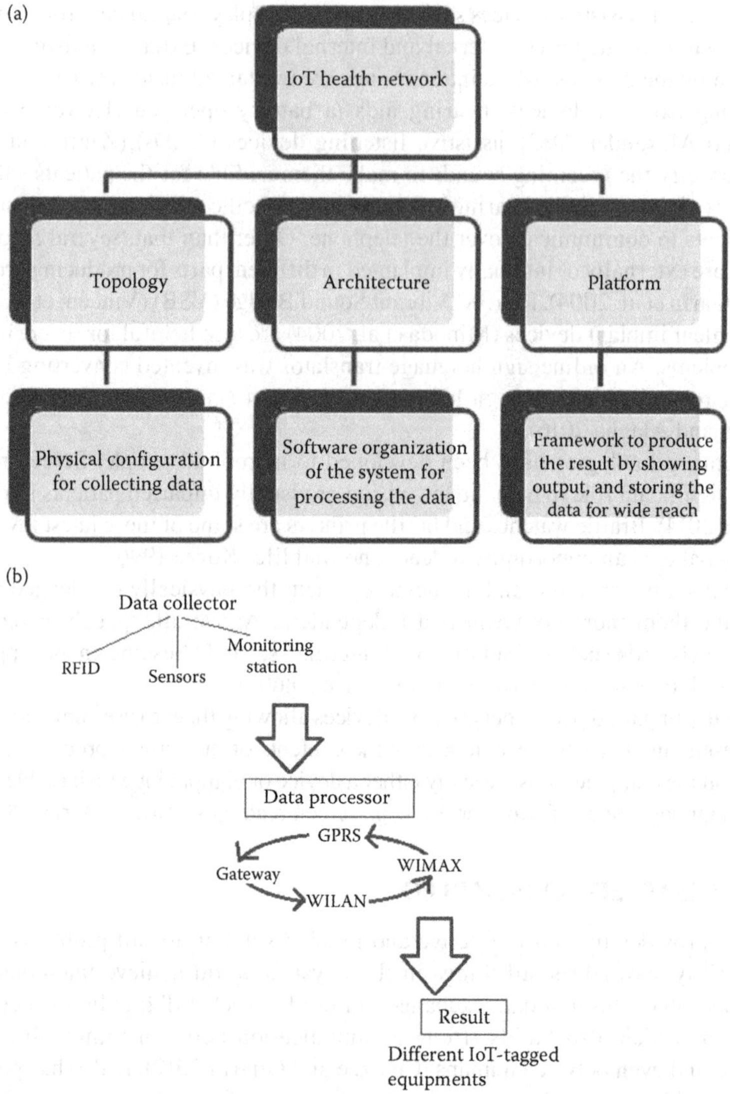

FIGURE 18.2 (a) The IoT health care network and its components, (b) The architecture of IoT equipments.

IoT-applied smart house or shopping scenarios helps the disabled person to carry out their daily activities. This also increases their self-confidence and autonomy. Smart houses enable the disabled person to move independently, handle objects, and also communicate effectively. Monitoring systems reduce or eliminate the dependency on caregivers by improving the autonomy of the individuals.

On the other side, disabled children can experience improved linguistic skills, emotional skills, and also social skills by using interactive learning and IoT (Hengeveld et al. 2009).

18.3.1 Physical and Dexterity

IoT plays a significant role in making the lives of physically challenged people easier, facilitating communications and helping them integrate into society.

Two equipments are known to play a major role, (1) body sensors, acutors, and neurochips, and (2) body sensors with RFID technology.

- Body sensors, acutors, and neurochips

Acutors are able to functionally reanimate paralyzed limbs. The sensors connected with the nerves are able to detect the user's urge to move a particular muscle, whereas acutors stimulate the movement of those muscles.

BIOnic Neurons (BIONs) (Domingo 2012) are wireless capsules injected in various parts of the body near the motor nerve. They detect voluntary command and stimulate a related appropriate response to the neuromuscular cells.

Electrocorticography (ECoG) (Yanagisawa et al. 2012) signals seem to control prosthetic arm controls and provide clinical motor restoration for patients.

Facial expression-based computer cursor control systems (Vasanthan et al. 2012) are also used commonly as an assistive device for the physically disabled persons. Various indicative expressions such as left and right cheek movement, eyebrow rising and lowering, and mouth open or shut controls various computer-related applications. Four illuminator stickers are placed in proper places to provide instructions, which are communicated to the computer through the BASIC STAMP microcontroller. The x-y coordinate changes are detected on the video image through movement marker and facial expression. The result is also represented by a binary number which is useful for controlling the cursor.

m-IoT (Istepanian et al. 2011) is a new concept introduced for helping the disabled elderly. It is a combination of IoT and m-health (mobile health i.e., wireless medical care). This technology synchronizes the functionalities of m-health and IoT for future and innovative applications. It connects IP-based communication technologies 6LoWPAN with 4G networks by introducing a new paradigm.

18.3.2 Visual

Visually impaired people also include people who suffer from partial blindness and even color blindness. People with visual disabilities face two main problems. First, they are not able to understand the pictorial or graphical information displayed on the screen. Second, the pointing devices (e.g., a mouse), which require eye-hand coordination, also create difficulties for them (Kavcic 2005).

The components introduced for visually impaired patients are different types of sensors such as body micro- and nanosensors, etc., as well as sensor assistive devices like RFID-based assistive devices. Schwiebert et al. discovered an artificial retina to restore a measure of vision to patients suffering from degenerative blindness due to retinitis pigmentosa and age-related diseases. In these two diseases, the photoreceptor cells of the outer retina start degrading but they do not affect the optic nerve, which is made of inner retinal ganglion nerve cells. Similarly, a camera attached to a pair of glasses is used to send data to the implanted device (body sensors) placed on the retina. This sensor made retina (Schwiebert et al. 2001) stimulates specified ganglion cells using electric impulses. The ganglion cells convert them into neurological signals and transfer the signals to the brain via the optical nerve.

Another important implant developed to solve the problem of visual impairment is the Bio-Retina (Domingo 2012). The Bio-Retina incorporates several nanosized components into the retinal implants and is used as a replacement for damaged photoreceptor cells. It converts the received natural light into electrical impulses, which stimulate neurons and transfers the image received by the Bio-Retina to the brain. The nanosized components are charged by a unique pair of activation eyeglasses.

RFID-based assistive devices are very useful for visually impaired people to navigate in unfamiliar envoronments. RFID tags are placed at the center of the sidewalks to prevent accidental falls. The Sesamonet System (Faggion and Azzalin 2011) provides safe and secure paths to allow visually challenged people to move and navigate independently. It is based on a path marked by passive low frequency RFID transponders. A unique identification number (UID) is assigned to every transponder and is stored in a database. The electronic stick reads the UID present on the road and transfers the data to a portable electronic device or smartphone where the Sesamonet software is opened. The software constantly maintains communication with saved local databases and relates the user's position to the important information of the surrounding environment (Figure 18.3).

Virtual leading blocks (Amemiya et al. 2004) is an effective tool for deaf-blind people. The wearable computer and ubiquitous environment play a role of virtual navigator. The navigator guides the user on the basis of position data detected by the RFID tags and the orientation of motion sensors in real time. Verbal instruction is translated by finger braille method, while the nonverbal information is translated into spatial, accentual, and numerical change of tactile stimuli.

Obstacle Recognition System (Nassih et al. 2012) is another sophisticated system proposed regarding this purpose. It consists of a can, which also has an RFID tag reader on it and a braille interface. It also possesses a mechanical battery, which provides full-time energy during the operation of the device.

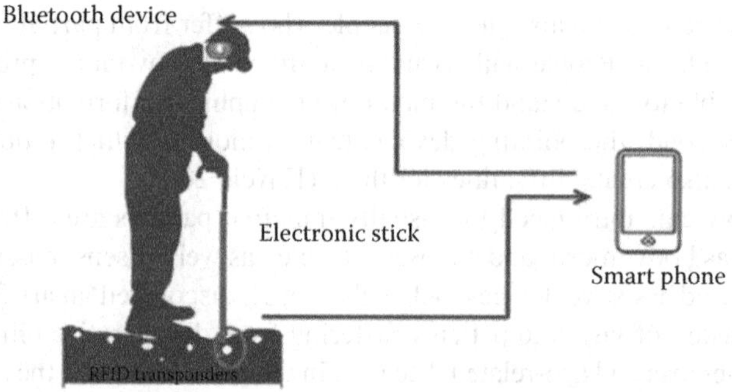

FIGURE 18.3 Electronic stick for visually impaired people.

Another important prototype "SmartVision" (Fernandes et al. 2013) has also been invented for this purpose. It uses an electronic walking stick enabling people to activate sensors on the floor, which are connected to form connected lines and clusters. These are able to guide the person by building a safe route to common locations of interest, such as a grocery store, book store, restaurant, and others. The software is able to run offline on a portable device using a local representation of the data presented in a geographic information system (GIS) stored in a tag identifier, which is a remote server. This uses corresponding geographic coordinates and tag ownership for tracing the location. After sensing the RFID, the software handles the inputs in two ways: navigating and scrolling. For navigation purposes, the software calculates the routes between the point of interest and the destination point, and each input from the location module triggers the navigation module to direct as well as keep the record on track. But for scrolling, the system picks the current coordinates and is able to find out all the relevant features of the surroundings using the GIS module.

Almeighder proposed a navigation system (Alghamdi et al. 2013) to help guide blind people in indoor locations where GPSs are not able to work properly. Indoor features are identified using RFID tags and users carry a mobile reader. The system helps to determine the shortest path to the destination. After calculating the shortest path, the path points are indicated by activated tags, which are scanned by a QR code and navigate the users to their destination.

18.3.3 Hearing

Deaf people have difficulties in detecting sounds and differentiating auditory information from the background. They communicate mostly through sign language (World Health Organization 1999). The World Health Organization (WHO) classified hearing disabilities into five categories. They are listed in Table 18.1 (Duthey 2013).

HandTalk (Sarji 2008) helps deaf people break the communication barrier. It is basically a pair of normal cotton gloves in which several flexor sensors are incorporated. They are normally embedded along the length of the each finger as

TABLE 18.1
Classification of Hearing Disabilities

Impairment Type/Grade	Audiometric Range (dbHL)	Description
No Impairment (Grade:0)	>25	Able to hear whisper
Slight Impairment (Grade:1)	26–40	Able to hear repeated word in normal voice within 1 meter
Moderate Impairment (Grade:2)	41–60	Able to hear repeated word in raised voice within 1 meter
Severe Impairment (Grade:3)	61–80	Able to hear the shouted voice
Profound Impairment (Grade:4)	<80	Not able to hear the shouted voice

well as the thumb. Normal gestures are sensed by sensors and transmitted to the BlueSentry software. This software converts the data into digital language in a range of 0 to 65,500. These numbers provide proper reflexes by using commercial off-the-shelf (COTS) elements. The architecture of these hand gloves is illustrated in Figure 18.4.

Language Acquisition Manipulatives Blending Early Childhood Research and Technology (LAMBERT) (Parton et al. 2010) allows the children, especially the deaf preschoolers, to play with objects that help them to learn the ASL language of most commonly used words via Multimedia Presentation. A 15–20 second multimedia presentation is prepared for each object. Various clip arts, coloring objects, and written descriptions help the children learn ASL.

18.3.4 Cognitive

Cognitive impairments include the impairments of thinking, memory, language, and perception. Several learning disabilities such as attention deficit disorders, dyslexia, and dementia fall into this category (Castaneda et al. 2008).

Brain computer interface (BCI) (Fazel-Rezai et al. 2011), (Lee et al. 2013) technology provides an electronic interface for communicating messages and commands to the computer from the human brain. BCI technology has the potential to improve the quality of life and increase its productivity. A commercially available product using this technology is the Nano Chip. This technology is also used for stroke rehabilitation (Ang and Guan 2013).

The Thought Translational Device (TTD) (Birbaumer et al. 2000) is a very innovative approach for helping mentally impaired patients. It is a device through

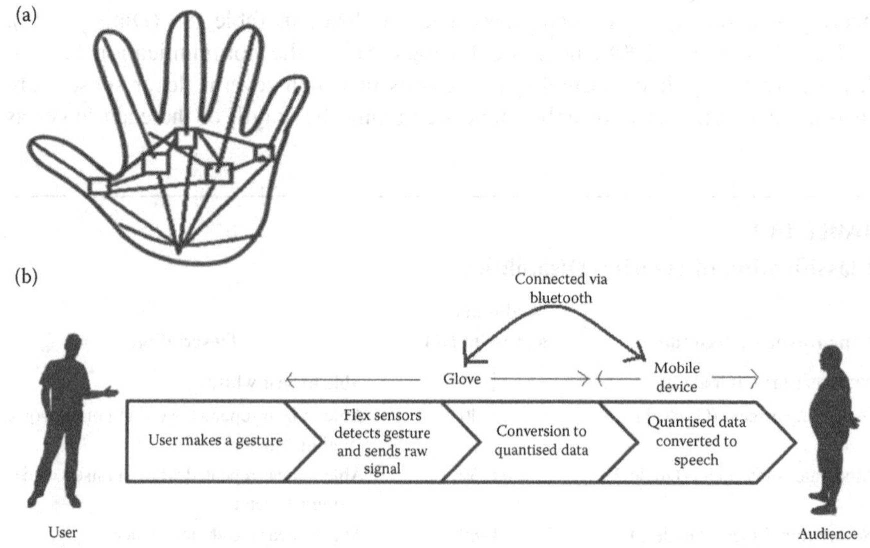

FIGURE 18.4 (a) Schematic model of HandTalk, (b) Working of HandTalk.

which patients are able to self-regulate the slow cortical potentials (SCPs) of their electroencephalograms (EEG). After learning the self-regulation process, patients are asked to select letters, words, pictograms, etc. from a computerized language support system, which finally help them describe and demonstrate their thoughts to others. As a result, mentally challenged people are able to establish a communicative network in society.

The assistive and augmentative communication (AAC) (Ann 2012) tool utilizes human computer interfacing to assist the cerebral palsy patients with their daily communications. Real-time emotions are recognized by the display and also send alerts to the caretakers through an SMS (Figure 18.5).

18.4 THE PATH AHEAD

In this chapter, we discussed the uses and advantages of IoT in improving the quality of life of disabled persons. But there are several drawbacks needing to be addressed before its worldwide adoption.

18.4.1 Accessibility

There are numerous companies and organizations exploiting IoT in some way or another. Accessibility should be maintained at each stage of the developmental process. All the parties related to this sector should maintain a universal design pattern (Alkhalil and Ramadan 2017).

IoT-related devices bring a huge profit to their manufacturing companies, but their accessibility is limited to a small part of society due to their excessive price tags. Companies should start the process of making the devices more cost effective to help the people in need.

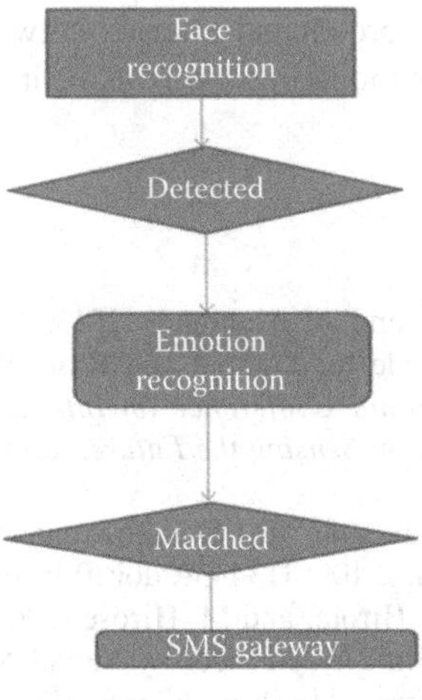

FIGURE 18.5 Working of assistive and augmentative communication.

18.4.2 Broadband Access

There are many patients who are not well accustomed with the Internet, as they are not able to access the web all the time. The broadband access must be increased by developing the infrastructure to realize the benefits of IoT.

18.4.3 Standardization

IoT has emerged with multidisciplinary applications in various fields, but there are problems with standardization of the product at present. Several organization have started working on it.

One such organization is the IEEE Standard Association (IEEESA), which has announced several parameters related to the standardization of IoT. These parameters are bluetooth connectivity, compatibility with Wi-Fi technology, and many more (Bandyopadhyay and Sen 2011).

18.4.4 Privacy

IoT collects data from the users' every single movement to understand their lifestyle patterns as well as their needs in a particular situation, which raises the question of privacy. Disabled people should be made aware of the data collection activities by providing proper disclaimer, and also providing an option for choosing the data types that help the user to decide which data they want to share (Zhang et al. 2014).

18.4.5 Security

IoT also creates a debate on security issues because large amounts of personal data are collected and stored by the software companies. This matter is discussed within companies and through private-public partnerships.

IoT is still in the developmental stages, and it will take years to become a part of our daily lives. IoT has to overcome several challenges, such as guidelines for standardization, extension of broadband Internet networks, private data protection, and many other before revolutionizing the lives of its users (van Kranenburg and Bassi 2012; Bekara 2014).

REFERENCES

Alghamdi, S., R. V. Schyndel, and A. Alahmadi. 2013. "Indoor Navigational Aid Using Active RFID and QR-Code for Sighted and Blind People." In *Proceedings of the 2013 IEEE 8th International Conference on Intelligent Sensors, Sensor Networks and Information Processing: Sensing the Future, ISSNIP 2013,* 1: 18–22. doi:10.1109/ISSNIP.2013.6529756.

Alkhalil, A., and R. A. Ramadan. 2017. "IoT Data Provenance Implementation Challenges." *Procedia Computer Science*, 109: 1134–39. doi:10.1016/j.procs.2017.05.436.

Amemiya, T., J. Yamashita, K. Hirota, and M. Hirose. 2004. "Virtual Leading Blocks for the Deaf-Blind: A Real-Time Way-Finder by Verbal-Nonverbal Hybrid Interface and High-Density RFID Tag Space." In *Proceedings—Virtual Reality Annual International Symposium*, 165–72. doi:10.1109/VR.2004.1310070.

Ang, K. K., and C. Guan. 2013. "Brain-Computer Interface in Stroke Rehabilitation." *Journal of Computing Science and Engineering* 7 (2): 139–46. doi:10.5626/JCSE.2013.7.2.139.

Ann, O. C. 2012. "Helping Disabled People: The Potentials of Biometric Information." *Interactive Technology and Smart Education* 9 (3): 153–70. doi:10.1108/17415651211258272.

Arrigo, M. 2005. "E-Learning Accessibility for Blind Students." *Recent Resaerch Development in Learning Technologies*, 1–5.

Bandyopadhyay, D., and J. Sen. 2011. "Internet of Things: Applications and Challenges in Technology and Standardization." *Wireless Personal Communications*, 58: 49–69. doi:10.1007/s11277-011-0288-5.

Bekara, C. 2014. "Security Issues and Challenges for the IoT-Based Smart Grid." *Procedia Computer Science*, 34: 532–37. doi:10.1016/j.procs.2014.07.064.

Birbaumer, N., A. Kübler, N. Ghanayim, T. Hinterberger, J. Perelmouter, J. Kaiser, I. Iversen, B. Kotchoubey, N. Neumann, and H. Flor. 2000. "The Thought Translation Device (TTD) for Completely Paralyzed Patients." *IEEE Transactions on Rehabilitation Engineering* 8 (2): 190–93. doi:10.1109/86.847812.

Castaneda, A. E., A. Tuulio-Henriksson, M. Marttunen, J. Suvisaari, and J. Lönnqvist. 2008. "A Review on Cognitive Impairments in Depressive and Anxiety Disorders with a Focus on Young Adults." *Journal of Affective Disorders.* 106(1–2): 1–27. doi:10.1016/j.jad.2007.06.006.

Coetzee, L., and G. Olivrin. 2012. "Inclusion Through the Internet of Things." *Assistive Technologies.* doi:10.5772/31929.

Cox, R. M., and G. C. Alexander. 1983. "The Abbreviated Profile of Hearing Aid Benefit." *British Journal of Audiology* 17 (1): 31–48. doi:10.1097/00003446-199504000-00005.

Domingo, M. C. 2012. "An Overview of the Internet of Things for People with Disabilities." *Journal of Network and Computer Applications* 35 (2). Elsevier: 584–96. doi:10.1016/j.jnca.2011.10.015.

Duthey, B. 2013. "Background Paper 6.21 Hearing Loss." *World Health Organization* 1 (February): 6.

Faggion, L., and G. Azzalin. 2011. "Low-Frequency RFID Based Mobility Network for Blind People: System Description, Evolutions and Future Developments." In *2011 IEEE International Conference on RFID-Technologies and Applications, RFID-TA 2011*, 364–69. doi:10.1109/RFID-TA.2011.6068663.

Fazel-Rezai, R., And Ahmad, W., C. Guger, G. Edlinger, and G. Krausz. 2011. *Recent Advances in Brain-Computer Interface Systems.* doi:10.5772/579.

Fernandes, H., V. Filipe, P. Costa, and J. Barroso. 2013. "Location Based Services for the Blind Supported by RFID Technology."*Procedia Computer Science*, 27: 2–8. doi:10.1016/j.procs.2014.02.002.

Garin, P., F. Genard, C. Galle, and J. Jamart. 2004. "The RetroX Auditory Implant for High-Frequency Hearing Loss." *Otology & Neurotology* 25 (4): 511–19. doi:10.1097/00129492-200407000-00019.

Hengeveld, B., C. Hummels, K. Overbeeke, R. Voort, H. van Balkom, and J. de Moor. 2009. "Tangibles for Toddlers Learning Language." In *Proceedings of the 3rd International Conference on Tangible and Embedded Interaction—TEI '09*, 161. doi:10.1145/1517664.1517702.

Istepanian, R. S. H., A. Sungoor, A. Faisal, and N. Philip. 2011. "Internet of M-Health Things 'M-IOT'." *IET Seminar on Assisted Living* 2011, 20–20. doi:10.1049/ic.2011.0036.

Kavcic, A. 2005. "Software Accessibility: Recommendations and Guidelines." In *EUROCON 2005—The International Conference on Computer as a Tool* 2: 1024–27. doi:10.1109/EURCON.2005.1630123.

Keating, E., and G. Mirus. 2004. "American Sign Language in Virtual Space: Interactions between Deaf Users of Computer-Mediated Video Communication and the Impact of Technology on Language Practices." *Language in Society* 32 (5): 693–714. doi:10.1017/S0047404503325047.

Kurze, M. 1996. "TDraw: A Computer-Based Tactile Drawing Tool for Blind People." In *Assets '96 Proceedings of the Second Annual ACM Conference on Assistive Technologies*, 131–38. doi:10.1145/228347.228368.

Lee, S., Y. Shin, S. Woo, K. Kim, and H.-N. Lee. 2013. "Review of Wireless Brain-Computer Interface Systems." In *Brain-Computer Interface Systems—Recent Progress and Future Prospects*, 215–37. doi:http://dx.doi.org/10.5772/56436.

Minoda, R., M. Izumikawa, K. Kawamoto, and Y. Raphael. 2004. "Strategies for Replacing Lost Cochlear Hair Cells." *Neuroreport* 15 (7): 1089–92. doi:10.1097/01.wnr.0000126216.79493.8d.

Nassih, M., I. Cherradi, Y. Maghous, B. Ouriaghli, and Y. Salih-Alj. 2012. "Obstacles Recognition System for the Blind People Using RFID." In *2012 Sixth International Conference on Next Generation Mobile Applications, Services and Technologies*, 60–63. doi:10.1109/NGMAST.2012.28.

Neto, R., and N. Fonseca. 2014. "Camera Reading for Blind People." *Procedia Technology* 16: 1200–1209. doi:10.1016/j.protcy.2014.10.135.

Ngu, A. H., M. Gutierrez, V. Metsis, S. Nepal, and Q. Z. Sheng. 2017. "IoT Middleware: A Survey on Issues and Enabling Technologies." *IEEE Internet of Things Journal* 4 (1): 1–20. doi:10.1109/JIOT.2016.2615180.

Parton, B. S., R. Hancock, and A. D. DuBusde Valempré. 2010. "Tangible Manipulatives and Digital Content: The Transparent Link That Benefits Young Deaf Children." In *Proceedings of IDC2010 The 9th International Conference on Interaction Design and Children*, 300–303. doi:10.1145/1810543.1810597.

Saleem, K., and I. Alagha. 2016. "System for People with Hearing Impairment to Solve Their Social Integration." In *2015 5th International Conference on Information and Communication Technology and Accessibility, ICTA 2015*. doi:10.1109/ICTA.2015.7426913.

Sarji, D. K. 2008. "HandTalk: Assistive Technology for the Deaf." *Computer* 41 (7): 84–86. doi:10.1109/MC.2008.226.

Schwiebert, L., S. K. S. Gupta, and J. Weinmann. 2001. "Research Challenges in Wireless Networks of Biomedical Sensors." In *Proceedings of the 7th Annual International Conference on Mobile Computing and Networking—MobiCom '01*, 151–65. doi:10.1145/381677.381692.

Sen, E., and S. Yurtsever. 2007. "Difficulties Experienced by Families with Disabled Children." *Journal for Specialists in Pediatric Nursing* 12 (4): 238–52. doi:10.1111/j.1744-6155.2007.00119.x.

Stucki, G. 2005. "International Classification of Functioning, Disability, and Health (ICF)." *American Journal of Physical Medicine & Rehabilitation* 84 (10): 733–40. doi:10.1097/01.phm.0000179521.70639.83.

van Kranenburg, R., and A. Bassi. 2012. "IoT Challenges." *Communications in Mobile Computing* 1 (1): 9. doi:10.1186/2192-1121-1-9.

Vasanthan, M., M. Murugappan, R. Nagarajan, B. Ilias, and J. Letchumikanth. 2012. "Facial Expression Based Computer Cursor Control System for Assisting Physically Disabled Person." In *Proceeding—COMNETSAT 2012: 2012 IEEE International Conference on Communication, Networks and Satellite*, 172–76. doi:10.1109/ComNetSat.2012.6380800.

Vincent, C., B. Fraysse, J. P. Lavieille, E. Truy, O. Sterkers, and F. M. Vaneecloo. 2004. "A Longitudinal Study on Postoperative Hearing Thresholds with the Vibrant Soundbridge Device." *European Archives of Oto-Rhino-Laryngology* 261 (9): 493–96. doi:10.1007/s00405-003-0669-9.

World Health Organization (WHO). 2003. "International Classification of Funcitoning, Disability and Health." *World Health Organization*. http://www.who.int/classifications/icf/icfchecklist.pdf?ua=1.

World Health Organization. 1999. "Future Programme Development for Prevention of Deafness and Hearing Impairment: Report of Second Informal Consultation." http://apps.who.int/iris/bitstream/10665/64994/1/WHO_PBD_PDH_99.6.pdf.

Yanagisawa, T., M. Hirata, Y. Saitoh, H. Kishima, K. Matsushita, T. Goto, R. Fukuma, H. Yokoi, Y. Kamitani, and T. Yoshimine. 2012. "Electrocorticographic Control of a Prosthetic Arm in Paralyzed Patients." *Annals of Neurology* 71 (3): 353–61. doi:10.1002/ana.22613.

Zanin, J., and G. Rance. 2016. "Functional Hearing in the Classroom: Assistive Listening Devices for Students with Hearing Impairment in a Mainstream School Setting." *International Journal of Audiology* 55 (12): 723–29. doi:10.1080/14992027.2016.1225991.

Zhang, Z. K., M. C. Y. Cho, C. W. Wang, C. W. Hsu, C. K. Chen, and S. Shieh. 2014. "IoT Security: Ongoing Challenges and Research Opportunities." In *Proceedings—IEEE 7th International Conference on Service-Oriented Computing and Applications, SOCA 2014*, 230–34. doi:10.1109/SOCA.2014.58.

19 Smart Analytical Lab

Subhrodeep Saha, Sourish Sen, Bharti Singh, and Shampa Sen

CONTENTS

19.1 INTRODUCTION

All analytical instruments, from a small pH meter to a large electron microscope, are integral parts of any biotechnological and chemical laboratory. Analytical labs play a major role in diagnostics, research, and development across all fields. Spectroscopy, colorimetric analysis, sensing and imaging techniques are a few analytical techniques having found widespread use in many fields. For example, in the field of medicine, magnetic resonance imaging (MRI) is used to take images of our anatomy without invasive procedures. In environmental sciences, or chemical and biotechnological industries these techniques are used to detect impurity, contamination, and quality and composition of any given sample.

However, the conventional analytical lab has certain challenges such as limited space and high cost of equipment, which limits the number of skills practiced in the lab. Also, handling most of the sophisticated diagnostic instruments requires a high level of expertise and experience, making the process of analysis very time consuming. This results in a long wait for the samples to be analyzed, which can result in the deterioration of sample quality, thus giving false results. Some of these problems can be tackled by a so-called "smart analytical lab."

Smart analytical lab is a concept based on Internet of things (IoT). IoT essentially puts the "smart" in the "smart analytical lab." It has the potential to change the way work is done in a lab by circumventing the challenges and limitations associated with conventional analytical labs (Alaa et al. 2017).

IoT is made up of three main parts:

1. The "things" (the instruments and smartphones)
2. The network connecting them, which allows data sharing
3. The computer systems (the "brain") controlling the flow of data from and to the "things" and interpret/process the data to give us coherent information (Stergiou et al. 2018).

As an important component of the IoT, analytical instruments can serve users effectively by communicating with various digital devices such as computers and smartphones (Alaa et al. 2017). This has become possible because automation in lab instruments has become increasingly sophisticated in recent years permitting them to gather and interchange data across different platforms. This can increase experimental data quality and efficiency, and reduces the time taken for performing experiments and processing data. It also allows experimentation, which generally would be impossible for reasons such as small sample sizes, lengthy procedures, etc. (Stergiou et al. 2018). Due to recent advancements in technology and the introduction of user-friendly operating systems and applications, smartphones have replaced laptops and desktop computers. Taking this fact into account, researchers have designed sensing systems more compatible with smartphones making them handy analytical devices (Kanchi et al. 2018). The processing power of microchips has doubled while their price has halved every few years for the last few decades (Moore's Law); modern smartphones are exponentially more powerful than the desktop computer of 1970s, thus it has become possible to utilize smartphones as handy biosensors and analytical instruments. Although, in most cases, smartphones don't work alone as lab devices, they can be augmented by peripherals, which give the smartphones theirs analytical features (Roda et al. 2016). The main reason behind this is that many sophisticated analytical instruments are not miniaturized enough to fit a smartphone.

The advancement of point-of-care (PoC) instruments, which can be used in nonlaboratory has enabled many tests to be performed at the point of need, outside the lab. A point-of-care instrument must have certain key features—portable, durable, cheap, and energy efficient (can be used on battery). These portable and cheap instruments could be utilized to run routine tests, which are right now performed by trained personnel utilizing research center instruments such as microscopes and spectrophotometers. This would offer huge potential for improving the detection and treatment of diseases, especially in remote regions (Erickson et al. 2014; Xu et al. 2015).

In the ideal version of a connected future, all appliances in a lab would communicate with one another seamlessly; this would require immense computing power and mass storage supported by mobile cloud computing (Alaa et al. 2017; Stergiou et al. 2018). It provides access to data and information from anywhere at any time. For example, integrating a blood glucose meter within a smartphone would help reduce the chances of a diabetic patient forgetting to carry their glucometer with them; it can also remind its users to take timely medication. In addition, with the use of IoT and cloud computing, the smartphone can autonomously store, process, and send blood glucose readings from the phone to a care provider, or cloud service, via the built-in wireless functionality (Bandodkar et al. 2018). Another example of this is the compact digital lens-free microscopy smartphone. This lightweight and compact microscope could be integrated in a simple smartphone; using lens-free holographic on-chip imaging, the setup can visualize microparticles of various size along with biological samples such as body fluids and microbes (Zhang and Liu 2016).

Further in this chapter, the feasibility and potential applications of a smart analytic laboratory are discussed in detail.

19.2 SMART ANALYTICS

In academia, smart analytics are defined as the process of collecting and analyzing data to help business make better decisions about their products and services. With the introduction of smart analytics in analytical labs, collection and interpretation of data using large sophisticated instruments have become easier and less time consuming. Moreover, advanced mobile phones furnished with high-resolution cameras are exploited to design bioelectronics and biosensors for optical imaging. Currently, with innovations in imaging nanoscale determinations can be achieved by smartphones with conservative gadgets. Following in the footsteps of microscopy imaging, fluorescent and colorimetric imaging have been improved by integration with advanced cameras of smartphones. Mainly, colorimetric analysis with advanced cell phone has been greatly detailed due to its broad potential applications in point-of-care diagnostics. Although smartphones are easy to understand, numerous explanatory systems still require the utilization of substantial and modern supplies.

19.2.1 Biosensors

A biosensor is essentially a device to detect a biological signal (for example urea in urine) and convert it into a measurable electrical signal, thus allowing us to find the "strength" of the biological signal (for example urea concentration). It requires the use of biological substances like nucleic acids or more commonly enzymes and proteins associated with the electrochemical transducers. This setup converts the biological signal into the electrical signal, for example antibody–antigen interaction; here, the antigen acts as the biological detector and the antibody as the analyte. The antigen is associated with the transducer and this method is used for disease diagnosis. A schematic representation of the working of a biosensor has been shown in Figure 19.1. A wide variety of biosensors are being used in the fields of medicine, food and cosmetics industry, and environmental impact assessment to detect and remove adulterants, contaminants, pathogens, impurities, etc. (Ali et al. 2017; Eggenstein et al. 2017).

For example, the glucose biosensor (glucometer) utilizes glucose oxidase to oxidize glucose in the sample, and this glucose reaction generates an electrical signal. The strength of this current is then measured to detect the quantity of glucose in the sample.

Time consuming qualitative and quantitative analysis of biological components can be greatly eased by the advent of "smart" biosensors in modern analytical labs.

Biosensors integrated with a smartphone have been researched through two main approaches where either the smartphone is modified to be a biosensor or the smartphone is used as an interface for the biosensor. These smartphone-based biosensors have tremendous potential as PoC devices considering they have most of the key features necessary for a PoC device (portable, cheap, energy efficient, etc.). Thus, they have use in health care, food safety, environmental monitoring, and biosecurity; especially in inaccessible and remote areas because a smartphone modified to function as biosensor would be far smaller and more portable than conventional biosensors (Roda et al. 2016). Smartphones can be useful even in cases where the biosensor cannot be integrated into a smartphone due to practical or technological limitations, as it can be used as an interface between the user and the biosensor. The biosensor could wirelessly transmit information to the smartphone and operated remotely, therefore the user does not need to go to the place where the biosensor is installed to take readings. This would be especially useful in inaccessible and/or hazardous zones. A smartphone-based biosensor would also be easy to operate, thus eliminating the need for expertise and laboratory conditions.

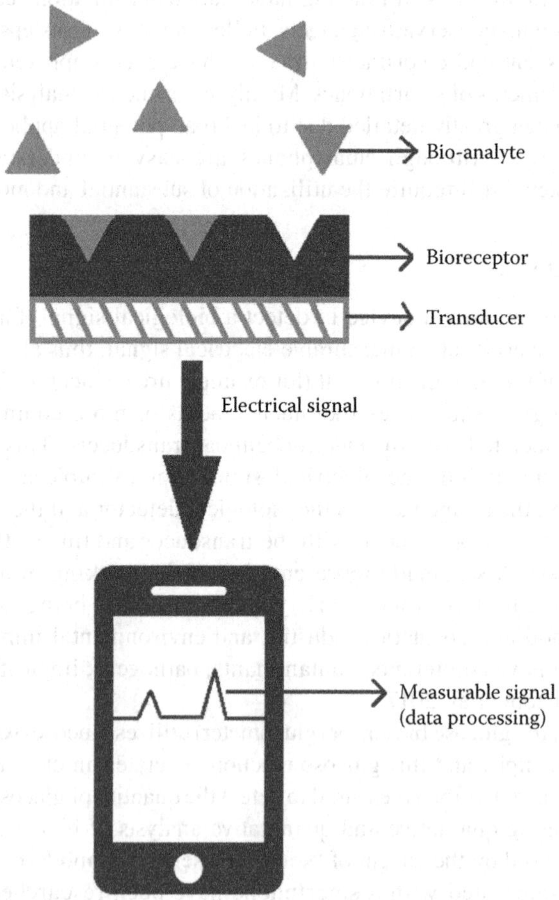

FIGURE 19.1 Biosensors.

19.2.2 Fluorescence Imaging

Fluorescence imaging is the process of visualizing molecular and cellular structures by labeling them with fluorescent dyes or markers. This technique has various applications in the fields of biotechnology, organic chemistry, and medicine because of easy implementation, innate sensitivity, inert nature, nonintrusiveness, and the commercial availability of many well-developed fluorescence dyes and labeling schemes for biological studies. A few cases of across the board use of fluorescence imaging are enzyme assays, capillary electrophoresis (CE), antibody detection and screening, bead-based immunoassays, DNA sequencing and fragment sizing or extraction, protein detection, food assay, cellular microscopy, and imaging detection.

The fluorescence microscope used in analytical labs is one of the most common imaging devices utilizing phosphorescence and fluorescence along with reflection and absorption to study properties of organic or inorganic substances. Here, the sample is illuminated with light of a particular wavelength (or wavelengths), which is absorbed by the fluorophores (a chemical compound that absorbs light and emits it as fluorescence), making them emit light of higher wavelengths (the emitted light has lower energy and different color). The light from the source is separated from the much weaker fluorescence of the sample using a spectral emission filter. Its major components are a light source (mercury-vapor lamp or xenon arc lamp are common; more advanced forms are high-power LEDs and lasers), the excitation filter, the dichroic beamsplitter, and the emission filter. The filters and the dichroic beamsplitter are selected to correspond to the spectral excitation and emission spectrum of the fluorophore used to label the sample. In this way, the fluorescence from the labels is imaged to find its distribution at a time. Multicolor images of numerous types of fluorophores are combinations of several single-color images. Epifluorescence microscopes are the most common type of fluorescence microscopes, here excitation of the fluorophore and fluorescence detection are done through the objective (i.e., straight path).

In the biological sciences fluorescence microscopy is a potent device for precise and sensitive staining of a sample to detect the distribution of biomolecules of interest. Numerous fluorescent stains have been produced for indirect labeling of our specimen, for instance nucleic acid stains such as Hoechst and DAPI bind to the minor groove of DNA, thus labeling the cell nuclei.

In smart analytical labs, smartphones can be used to detect fluorescence and obtain the image. Ochratoxin A (OTA) concentration can be measured using a smartphone as a fluorescence device. Here ultraviolet (UV) light is used to excite the sample. A lens is used to focus the fluorescence from the excited sample to the smartphone camera. Finally, the data is transmitted to a personal computer. The fluorescence image data from the smartphone camera is analyzed by the personal computer and represented in its red, green, and blue (RGB) components. OTA is naturally fluorescent and hence when the light is emitted through a solution with OTA present a blue fluorescence is emitted. Thus, the image is captured with the smartphone camera and no fluorescence is observed for a blank solution (Bueno et al. 2016). This technique can be used for various other bio molecules with appropriate excitation and emission spectrum.

19.2.3 Colorimetric Analysis

Colorimetric analysis is used to determine the concentration of the analyte in a colored solution. The colorless analyte can be transformed into a visible color by the addition of a shading reagent. It is significant to both inorganic and natural mixes and might be used by means of enzymatic or nonenzymatic steps. This approach is normally used as a part of therapeutic labs and for mechanical purposes like examinations related to water treatment. The types of apparatus required for colorimetric examination are a colorimeter, an appropriate shading reagent, and some cuvettes. The method might be automated, for example, by using an auto analyzer or by flow infusion examination.

Colorimetric examination can be classified into three types of techniques: enzymatic strategies, nonenzymatic techniques, and ultraviolet techniques.

1. Nonenzymatic techniques: In these techniques, the complex of the test substance with the shading reagent is formed without the use of biocatalysts/enzymes. For example, calcium and copper are prepared for colorimetric analysis by the methods shown below.

Calcium + o-cresolphthalein complexone → Hued complex

(Ray Sarkar and Chauhan 1967)

Copper + Bathocuproin disulfonate → Hued complex (Zak 1958)

2. Enzymatic techniques: It require the use of an enzymatic catalyst to form a stable complex between the substance to be tested and the shading reagent. Some examples are illustrated below:
   a. *Glucose (GOD-Perid technique)* (Werner et al. 1970)

$$\text{Glucose} + \text{Oxygen} + \text{Water} \xrightarrow{\text{Glucose oxidase}} \text{Gluconate} + \text{Hydrogen Peroxide}$$

$$\text{Hydrogen peroxide} + \text{ABTS} \xrightarrow{\text{Peroxidase}} \text{Colored complex}$$

   b. *Urea* (Fawcett and Scott 1960)

$$\text{Urea} + \text{Water} \xrightarrow{\text{Urease}} \text{Ammonium carbonate}$$

$$\text{Ammonium carbonate} + \text{Phenol} + \text{Hypochlorite} \rightarrow \text{Coloured complex}$$

19.2.3.1 Ultraviolet (UV) Method

No visible change in color is observed with the UV method, but the mechanism is exactly the same, that is, the measurement of an alteration in the absorbance of the solution. UV methods generally measure the disparity in absorbance at the 340 nm wavelength between nicotinamide adenine dinucleotide (NAD) and its reduced form (NADH).

However, these strategies are less exact and tedious. A versatile location framework, named Smart Forensic Phone, has been created, has empowered us to quickly and definitely perform colorimetric investigations of a bloodstain for age estimation. In this strategy, blood is set on five distinct materials (backdrop, texture, glass, wood, and A4 paper), the RGB values per pixel of the bloodstain picture at 6-h interims are observed, and the inexact age of the bloodstain was evaluated utilizing a cell phone application. The RGB values are changed over into the V estimation of HSV (tint, immersion, and brilliance). A quick decrease in RGB and V values is seen over the initial 42 h and from that point remains moderately unaltered. The age of the bloodstain is determined from the plot of V versus time. This offers a novel technique to estimate the time since the victim's demise technique (Kanchi et al. 2018).

Additionally, an advanced cell phone-based optical platform is utilized for colorimetric investigations of blood utilizing a dispensable miniaturized scale fluidic gadget in an advanced scientific lab. Using an incorporated camera in a personal digital assistant (PDA), pictures of human blood are taken in the smaller scale channel and broken down by a versatile application. To avoid image burning and ambient light effect, a special light-diffusing system consisting of a microfluidic device enclosed inside a white acrylic box is used. With the picture software program on the PDA, the created gadget is effectively connected to decide different haematocrit levels of human blood from 10% to 65% (Kim et al. 2017).

Also, a quick and direct technique is being produced for colorimetric detection of smelling salts, which release ammonia, utilizing advanced mobile phones. This system depends on the control of surface plasmon bands of silver nanoparticles (AgNPs) by forming $Ag(NH_3)^{2+}$ complex. Hence, there is a reduction in AgNP concentration, and thus, the color of the suspension diminishes. The variation in color intensity can not only be measured by the standard UV-Vis spectrometer but it can also be done using a smartphone by RGB analysis. This was verified by detecting ammonia in the range of 10–1000 mg L^{-1}, by UV-Vis spectrometry and the developed smartphone colorimetric technique. It was seen that the cell phone-based colorimetric technique can be used as reliably as spectrophotometer for quantification of ammonia in the sample (Amirjani and Fatmehsari 2018).

With the specific goal of understanding the location and visual recognition of antimicrobial compounds, a PoC Testing technique is built by utilizing computerized picture colorimetry based on advanced smartphone. To test this, streptomycin aptamer was taken as the analyte model of antitoxins. The streptomycin aptamer interacted with the analyte, and the unreacted aptamer was hybridized with the correlative DNA to form dsDNA. SYBR Green I was joined with the dsDNA to emit clear green fluorescence. The fluorescence diminishes with the increase of streptomycin fixation. This was detected once again by using the smartphone-based technique. In this example, the phone camera takes the pictures of the fluorescence and the Touch Color application is used to perform the RGB estimations of the pictures. This strategy has the upsideof good selectivity, straightforward operation, and on location representation (Lin et al. 2018).

19.2.4 Spectroscopy

Spectroscopy is the study of the interaction between electromagnetic radiation and matter. Spectral measurement devices are otherwise called spectrometers, spectrophotometers, spectrographs, or spectral analyzers. To put it plainly, the succession of events in an advanced spectrophotometer is as follows:

1. The beam of light is passed through a monochromator, which diffracts it into its components and allows only the required monochromatic light to pass through it, and split into two beams. It is then incident over the sample and the reference arrangements.
2. Only a portion of the incident wavelengths reflects from, or transmits through the sample and the reference.
3. The resultant light strikes the photo detector, which extrapolates the relative power of the two beams.
4. Computerized systems convert the relative currents into absorbance values, which allow us to determine the sample concentration.

These days handheld mobile phone-based spectrometers have been created, which work in both assimilation and discharge modes. The device, named Spectrophone, includes modules made in medium-density fibreboard (MDF), an embedded light source intended for retention mode, a digital video disc (DVD) for the diffraction grinding, and a cell phone to process the picture information obtained. No significant difference was observed when compared to the values obtained from conventional measurement methods. The figures of merit of the Spectrophone are decent, with cutoff points of measurement of 70 g/L for absorption and 60 g/L for emission modes, while showing a high linearity ($R^2 > 0.9995$). High precision, simplicity, and an ease of use of the application, while requiring minimal efforts make this handheld device an attractive option for absorption/emission estimations (de Oliveira et al. 2017).

An advanced mobile phone-based spectrometer design, which is independent and held on a remote stage can be utilized for quick, nondamaging testing of organic products. This gadget is inherently low-cost and its power consumption is negligible, making it convenient to do a range of studies in the field. Every single basic component of the device like the light source, spectrometer, channels, microcontroller, and remote circuits are integrated into a device of small dimensions (88 × 37 × 22 mm) weighing only 48 g. The resolution of the spectrometer is 15 nm, conveying exact and repeatable estimations. The device has a dedicated application interface on the PDA to receive, communicate, analyze, and plot spectral data. The execution of the advanced cell spectrometer is practically identical to existing bench top spectrometers regarding soundness and wavelength determination. Validations of the device are carried out by exhibiting nondangerous readiness testing in organic product tests. During the ripening process, ultra-violet (UV) fluorescence from chlorophyll present in the skin are measured across various apple varieties and correlated with destructive firmness tests. An acceptable correlation is seen among ripeness and fluorescence signals (Das et al. 2016).

19.3 CLOUD COMPUTING

Cloud computing is a system which allows wireless access to a collective pool of electronic resources, as demonstrated in Figure 19.2. These computing services can be easily managed and released with minimal effort or human interaction. It has brought a paradigmatic shift in providing data storage as well as computing resources (Ravi et al. 2018).

In simple terms, cloud computing basically provides access to data and information from anywhere at any time, all the while reducing the need for bulky hardware. More specifically, mobile cloud computing can be defined as the incorporation of mobile devices (such as smartphones) with cloud computing technology in order to make these devices more resourceful by increasing their storage capacity, computational power, energy efficiency, and context awareness ("smartness") (Stergiou et al. 2018).

The modern smartphone and smart analytics necessitate the use of cloud computing to manage their data and resources. Cloud computing has become a versatile and powerful platform to achieve large-scale and complex computing. Some of the benefits of cloud computing can be listed as scalable data storage, virtualized resources, data/information security, and parallel processing. Virtualized resources saves space, parallel processing reduces time consumption, data security is essential in an easily accessible network, and scalable data storage makes sure that storage space is plentiful.

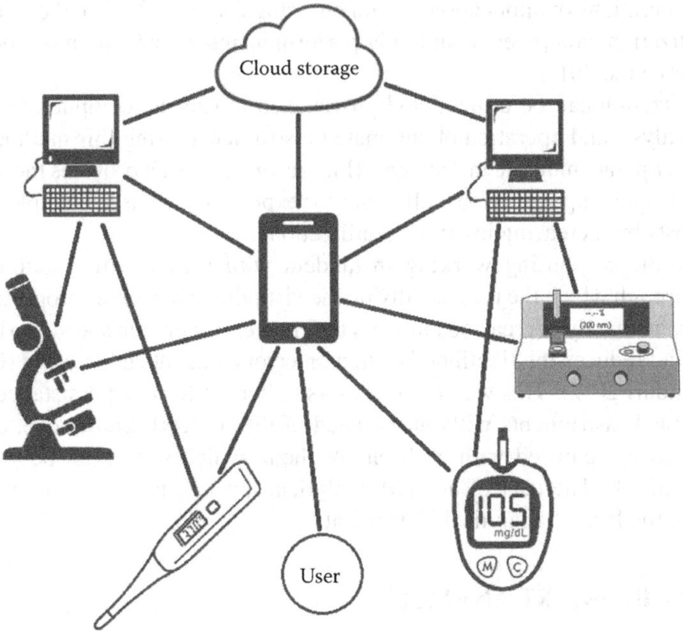

FIGURE 19.2 Interconnected devices in a smart lab.

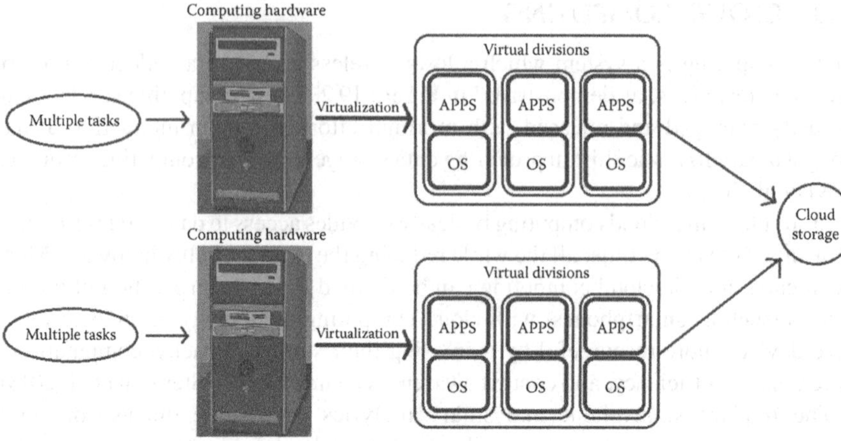

FIGURE 19.3 Virtualization in cloud computing.

Apart from the previously mentioned advantages, cloud computing helps a smart lab meet major challenges faced by a conventional analytical lab by increasing efficiency and decreasing operational cost by restricting its owned software and hardware, as cloud services can be readily outsourced (Ravi et al. 2018). Virtualization is the key technology behind cloud computing, it makes cloud computing possible. It is a method of logically dividing the computer's processing power to allow for multiple operations or applications to run on a single system. Before the introduction of virtualization, computers could only perform a single application or operation at a time (Graziano 2011).

Virtualization can be utilized and provisioned to execute computing tasks such as data analysis and operation of automated instruments; using this method a single system can operate multiple instruments (Figure 19.3). Thus, it provides the versatility required to speed up operations, allowing for exponentially larger sample sizes and cutting costs by increasing instrument utilization.

Autonomic computing working in tandem with virtualization automates the process through which the user can divide the virtualized resources proportionally to the requirements. By decreasing human interference, automation speeds up the whole process, and reduces the likelihood of human errors and cuts labor costs (Hamdaqa and Tahvildari 2012). This would allow users to benefit from the assistance of all of the analytical instruments without the need of the in-depth knowledge, expertise, and proficiency required with each one of them, which would not be possible in a traditional lab. The user of a smart analytical lab will not need to compromise because of the lack of analytical instruments.

19.4 IoT IN SMART ANALYSIS

The twenty-first century is the era of IoT, which has become greatly intertwined in our lives in the form smartphones and other smart devices. Generally speaking, IoT refers to the communication between everyday electronic objects, which are usually equipped with ubiquitous intelligence; in case of smart analytical labs the "objects with ubiquitous intelligence" are the computers, smartphones, and smart analytical lab equipment. IoT increases the omnipresence of the Internet by connecting all the objects to it via embedded (rooted) systems (as shown in Figure 19.3), which forms a widespread web of devices interacting with its users as well as other devices (León et al. 2010). This concept can be essentially used in a smart analytical lab where computers, smartphones, and analytical instruments would form a network and complement each other and their user; labs in different cities could share their data and resources easily, which would greatly aid and improve the quality and quantity of research. The IoT can connect all the instruments in the lab and turn it into a "smart," cohesive, and coherent system, which can function with minimal intervention and caters to the users' needs.

IoT and cloud computing are two sides of the same coin; cloud computing requires and enables IoT, as IoT is basically a network of devices, which is the backbone of cloud computing. It enables communication between different elements of the smart analytical lab and provides an easy interface for users to operate the analytical instruments; this eliminates the need for specialized expertise and experience, which is a limitation in a conventional analytical lab. Using IoT it would be possible to operate all analytical lab instruments, small to big, simple to complex, using specialized applications present in our smartphones.

The ideal smart analytical lab would essentially be an environment with minimal manual interference; samples would be analyzed and data would be obtained and interpreted automatically. Due to the connectivity provided by IoT, lab resources could be accessed and used remotely. The smartphone would essentially be a PDA, which would act as a guide, doorway, or interface between the user and the smart analytical lab.

REFERENCES

Alaa, M., A. A. Zaidan, B. B. Zaidan, M. Talal, and M. L. M. Kiah. 2017. "A review of smart home applications based on internet of things." *Journal of Network and Computer Applications*, 97, 48–65. Elsevier Ltd. https://doi.org/10.1016/j.jnca.2017.08.017.

Ali, J., J. Najeeb, M. A. Ali, M. F. Aslam, and A. Raza. 2017. "Biosensors: Their fundamentals, designs, types and most recent impactful applications: A review." *Journal of Biosensors & Bioelectronics*, 8(1), 1–9. https://doi.org/10.4172/2155-6210.1000235.

Amirjani, A. and D. H. Fatmehsari. 2018. "Colorimetric detection of ammonia using smartphones based on localized surface plasmon resonance of silver nanoparticles." *Talanta*, 176, 242–246.

Bandodkar, A. J., S. Imani, R. Nuñez-Flores, R. Kumar, C. Wang, A. M. Vinu Mohan, J. Wang, and P. P. Mercier. 2018. "Re-usable electrochemical glucose sensors integrated into a smartphone platform." *Biosensors and Bioelectronics*, 101, 181–187 (July 2017). Elsevier B.V. https://doi.org/10.1016/j.bios.2017.10.019.

Bueno, D., R. Muñoz, and J. L. Marty. 2016. "Fluorescence analyzer based on smartphone camera and wireless for detection of Ochratoxin A." *Sensors and Actuators B: Chemical*, 232, 462–468.

Das, A. J., A. Wahi, I. Kothari, and R. Raskar. 2016. "Ultra-portable, wireless smartphone spectrometer for rapid, non-destructive testing of fruit ripeness." *Scientific reports*, 6, 32504.

de Oliveira, H. J. S., P. L. de Almeida, B. A. Sampaio, J. P. A. Fernandes, O. D. Pessoa-Neto, E. A. de Lima, and L. F. de Almeida. 2017. "A handheld smartphone-controlled spectrophotometer based on hue to wavelength conversion for molecular absorption and emission measurements." *Sensors and Actuators B: Chemical*, 238, 1084–1091.

Eggenstein, C., M. Borchardt, C. Diekmann, B. Gründig, C. Dumschat, K. Cammann, M. Knoll, and F. Spener. 2017. "A disposable biosensor for urea determination in blood based on an ammonium-sensitive transducer." *Biosensors & Bioelectronics*, 14, 33–41 (January 01). Accessed January 06, 2018. https://www.ncbi.nlm.nih.gov/pubmed/10028647.

Erickson, D., D. O'Dell, L. Jiang, V. Oncescu, A. Gumus, S. Lee, M. Mancuso, and S. Mehta. 2014. "Smartphone technology can be transformative to the deployment of lab-on-chip diagnostics." *Lab on a chip*, 14, 3159–3164 (September 07). https://www.ncbi.nlm.nih.gov/pmc/articles/PMC4117816/.

Fawcett, J. K. and J. E. Scott. 1960. "A rapid and precise method for the determination of urea." *Journal of Clinical Pathology*, 13(2), 156–159. https://doi.org/10.1136/jcp.13.2.156.

Graziano, C. D. 2011. "A Performance Analysis of Xen and KVM Hypervisors for Hosting the Xen Worlds Project." *Master's Thesis*. http://lib.dr.iastate.edu/etd/12215/.

Hamdaqa, M. and L. Tahvildari. 2012. "Cloud computing uncovered: A research landscape." *Advances in Computers*, 86, 41–85. doi:10.1016/b978-0-12-396535-6.00002-8.

Kanchi, S., M. I. Sabela, P. S. Mdluli, Inamuddin, and K. Bisetty. 2018. "Smartphone based bioanalytical and diagnosis applications: A review." *Biosensors and Bioelectronics*, 102, 136–149 (August 2017). Elsevier B.V. https://doi.org/10.1016/j.bios.2017.11.021.

Kim, S. C., U. M. Jalal, S. B. Im, S. Ko, and J. S. Shim. 2017. "A smartphone-based optical platform for colorimetric analysis of microfluidic device." *Sensors and Actuators B: Chemical*, 239, 52–59.

León, O., J. Hernández-Serrano, and M. Soriano. 2010. "Securing cognitive radio networks." *International Journal of Communication Systems*, 23(5), 633–652. https://doi.org/10.1002/dac.

Lin, B., Y. Yu, Y. Cao, M. Guo, D. Zhu, J. Dai, and M. Zheng. 2018. "Point-of-care testing for streptomycin based on aptamer recognizing and digital image colorimetry by smartphone." *Biosensors and Bioelectronics*, 100, 482–489.

Ravi, K., Y. Khandelwal, B. S. Krishna, and V. Ravi. 2018. "Analytics in/for cloud-an interdependence: A review." *Journal of Network and Computer Applications*, 102, 17–37 (August 2017). https://doi.org/10.1016/j.jnca.2017.11.006.

Ray Sarkar, B. C. and U. P. S. Chauhan. 1967. "A new method for determining micro quantities of calcium in biological materials." *Analytical Biochemistry*, 20(1), 155–166. https://doi.org/10.1016/0003-2697(67)90273-4.

Roda, A., E. Michelini, M. Zangheri, M. Di Fusco, D. Calabria, and P. Simoni. 2016. "Smartphone-based biosensors: A critical review and perspectives." *TrAC—Trends in Analytical Chemistry* 79, 317–25. Elsevier B.V. https://doi.org/10.1016/j.trac.2015.10.019.

Stergiou, C., K. E. Psannis, B. G. Kim, and B. Gupta. 2018. "Secure integration of IoT and cloud computing." *Future Generation Computer Systems* 78, 964–75. Elsevier B.V. https://doi.org/10.1016/j.future.2016.11.031.

Werner, W., H. G. Rey, and Wielinger, H. 1970. Über die Eigenschaften eines neuen Chromogens für die Blutzuckerbestimmung nach der GOD/POD-Methode. *Fresenius' Zeitschrift Für Analytische Chemie*, 252(2–3), 224–228. https://doi.org/10.1007/BF00546391.

Xu, X., A. Akay, H. Wei, S. Wang, B. Pingguan-Murphy, B. E. Erlandsson, X. Li et al. 2015. "Advances in smartphone-based point-of-care diagnostics." *Proceedings of the IEEE*. July 15.

Zak, B. 1958. "Simple procedure for the single sample determination of serum copper and iron." *Clinica Chimica Acta*, 3(4), 328–334. https://doi.org/10.1016/0009-8981(58)90021-4.
Zhang, D. and Q. Liu. 2016. "Biosensors and bioelectronics on smartphone for portable biochemical detection." *Biosensors and Bioelectronics* 75, 273–284. Elsevier. https://doi.org/10.1016/j.bios.2015.08.037.

20 Crop and Animal Farming IoT (CAF-IoT)

Neha Agnihotri, Soumyadipto Santra, and Shampa Sen

CONTENTS

20.1 INTRODUCTION

The internet of things (IoT) is an innovation which gathers and trades information from physical things utilizing electronic sensors and the Internet. The (IoT) innovation is presently enabling ranchers to interface with devices on the web to enhance agriculture.

The Agriculture industry must defeat expanding water scarcity, constrained accessibility of fields, increasing costs, while meeting the expanding utilization needs of a worldwide population estimated to increase 70% by 2050 (FAO 2016).

While applying IoT to the farming industry, sensors collect billions of readings on ecological information such as temperature, moisture, drinking water, sustaining rates, CO_2 fixation, sodicity, and pH levels. This information gathered will have applications with regards to agricultural administration, profitability, and productivity, while even filling in as a ready framework for agriculturists.

Apart from farming, IoT innovations can significantly facilitate animal husbandry, where ranchers can now insert web-associated sensors inside their animals without any discomfort (Neethirajan 2017). Utilizing data from these sensors, agriculturists can screen their general health parameters and help farmers locate their animals and check for any imminent danger to them.

IoT innovations enables ranchers to screen their gear. Sensors can be incorporated into tractors to decide whether it is running proficiently. A notice will be sent to the rancher before repairs should be made. This will help avoid sudden breakdowns enabling it to remain in the field longer to expand efficiency. Adopting a proactive strategy can enable hardware to be utilized longer and decrease costs.

20.2 IoT IN CROP FARMING

20.2.1 Traditional Farming Practices

After the two world wars, customary agribusiness, prospered around the globe and became the most rehearsed type of agriculture as science and technology progressed simultaneously. Artificial chemical products are used to treat the soil and plants and are in general harmful to the ecosystem. Their purpose is to help prevent diseases or bugs in plant and form the backbone of customary farming.

The outcome is that these chemicals—and the same applies to items utilized as a part of natural agriculture—end up in our food. It has been proven these substances gather in our tissues and when their levels are high they increase the incidence of diseases that is like cancers, as these chemicals can act as carcinogens.

The examination found that agriculturists are quite happy to give up conventional cultivating techniques and move to better sustainable and fruitful techniques. Customary cultivating devices, for example, the techniques to get land ready for agriculture purpose. Kraal compost is the principle soil ripeness technique sought after by agriculturists.

20.2.2 Modern Farming Technologies

20.2.2.1 Problems with Traditional Farming Practices

Indian agricultural sector is in a troublesome state because of the absence of motorization and the shortage of mechanical progress. The agriculture in India

- Needs significant specialized learning
- Produces from low grades of seeds
- Does not utilize composts and pesticides judiciously
- Needs a satisfactory water system foundation
- Needs adequate credit of capital for development

Alongside this, the modern agricultural practices like crop rotation are not acknowledged in India. Majority of Indian ranchers plant one crop in the same field for a considerably long time. This prompts exhaustion of the soil with a consequential reduction in harvest yield. The goal of current farming practices is to keep up the richness of the soil throughout the year.

20.2.2.2 What is Modern Farming Technology

Current cultivating innovations are aimed to enhance traditional cultivating practices. It makes utilization of best quality seeds, better water management systems, improved modern equipments, composts and pesticides.

Present day agrarian practices utilize automated hardware for watering systems, working, and collecting alongside crossbred seeds. It also involves the incorporation of big data for smart farming (Wolfert et al. 2017). Majority of India's population depends on farming as a means of living, hence any innovations in this sector will impact the economy of the country. The agricultural land in India are distributed sparsely, thus making communication and subsequent automation troublesome.

On the other hand, land for modern agriculture is being consolidated into larger farms. This leads to easy access for mechanized equipment and the use of hybrid seeds for increased productivity and disease resistance (Schmitz and Moss 2015). The capital invested behind present day agribusiness is significantly more than what is invested in conventional cultivating advancements. Current agriculture picks up from endowments on vitality, water system, seeds and manures.

Modern farming has developed with innovations such as irrigation via tube wells, sprinklers, and dribbling systems. Modern agriculture is far more advanced and big in terms of capacity due to the use of tractors, motorized hardware for working, furrowing and reaping. Correspondingly, with the development of better temperature and moisture controlled storage rooms, it is now possible to shield from water, creepy crawly bugs and also provides additional warmth.

20.2.3 IoT and Smart Agriculture

Agriculture assumes an imperative role in the advancement of the nation. In India, around 70% of the population relies on cultivating, and 33% of the country's capital originates from cultivation. Problems concerning agriculture have continually hindered the improvement of the nation. The solution to these issues is modernizing current conventional agribusiness techniques. A schematic representation of the working of agricultural IoT has been demonstrated in Figure 20.1.

IoT has many applications in agriculture, some of which are mentioned below:

- *Managing pest*: Frequently, the ranchers hard work is annihilated by predators or vermin bringing about enormous misfortune to the farmers. To anticipate these circumstance, IoT enables recognizing the movement of predators or vermin utilizing passive infrared (PIR) sensors. This data can be utilized by the ranchers to prevent destruction by predators.
- *Managing crop water*: Farmers need water to irrigate the crop. This may bring about wastage of water and possible harm to the harvests as well. Therefore, IoT can play an essential role by sending warnings to agriculturists if there should be too much or too little irrigation.

20.2.3.1 Soil Moisture Sensor

A sensor that detects the dampness in the soil, is called a soil moisture sensor. It sends a warning message to the rancher in order to keep the land and harvest from being harmed (Eller and Denoth 1996).

20.2.3.2 PIR Sensors

PIR stands for passive infrared sensor. A PIR-based movement finder is utilized to detect the movements of predators, creatures, people, or different things. It would keep away vermin from causing damage to the crops or herds. Thus, installing these sensors in the field can reduce the need for farmers to constantly monitor them in order to maximize production. But the limitation is here is, PIR sensor cannot be used on all soil types hence the soil needs to be first assessed before using (Frankiewicz and Cupek 2013).

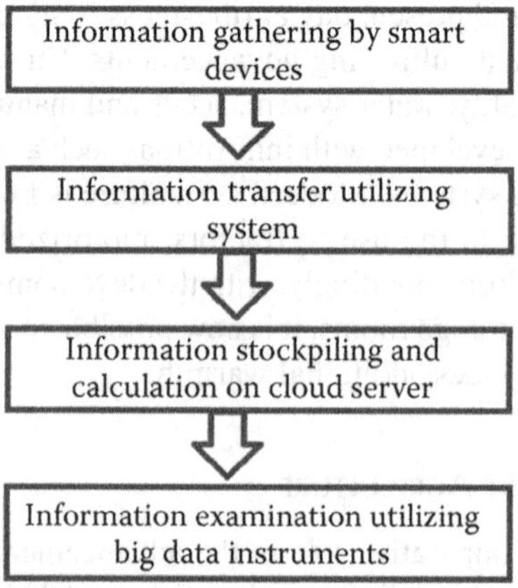

FIGURE 20.1 Flowchart of the agricultural IoT.

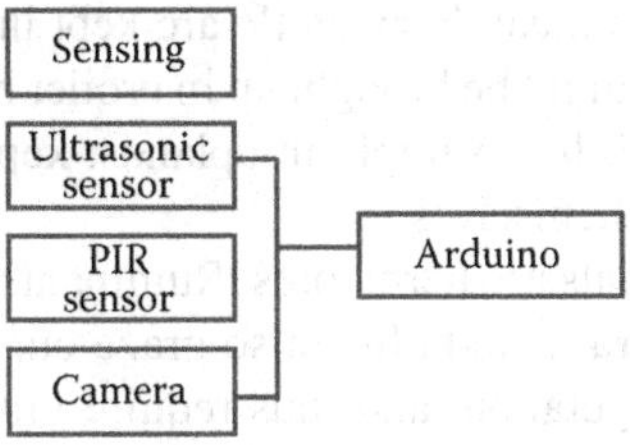

FIGURE 20.2 Block diagram of Arduino.

20.2.3.3 Arduino

Both soil moisture sensors and PIR sensors are associated with Arduino to to carry out the activity (Figure 20.2). Arduino will send the information to the database utilizing an ethernet shield, and if there is a crisis it will likewise send a message to the client utilizing a GSM module.

20.2.3.4 IoT-Based Smart Farming Stick Using Arduino and Cloud Computing

In view of IoT (Internet of things) innovation, it is a cultivating stick which has filled up the flaws of traditional agriculture and saved farmers lives. The aim is to build a novel IoT-based agribusiness stick able to collect live data (temperature, soil moisture, and smoke detection) for effective condition sensing, which will help them to do smarter cultivating and increase yields. The agriculture stick also incorporates the Arduino innovative motherboard fitted with different sensors. Thus, this stick combines all the features essential to protect the land as well as harvest. If we look into its make-up it combines the qualities of a soil moisture sensor, a PIR sensor, etc. thus maintaining the overall expense and reducing the load on farmers and machines (Moghe, Lambert, and Divan 2012).

20.3 IoT FOR ANIMAL HUSBANDRY

20.3.1 Traditional Animal Husbandry

20.3.1.1 Introduction

Animal husbandry concerns raising animals for meat, fiber, dairy, eggs, and so on. It incorporates everyday care, specific rearing, and the raising (with proper care) of animals. The history of domesticated animals begins around 13,000 BC during the Neolithic, preceding the cultivation of the first harvests. An extensive variety of different species, such as the horse, water bison, lama, rabbit, and guinea pig are utilized as domesticated animals around the world. The aquaculture of fish, mollusks, and the keeping of honeybees or silkworms is boundless. Insect raising for human consumption takes place in several nations; for instance, cricket farming is a productive industry in Thailand.

Modern animal husbandry relies on production system adapted to the type of land available. Subsistence farming is now being superseded by intensive animal farming, where for instance meat dairy cattle are kept in high density feedlots, and large numbers of chickens might be brought up in broiler houses or batteries (Kuhnen et al. 2015). On poorer soil, for example in uplands, kept in semi-captivity,and are allowed to wander to feed themselves.

Most domesticated animals are herbivores. Ruminants like dairy cattle and sheep are adapted to feed upon grass, and often also graze outside. Pigs and poultry can't process the cellulose in vegetation, and thus require grains and other high-vitality nourishments.

20.3.1.2 Husbandry

20.3.1.2.1 Systems

Traditional animal husbandry was a source of sustenance for the farmer and his/her family. Killing animals for nourishment was a secondary option, and whenever possible its fleece, eggs, milk, and blood were reaped while the creature was still alive. In the customary cycle of transhumance, individuals and animals moved between summer and winter pastures; the midyear pastures were up in the mountains, and the winter rangelands were in the valleys (Blench 2001). Animals can be kept in free ranges or in intensive farming structures. Broad frameworks includes extensive system that involves animals roaming at will, or under the supervision of herdsman. Farming in western United States includes animals roaming freely at will or under the supervision of herdsmen in open as well as private lands (Starrs 2000).

20.3.1.2.2 Feeding

Herbivores are typically classified into "think selectors" (who specifically eat seeds, leafy foods, and nutritious young foliage), "slow eaters" who mostly eat grass, and "middle of the road feeders" who feed from an entire scope of accessible plant materials. Cows, sheep, goats, deer, and impalas are ruminants; they process sustenance in two stages, biting and gulping in the ordinary way, and afterward spewing the semidigested cud to bite it again and in this way they remove the utmost nourishment value (Science and Dryden 2008).

The dietary needs of these animals are for the most part met by eating grass. Grasses develop from the base of the leaf-sharp edge, enabling it to flourish even when vigorously nibbled or cut ("The Living Planet: A Portrait of The Earth by David Attenborough," n.d.).

In many climates grass growth is seasonal, for instance in the calm summer or tropical stormy season. Other scavenge crops develop as well and a significant number of these, and also crop residues, can be used to provide for the nutritious needs of domesticated animals in the lean season. Animals raised on extended ranges may subsist completely on grazing, yet more intensively kept domesticated animals will require more vigorous and protein-rich nourishments. Nutrition is for the most part found in grains and cereal by products, fats and oils, and sugar-rich nourishments, while proteins may originate from fish or meat, dairy items, vegetables, and other plants, frequently the products of vegetable oil extraction (Science and Dryden 2008). Since pigs and poultry are nonruminants they are unfit to process the cellulose in grass and different rummages, so they are fed completely on grains and other high-nutrition foodstuffs. The elements for the animals' feed can be procured on the homestead or can be purchased, as commercially available feeds are targeted for the distinctive classes of domesticated animals, their developmental stages, and their particular nourishing necessities. Vitamins and minerals are added accordingly to adjust their diet. Farmed fishes are usually fed pelleted food.

20.3.1.2.3 Breeding

The breeding of ranch animals seldom occurs spontaneously but is managed by farmers with a view of encouraging certain traits that are seen as desirable. These incorporate resistance, productivity, breeding capacities, quick development rates, low bolster (feed) utilization per unit of development, better body features, higher yields, better fiber qualities, and other attributes. Unwanted attributes are, for example, health defects, forcefulness or absence of submission are chosen against (Rodenburg, Breeding, and Centre, n.d.).

Rearing in particular has seen huge increases in efficiency. In 2007, a run of the mill two-month-old oven chicken was 4.8 times larger than in 1957. In the 30 years prior to 2007, the normal yield of a dairy bovine in the United States almost doubled (Rodenburg and Turner 2015). Techniques, such as manual sperm injection and fetus exchange are commonly used today, not just as strategies to ensure females breed routinely, but additionally to help enhance breed hereditary qualities.

20.3.1.2.4 Animal Health

Great cultivation, appropriate feeding, and cleanliness are the fundamental principles of animal wellbeing on the homestead, bringing monetary advantages through increased production. When, despite these safety measures animals become sick they are treated with veterinary pharmaceuticals by the agriculturist and the veterinarian. In the European Union, when agriculturists treat their own animals they are required to follow treatment guidelines and to keep records of the medicines given (Kools, Moltmann, and Knacker 2008). Herds are vulnerable to various diseases and conditions, which may influence their wellbeing. A few known swine fever and scrapie are particular to one sort of stock, while others, similar to foot-and-mouth illness influence all cloven-hoofed animals. For this very reason the body temperature of the animals must be measured and monitored closely (Sellier, Guettier, and Staub 2014). Governments are especially worried about zoonoses—ailments people may contract from animals. Wild animals may harbor diseases able to infect domesticated animals, which may be due to inadequate biosecurity. An episode of Nipah infection, in Malaysia in 1999, was traced back to pigs ending up sick coming in contact with organic product eating flying foxes (*Pteropus hypomelanus and Petropus vampyrus*). The pigs thus passed on the contamination to humans. Avian influenza H5N1 is present in wild bird populations and can be conveyed over vast distances by migrating birds. This infection is highly transmissible to local poultry and people living close to them. Different contagious illnesses affect wild creatures, domesticated animals, and people such as rabies, leptospirosis, brucellosis, tuberculosis, and trichinosis.

20.3.2 Technological Developments

Domesticated animals revenues are invaluable for the wealth of a nation like India. In India, animal farming is no longer a side-revenue to agriculture or a assistance to farming (using cattles to plow the field). Animal farming has transformed into an industry and the most recent reports state that its contribution to the GDP of the country is significantly higher than ever before. Animal husbandry offers a superior degree of financial security for small ranchers whose wages from agriculture suffer because of storms, loss of landholdings, insects, poor planning, and so forth. Despite the fact that the development of the domesticated animal industry is exceptionally encouraging, keeping in mind the end goal to make India a worldwide leader in animal farming, it is logical to integrate in this industry the advances made in different fields. The advancements in Information Technology over the course of recent decades are colossal and offer extraordinary potential in enhancing animal productivity through different measures like eradication of dangerous illnesses, quick and precise infection detection, and so on.

20.3.2.1 Geo-Informatics Technologies

Geographic Information System (GIS) (QGIS Development Team 2015) is an electronic database administration system for capturing, storing, checking, coordinating, controlling, dissecting, and showing information tied to a geographical area. Early identification of an irresistible sickness episode is a vital initial move towards actualizing powerful ailment mediations and diminishing coming about mortality and dismalness. Both topographical and regular spreads of numerous irresistible maladies through the atmosphere, has evoked the use of occasional atmosphere figures as prescient pointers in ailment early cautioning frameworks (EWS) for quite some time. GIS, Remote Sensing (RS), and GPS are the three commonly utilized veterinary geo-informatics tools utilized with this type of data for fast global mapping of information to enable management of animal infections.

20.3.2.2 Remote Sensing

RS has changed the way researchers handle and break down geographic information. Remote detecting alludes to the securing of geographic information without reaching the zone of study. Today, remote detecting by and large rests on satellite pictures and airborne photography/imaging, both of which are sensitive to portions of the electromagnetic spectrum not noticeable to the human eye or to most cameras (e.g., infrared, small-scale waves) (Toth and Jóźków 2016). The capacity to determine otherwise undetectable vitality and to record precise measurements remotely, shows the capability of the apparatus.

20.3.2.3 ShearEzy

The ShearEzy is a result of Australian Wool Innovation (AWI) interest in the Upright Posture Shearing Platform (UPSP) innovations. The product turned out to be financially accessible in late 2007, with the latest adaptation released in June 2012. ShearEzy's benefits include:

- Diminished physical exertion and strain from shearing.
- Can be more beneficial than "on the board" shearing, particularly for the novice or normal shearer.
- Simpler blow patterns and effective animal restraints.
- Simple detachment of crutching's and lower quality fleeces.

Many sheep and rams are effortlessly and safely taken care of without sedation. ShearEzyis an air-worked system, comprising of a shearing stage with an articulated parallel sheep loader. Sheep stroll into the loader on a slanted, nonreinforced incline, and are kept there until prepared to be shorn (De Boos 2001).

20.3.2.4 Information Technology in Molecular Biology

A standout among the most vital examinations for animal wellbeing is the examination of their genomes. Results of genomic examination are always correct and appropriate and thus helps in better understanding of the disease. To create quality tests, the whole genome must be mapped. After mapping, the unique areas have to be identified comparing it with the genome of other infectious agents. Despite the fact that the genome is little in measure, the quantity of bases in every genome is past manual correlation. Numerous products are accessible that can contrast the genome and existing genomes in a quality bank to recognize special territories. Such programming additionally gives degree to building up a phylogenetic tree, atomic clock and for creating preliminaries for PCR response. The phylogenetic tree and sub-atomic clock are basic for sub-atomic scientists to land at conceivable sources of the infection. This discovering acquires essentialness if the infection strains are immunologically particular.

20.3.2.5 Information Technology in Simulation Studies

Another area where IT can contribute significantly is the reconstruction of in vivo conditions. Certain advanced graphics based softwares offer best solution for this. Trafficking of etiological operators, proteins and so forth in the middle of the cells can be considered utilizing this product. Concentrate on these trafficking designs is fundamental to create immunizations for cell related infections and intra cell microorganisms. The part of designs based programming in concentrate the three dimensional structures of antigen and immune response is additionally imperative to dissect the cooperation's between them in vivo. Cooperation of antigen and immune response is fundamental to think about the pathogenesis of any irresistible ailments.

20.3.2.6 Sperm Analyzer (Aidmics Biotechnology)

The iSperm analyzer enables animal husbandry specialists to complete swine sperm assessment quickly and decrease their costs of equipment by about 90%. The iSperm can associate with cell phones and laborers can examine the sperm's morphology, quantity, and motile rating. The information can likewise be stored to enable breeders to keep records and build up a precise database.

Thus, as compared with the traditional process of sperm collection (checking motility rate, concentration, and manually recording all the data), which is time consuming, tedious, and requires labor, this sperm analyzer reduces the overall time spent and also skips the step of manual recording by replacing it with an automatic recording system. Thus, this can be immediately followed by the artificial insemination if desired.

20.3.3 Automation and IoT

20.3.3.1 Biosensors

Early identification of sicknesses utilizing biosensors enables can prevent their death or spread the illness thus, decreasing the impact of the illness and related social and financial consequences. In order to diagnose a disease early in animals we need a continuous monitoring biosensor to predict the disease at early stages (Diamond, Zhao, and Li 2017).

The utilization of biosensors and wearable innovations is becoming progressively imperative for animal wellbeing management. These devices, if fabricated correctly and utilized effectively, can determine infections in the long run and diminish their financial impact. Such devices are especially valuable for dairy cows and poultry ranches. Rather than depending exclusively on the ranchers' faculties and knowledge, nearby sensors can give solid information about the physical state of the herds. Because of the unrivaled progresses of wearables and sensors, they can produce a leap forward in domesticated animals care, and are guaranteed to be one the most impactful and practicable innovations in the animal farming market. New wearable advancements are being aimed at addressing the issues of pets and farm animals. Items such as solution patches, tracking collars, and sitting posture corrector are being acquired at higher rates and outfit for the more advantageous childhood of homestead creatures. These wearable advances are multifunctional and productive, enabling animal owners to accomplish more in less time. Worldwide development of this sector—in the next 10 years—has been anticipated to take off from $0.91 billion to $2.6 billion.

Sensor and wearable innovations can be embedded on animals to analyze their sweat constituents (McCaul et al. 2017; Luong, Male, and Glennon 2008), measure their body, monitor their behaviors (Sellier, Guettier, and Staub 2014), and develop characteristic sounds (Lee et al. 2015; Kucharzyk et al. 2015), identify pH (Zhang et al. 2017), avert illness, recognize analytes, and recognize the presence of infections and pathogens. Wearable sensors enable ranchers to indentify ailments early, and avoid spread to other animals. Agriculturists can likewise separate sick creatures so as to keep the spread of an ailment contained. Aside from gathering helpful information on animal wellbeing, general homestead work can likewise be made less demanding and more dependable by utilizing biosensors coordinated with cell phones and handheld devices, rather than regular techniques such as redacting notes, keeping a ranch journal, or utilizing basic gear without information sharing capacities.

It's a major challenge to produce great quality and safe meat products to meet the expanding worldwide demand for such items. With greater vested interests comes increasing concerns for managing farm animal wellbeing. Devices placed under their skin, or in their stomachs give animal farmers access to helpful data with respect to their herd's conduct and medicinal conditions. These electronic devices are relied upon for the therapeutic treatment of animals, the recognition of warming and cooling needs, iontophoretic sedate conveyances, and even the protection of wild species. Another critical utilization of biosensors is antimicrobial recognition. With the intensive and successive use of anti-infection agents in the animal business, antitoxin protection has become a noteworthy risk for agriculturists. Environmental precariousness is caused by the uncontrolled utilization of subrestorative antitoxins in concentrated animal feeding operation, which promotes anti-infection protection in creatures. There is a critical requirement for agriculturists to change to other options to maintain a strategic distance so that animals do not become invulnerable to antitoxin treatment. The dosage of anti-infection agents in the blood serum and muscles of ranch animals ought to be kept within a specific range, and there should be legitimate guidelines to test their antimicrobial levels (Mungroo and Neethirajan 2014). It is difficult to put a restriction on the use of anti-infection agents in domesticated animals, as antitoxins help cure the most widely recognized infirmities, such as enteric and respiratory contaminations. The utilization of anti-infection agents to protect ranch creatures against microbial infection is being perceived likewise all around the globe. To address this issue, the European Union set up a standard to keep a check on antimicrobial protection. This guideline, which has been recommended as a prudent step, concentrates on forbidding certain antimicrobial development products. Most extreme Residue Limits (MRLs) have been set up to limit those antimicrobials still permitted for animal use in the United States and European nations. MRL is that measure of pharmacologically dynamic substances and their inferred metabolites, which is legitimately allowed. Biosensors have been useful in such matters; they can easily recognize antimicrobial levels and caution the ranchers if antitoxins levels surpass a certain range.

In the following decade, the global market for wearable innovations aimed at animals is expected to generate around $1 billion to $2.5 billion, expanding more than 2.5 times. The leader for these exceptional innovations is China, which is producing these items at an extremely low cost, followed by the United States. A lot of cash is spent each year on agricultural research and animal wellbeing management. Be that as it may, it improves profitability to increase animal wellbeing. As a general rule, the financing is intended to give more up to date answers for the issues, as opposed to crossing over any barrier amongst research and industry. Banhazi and Black (Banhazi et al. 2012) have recommended that a thorough system be set up to guarantee farming practices are adequate and promote the flow of learning and research discoveries (Banhazi et al. 2012). In order to know the exact standard of animals' health well being and to counter the current high disease rates. it is evident that a new way should be developed to identify ailments. This move includes supplanting off-site lab tests for homesteads with fast test findings on the ranch itself. The world association for animal wellbeing (OIE) has cautioned that the zoonotic illnesses from intensive farming of domesticated animals can impact their general wellbeing if there is overflow from the cultivated creature repository, and the domesticated animals industry is under overwhelming strain to enhance its biosecurity conventions and improve creature traceability and welfare. Consequently, the animal business and sustenance wellbeing review offices are looking for new instruments and innovations, to empower fast, continuous and on-cultivate checking of sicknesses and record keeping.

Biosensing improvements guarantee enhancements to the execution, cost, and efficiency in animal disease management. Advancement and promotion of dependable, fast tests will permit a prior and more targeted treatment of diseases, conceivably bringing about decreases in antimicrobial use and enhancements in animal welfare. The biosensors and early detecting tools help in designing the medicines or drugs even before the sickness kicks in and this could shape the administration segments of the coordinated ranch assessment and proAction display (Awasthi et al. 2016). Figure 20.3 shows the schematic working of a biosensor.

The combination of novel indicators and illness identifiers utilizing biosensors would keep animals and the farming industry one step ahead of undetectable infections through the use of satellites and cell phones. Smart and accurate domesticated animal farming and wellbeing management will keep on growing in importance to tackle the expanding demand for nourishment and guarantee sustainability. The biosensing advances—channeled through the IoT worldwide—will advance fast, establishing constant on-site farm monitoring of animal farming infections. The continuous spread of information gathered from the homesteads through these biosensors will have an impact beyond the ranch too; permitting entrance to this data that will demonstrate fundamental to the social permit issues confronting our rural area and will be a key to our proceeded with worldwide aggressiveness.

20.3.3.2 IoT-Based Beef Cattle Breeding System

Beef cattle breeding industry plays an important role in animal husbandry however the traditional methods are unable to meet the growing demands of beef. To deal with the problems of beef cattle breeding industry in China an IOT based beef cattle breeding system that take RFID technology into beef cattle was designed and combined with B/S and C/S structures. The framework gives the help of gathering and figuring out how to the cowshed and steers data, nourishing data, ailment treatment data, pandemic avoidance data, weight data and client data. In the meantime, the settled RFID peruser is utilized with the electronic scale to understand the precise record of weight change amid the procedure of hamburger cows rearing. The planned IOT based beef cattle breeding system has been applied and in use for several years in Lincang, Yunnan Province, China. The running outcomes demonstrated that it

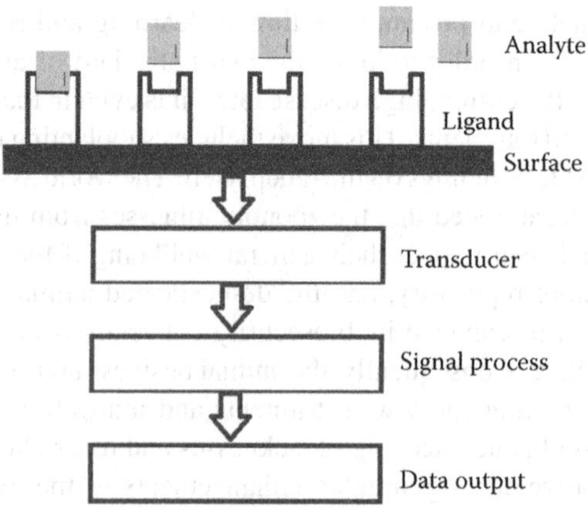

FIGURE 20.3 Biosensor.

is steady and solid, and the genuine needs likewise can be met by each capacity. Generally speaking, the framework is reasonable for fine hamburger steers rearing and can enable the dairy cattle to cultivate understanding the advanced and traceable logical reproducing strategy (Wolfová et al. 2005).

20.4 CONCLUSION

The advantages of IoT innovations in farming operations include, yet are not restricted to, diminishing waste, better pest and animals management, and expanded profitability. New creative IoT applications are addressing these issues and expanding the quality, amount, sustainability, and cost viability of rural production. The present large and small farms can benefit from IoT innovations to screen soil moisture, guide crop management, improve herd nutrition levels, improve pest management system and water system infrastructures, and rapidly analyze operational information. Combined with other sources of data, such as climate predictions, this available knowledge can additionally enhance basic farming practices. On account of headway made by IoT, another era of farming is upon us soon.

REFERENCES

Awasthi, A., A. Awasthi, D. Riordan, and J. Walsh. 2016. "Non-Invasive Sensor Technology for the Development of a Dairy Cattle Health Monitoring System." *Computers* 5 (4): 23. doi: 10.3390/computers5040023.

Banhazi, T. M., H. Lehr, J. L. Black, H. Crabtree, P. Schofield, M. Tscharke, and D. Berckmans. 2012. "Precision Livestock Farming: An International Review of Scientific and Commercial Aspects." *International Journal of Agricultural and Biological Engineering.* 5(3): 1 doi: 10.3965/j.ijabe.20120503.00?

Blench, R. 2001. "'You Can't Go Home Again' Pastoralism in the New Millennium." *Odi* no. May 2001: 106. http://www.odi.org/sites/odi.org.uk/files/odi-assets/publications-opinion-files/6329.pdf. www.org.odi.uk.

De Boos, A. 2001. "Australian Wool - Competitive through Innovation." *International Textile Bulletin* 47 (2): 8–21. https://www.scopus.com/inward/record.uri?eid=2-s2.0-0035005472&partnerID=40&md5=a72c949083a03fcb767137d23231dfa8.

Diamond, D., C. Zhao, and L. Li. 2017. "5th International Symposium on Sensor Science Session Chairs," September 27–29, 2017, Barcelona, Spain.

Eller, H. and A. Denoth. 1996. "A Capacitive Soil Moisture Sensor." *Journal of Hydrology* 185 (1–4): 137–46. doi: 10.1016/0022-1694(95)03003-4.

FAO. 2016. "Food and Agriculture Organization of United Nations." *The Food and Agriculture Organization of the United Nations.* http://faostat.fao.org.

Frankiewicz, A. and R. Cupek. 2013. "Smart Passive Infrared Sensor—Hardware Platform." In *IECON Proceedings (Industrial Electronics Conference)*, 7543–47. doi: 10.1109/IECON.2013.6700389.

Kools, S. A. E., J. F. Moltmann, and T. Knacker. 2008. "Estimating the Use of Veterinary Medicines in the European Union." *Regulatory Toxicology and Pharmacology* 50 (1): 59–65. doi: 10.1016/j.yrtph.2007.06.003.

Kucharzyk, K. H., M. A. Deshusses, K. A. Porter, and H. Hsu-Kim. 2015. "Relative Contributions of Mercury Bioavailability and Microbial Growth Rate on Net Methylmercury Production by Anaerobic Mixed Cultures." *Environmental Science: Processes & Impacts.* 17(9): 1568–77. The Royal Society of Chemistry. doi: 10.1039/C5EM00174A.

Kuhnen, S., R. B. Stibuski, L. A. Honorato, and L. C. P. M. Filho. 2015. "Farm Management in Organic and Conventional Dairy Production Systems Based on Pasture in Southern Brazil and Its Consequences on Production and Milk Quality." *Animals* 5 (3): 479–94. doi: 10.3390/ani5030367.

Lee, J., B. Noh, S. Jang, D. Park, Y. Chung, and H. H. Chang. 2015. "Stress Detection and Classification of Laying Hens by Sound Analysis." *Asian-Australasian Journal of Animal Sciences* 28 (4): 592–98. doi: 10.5713/ajas.14.0654.

Luong, J. H. T., K. B. Male, and J. D. Glennon. 2008. "Biosensor Technology: Technology Push versus Market Pull." *Biotechnology Advances.* doi: 10.1016/j.biotechadv.2008.05.007.

McCaul, M., A. Porter, T. Glennon, R. Barrett, S. Beirne, G. G. Wallace, P. White, F. Stroiescu, and D. Diamond. 2017. "Wearable Sensor for Real-Time Monitoring of Electrolytes in Sweat." *Proceedings* 1 (10): 724. doi: 10.3390/proceedings1080724.

Moghe, R., F. C. Lambert, and D. Divan. 2012. "Smart Stick-on Sensors for the Smart Grid." *IEEE Transactions on Smart Grid* 3 (1): 241–52. doi: 10.1109/TSG.2011.2166280.

Mungroo, N. A. and S. Neethirajan. 2014. "Biosensors for the Detection of Antibiotics in Poultry Industry-A Review." *Biosensors* 4 (4): 472–93. doi: 10.3390/bios4040472.

Neethirajan, S. 2017. "Recent Advances in Wearable Sensors for Animal Health Management." *Sensing and Bio-Sensing Research* 12: 15–29. The Author. doi: 10.1016/j.sbsr.2016.11.004.

QGIS Development Team. 2015. "QGIS Geographic Information System." *Open Source Geospatial Foundation Project.* doi: http://www.qgis.org/.

Rodenburg, T. B., A. Breeding, and G. Centre. n.d. "The Role of Breeding and Genetics in Animal Welfare".

Rodenburg, T. B. and S. P. Turner. 2015. "The Role of Breeding and Genetics in Animal Welfare," 38.

Schmitz, A. and C. B. Moss. 2015. "Mechanized Agriculture: Machine Adoption, Farm Size, and Labor Displacement." *AgBioForum* 18 (3): 278–96.

Science, Animal Nutrition, and Gordon Mcl Dryden. 2008. "DOWNLOAD http://bit.ly/1J5XwKE http://goo.gl/RwOeh http://en.wikipedia.org/w/index.php?search=Anima l+Nutrition+Science" 16.

Sellier, N., E. Guettier, and C. Staub. 2014. "A Review of Methods to Measure Animal Body Temperature in Precision Farming." *American Journal of Agricultural Science and Technology.* doi: 10.1016/j.sbsr.2016.11.00410.7726/ajast.2014.1008.

Starrs, P. F. 2000. "Let the Cowboy Ride: Cattle Ranching in the American West," no. March: 386. doi: 10.2307/2649639.

The Living Planet: A Portrait of The Earth by David Attenborough. n.d.

Toth, C. and G. Jóźków. 2016. "Remote Sensing Platforms and Sensors: A Survey." *ISPRS Journal of Photogrammetry and Remote Sensing.* doi: 10.1016/j.isprsjprs.2015.10.004.

Wolfert, S., L. Ge, C. Verdouw, and M. J. Bogaardt. 2017. "Big Data in Smart Farming—A Review." *Agricultural Systems.* doi: 10.1016/j.agsy.2017.01.023.

Wolfová, M., J. Wolf, J. Přibyl, R. Zahrádková, and J. Kica. 2005. "Breeding Objectives for Beef Cattle Used in Different Production Systems: 1. Model Development." *Livestock Production Science* 95 (3): 201–15. doi: 10.1016/j.livprodsci.2004.12.018.

Zhang, Q., J. R. Castro Smirnov, R. Xia, J. M. Pedrosa, I. Rodriguez, J. Cabanillas-Gonzalez, and W. Huang. 2017. "Highly pH-Responsive Sensor Based on Amplified Spontaneous Emission Coupled to Colorimetry." *Scientific Reports* 7: 1–6. doi: 10.1038/srep46265.

Index

A

F

G

H

I

N

O

P

Q

R

T

U

V

W

Y

Z